T0325156

Lattice Basis Reduction

An Introduction to the LLL Algorithm and Its Applications

PURE AND APPLIED MATHEMATICS

A Program of Monographs, Textbooks, and Lecture Notes

MONOGRAPHS AND TEXTBOOKS IN PURE AND APPLIED MATHEMATICS

Recent Titles

Santiago Alves Tavares, Generation of Multivariate Hermite Interpolating Polynomials (2005)

Sergio Macías, Topics on Continua (2005)

Mircea Sofonea, Weimin Han, and Meir Shillor, Analysis and Approximation of Contact Problems with Adhesion or Damage (2006)

Marwan Moubachir and Jean-Paul Zolésio, Moving Shape Analysis and Control: Applications to Fluid Structure Interactions (2006)

Alfred Geroldinger and Franz Halter-Koch, Non-Unique Factorizations: Algebraic, Combinatorial and Analytic Theory (2006)

Kevin J. Hastings, Introduction to the Mathematics of Operations Research with *Mathematica®*, Second Edition (2006)

Robert Carlson, A Concrete Introduction to Real Analysis (2006)

John Dauns and Yiqiang Zhou, Classes of Modules (2006)

N. K. Govil, H. N. Mhaskar, Ram N. Mohapatra, Zuhair Nashed, and J. Szabados, Frontiers in Interpolation and Approximation (2006)

Luca Lorenzi and Marcello Bertoldi, Analytical Methods for Markov Semigroups (2006)

M. A. Al-Gwaiz and S. A. Elsanousi, Elements of Real Analysis (2006)

Theodore G. Faticoni, Direct Sum Decompositions of Torsion-Free Finite Rank Groups (2007)

R. Sivaramakrishnan, Certain Number-Theoretic Episodes in Algebra (2006)

Aderemi Kuku, Representation Theory and Higher Algebraic K-Theory (2006)

Robert Piziak and P. L. Odell, Matrix Theory: From Generalized Inverses to Jordan Form (2007)

Norman L. Johnson, Vikram Jha, and Mauro Biliotti, Handbook of Finite Translation Planes (2007)

Lieven Le Bruyn, Noncommutative Geometry and Cayley-smooth Orders (2008)

Fritz Schwarz, Algorithmic Lie Theory for Solving Ordinary Differential Equations (2008)

Jane Cronin, Ordinary Differential Equations: Introduction and Qualitative Theory, Third Edition (2008)

Su Gao, Invariant Descriptive Set Theory (2009)

Christopher Apelian and Steve Surace, Real and Complex Analysis (2010)

Norman L. Johnson, Combinatorics of Spreads and Parallelisms (2010)

Lawrence Narici and Edward Beckenstein, Topological Vector Spaces, Second Edition (2010)

Moshe Sniedovich, Dynamic Programming: Foundations and Principles, Second Edition (2010)

Drumi D. Bainov and Snezhana G. Hristova, Differential Equations with Maxima (2011)

Willi Freeden, Metaharmonic Lattice Point Theory (2011)

Lattice Basis Reduction

An Introduction to the LLL Algorithm and Its Applications

Murray R. Bremner

University of Saskatchewan

Saskatoon, Canada

CRC Press
Taylor & Francis Group
Boca Raton   London   New York

CRC Press is an imprint of the
Taylor & Francis Group, an **informa** business
A CHAPMAN & HALL BOOK

The author is very grateful to Chris Applegate for the cover art, which is a photograph of an Islamic mosaic at the Alcazar in Seville.

CRC Press
Taylor & Francis Group
6000 Broken Sound Parkway NW, Suite 300
Boca Raton, FL 33487-2742

© 2012 by Taylor & Francis Group, LLC
CRC Press is an imprint of Taylor & Francis Group, an Informa business

No claim to original U.S. Government works

Printed in the United States of America on acid-free paper
Version Date: 20110615

International Standard Book Number: 978-1-4398-0702-6 (Hardback)

Visit the Taylor & Francis Web site at
http://www.taylorandfrancis.com

and the CRC Press Web site at
http://www.crcpress.com

Contents

List of Figures

Preface

This book is intended to be an introductory textbook on lattice algorithms for advanced undergraduate students or beginning graduate students. It is designed to be either the principal text for a one- or two-semester course on lattice basis reduction, or a secondary reference for courses on computer algebra, cryptography, and computational algebraic number theory. It could also be used as a source of topics for presentations in an undergraduate seminar. The book will be useful for graduate students and researchers in many branches of pure and applied mathematics who need a user-friendly introduction to lattice algorithms, suitable for self-instruction, in order to apply these algorithms in their own work. This book is not intended as a state-of-the-art research monograph for experts in the field. All of the algorithms and theorems presented in this book are at least a few years old, and most of them were first published in the the last two decades of the 20th century. Most of these algorithms were originally published in an "unstructured" form; I have rewritten them without the use of "goto" statements.

The goal of the book is to present the essential concepts that should be familiar to all users of lattice algorithms. The book is based primarily on a number of fundamental papers in the area, including of course the paper of Lenstra, Lenstra and Lovász which introduced the LLL algorithm. I have developed the topic following these papers, but using consistent notation, providing numerous computational examples (primarily using Maple), and including suggested projects and exercises.

The most important prerequisite for the book is a knowledge of basic linear algebra; the essential facts are reviewed briefly in Chapter 1. Some knowledge of elementary number theory would also be helpful, but not essential. For the chapters on polynomial factorization, some familiarity with basic abstract algebra (especially the theory of polynomial rings) is assumed.

In the following paragraphs I give a summary of the contents chapter by chapter; many of these topics have not previously appeared in a textbook.

Chapter 1 is introductory; it first recalls basic facts about Euclidean vector spaces, then introduces the concepts of lattice, sublattice, and lattice basis, and concludes by summarizing without proof some results from the geometry of numbers which are necessary for an understanding of lattice algorithms.

Chapter 2 presents a detailed analysis of the Gaussian algorithm (attributed by some authors to Lagrange) for lattice basis reduction in two dimensions; an understanding of this algorithm is essential for all the other topics discussed later in the book. Chapter 3 provides details of the Gram-

Schmidt orthogonalization process: the main theorem is proved in detail, and the complexity of the algorithm is analyzed. Chapter 4 presents a detailed exposition of the original paper by Lenstra, Lenstra and Lovász [88] which used lattice basis reduction to provide an efficient algorithm for polynomial factorization; this material depends fundamentally on the results of the previous two chapters. (The application to polynomial factorization is postponed to the last chapter of the book.) Chapter 5 discusses a modification of the LLL algorithm which uses "deep insertions"; Chapter 6 discusses the version of the LLL algorithm which allows the input vectors to be linearly dependent.

Chapter 7 introduces the first application of lattice basis reduction to cryptography: it shows how the LLL algorithm can be used to break a knapsack cryptosystem. Chapter 8 discusses the second application to cryptography: the famous algorithm of Coppersmith which uses the LLL algorithm to find small roots of a modular polynomial, and which has important consequences for the security of RSA cryptosystems. Chapter 9 presents a brief discussion of an application of lattice basis reduction to a problem in algebraic number theory, namely simultaneous Diophantine approximation.

The LLL algorithm has polynomial running time, and produces a good basis for the input lattice, but unfortunately it does not in general produce the best basis, or even a basis containing a shortest nonzero vector. Chapter 10 presents the Fincke-Pohst algorithm which performs an exhaustive search to find a shortest nonzero vector in the input lattice. Chapter 11 discusses Kannan's algorithm, which recursively calls an exhaustive search procedure in order to produce (in exponential time) a very good basis for the input lattice, which is guaranteed to contain a shortest nonzero lattice vector, and which is "reduced" in a much stronger sense than the output of the LLL algorithm. Chapter 12 presents Schnorr's hierarchy of algorithms, which use the concept of block reduction to modify Kannan's algorithm in such a way that polynomial-time complexity is restored.

The algorithms in Chapters 10 to 12 compute a strongly reduced lattice basis, which in particular contains a nonzero lattice vector which is as short as possible with respect to the Euclidean norm on $\mathbb{R}^n$. Chapter 13 considers instead the max norm, and presents the proof by van Emde Boas that the problem of finding a shortest nonzero vector in a lattice with respect to the max-norm (not the Euclidean norm) is NP-complete.

Chapter 14 forms a short course on algorithms for the Hermite normal form of an integer matrix. It begins by reviewing Gaussian elimination over a field, and the generalization to matrices with integer entries, and then shows how the LLL algorithm can be used to find a basis consisting of relatively short vectors for the nullspace lattice of an integer matrix. The remaining sections of the chapter present the algorithm of Havas, Majewski and Matthews which uses linear algebra to efficiently compute the greatest common divisor of a set of integers, and then uses this algorithm to efficiently compute the Hermite normal form of an integer matrix. (The author's own interest in lattice basis reduction originated in computational problems related to polynomial identities for nonassociative algebras. One can represent such identities as the nonzero vectors in the nullspace of a large integer matrix. Finding the simplest identities is equivalent to finding the shortest vectors in the nullspace.)

Chapter 15 forms a short course on polynomial factorization. It first presents the necessary background material on polynomial factorization over finite fields, and on Hensel lifting to a p-adic factorization. The remaining sections of the chapter apply these results to give the polynomial-time algorithm (from the original paper by Lenstra, Lenstra, and Lovász) for factoring polynomials with rational coefficients.

Each chapter includes two supplements: a section of Projects which would be suitable as substantial programming assignments or as topics for written reports and class presentations; and a section of Exercises which would be suitable for inclusion in problem sets. Roughly 60% of the material in the book has been classroom-tested in graduate courses on computer algebra and lattice basis reduction that I have taught at the University of Saskatchewan.

There are a number of other textbooks on various aspects of the theory of lattices, but they are all fall into two categories: theoretical monographs on the pure mathematical theory, or research-level monographs on a special topic in the computational theory. Examples of the former category are the classical texts on the geometry of numbers by Cassels [22] and Lekkerkerker [87] (see also Gruber and Lekkerkerker [52]), and the more recent book by Martinet [93]. Also worth mentioning is the book by Conway and Sloane [27], which deals with the interplay between the geometry of numbers and finite simple groups. The primary example of the latter category is the book by Micciancio and Goldwasser [100] which gives a cryptographic perspective on the complexity of lattice algorithms. Two other monographs which deserve special mention, since one of the authors is a co-discoverer of the LLL algorithm, are Lovász [91] and Grötschel et al. [51]. Many textbooks on computer algebra contain at least one chapter on lattice algorithms; this author's personal favorites are von zur Gathen and Gerhard [147], Cohen et al. [25], and Cohen [26]. There is also a very recent conference proceedings edited by Nguyen and Vallée [110] , which contains many articles by experts on special topics in the theory of lattice algorithms and their applications. However, none of these references does what I have attempted to do in the present book: to survey the entire topic of lattice basis reduction at a level suitable to anyone with a good background in undergraduate linear algebra.

This book concentrates on the computational aspects of the theory of lattices, and seems to be the only complete introduction for non-specialists. I hope that this textbook will fill a gap in the existing literature on lattice algorithms, and encourage many more people to learn about lattice basis reduction and its applications throughout pure and applied mathematics. This seems to be the first book that attempts to give a broad survey of the field, covering the essential topics, but not developing the material in the specialized detail that would be expected from a research monograph.

As usual in a project of this nature, many people made indirect but important contributions. The anonymous referees of the original proposal gave me the idea to include the chapters on Coppersmith's algorithm and on NP-completeness. The students in my graduate classes on computer algebra pointed out a number of errors and inconsistencies. My research collaborators Luiz Peresi (University of São Paulo, Brazil) and Irvin Hentzel (Iowa State University, USA), and my graduate students Hader Elgendy and Jiaxiong Hu, assisted me with the application of the LLL algorithm to problems in nonassociative algebra. Cristina Draper and her colleagues at the University of Málaga in Spain kindly invited me to give a short course on lattice basis reduction in the summer of 2009. Earl Taft of Rutgers University was very encouraging at an early stage of the project. The staff at CRC Press responded promptly to my questions; in particular, I would like to thank David Grubbs, Bob Stern, Amber Donley, and Shashi Kumar.

No doubt there remain some errors in the book, either typographical or otherwise, for which I take full responsibility; I hope that they are not numerous. I would be very happy to receive comments, suggestions, and corrections from readers, by email at the address below.

Murray R. Bremner (March 2011)

bremner@math.usask.ca

Department of Mathematics and Statistics
University of Saskatchewan
McLean Hall, Room 142
106 Wiggins Road
Saskatoon, Saskatchewan
Canada S7N 5E6

(306) 966-6122

About the Author

Murray R. Bremner received a Bachelor of Science from the University of Saskatchewan in 1981, a Master of Computer Science from Concordia University in Montreal in 1984, and a Doctorate in Mathematics from Yale University in 1989. He spent one year as a Postdoctoral Fellow at the Mathematical Sciences Research Institute in Berkeley, and three years as an Assistant Professor in the Department of Mathematics at the University of Toronto. He returned to the Department of Mathematics and Statistics at the University of Saskatchewan in 1993 and was promoted to Professor in 2002. His research interests focus on the application of computational methods to problems in the theory of linear nonassociative algebras, and he has had more than 50 papers published or accepted by refereed journals in this area.

1

Introduction to Lattices

CONTENTS

In this chapter we begin with a review of elementary linear algebra, and in particular the geometry of Euclidean vector space $\mathbb{R}^n$. The main purpose of this first section is to fix our conventions on notation and terminology. We then introduce the concept of a lattice, the main object of study throughout this book, and prove some basic lemmas about these structures. The last section of the chapter recalls some essential facts from the geometry of numbers, by which is meant the interplay between Euclidean geometry and the theory of numbers. Throughout this book we will use the following standard notation:

$\mathbb{Z}$ the domain of integers

$\mathbb{Q}$ the field of rational numbers

$\mathbb{R}$ the field of real numbers

$\mathbb{C}$ the field of complex numbers

$\mathbb{F}_p$ the field of congruence classes modulo the prime number p

1.1   Euclidean space $\mathbb{R}^n$

We regard n-tuples of elements from a field $\mathbb{F}$ as either column vectors or as row vectors, and denote them by boldface roman letters:

$$\mathbf{x} = \begin{bmatrix} x_1 \\ x_2 \\ \vdots \\ x_n \end{bmatrix} \in \mathbb{F}^n, \qquad \mathbf{x} = \begin{bmatrix} x_1, x_2, \cdots, x_n \end{bmatrix} \in \mathbb{F}^n.$$

We use the column format when we consider an $n \times n$ matrix acting as a linear operator on $\mathbb{R}^n$ by left multiplication on column vectors. However, we will be primarily concerned with operations on a basis of $\mathbb{R}^n$, and for this reason it is convenient to represent the basis vectors $\mathbf{x}_1, \mathbf{x}_2, \ldots, \mathbf{x}_n$ as the rows of an $n \times n$ matrix X. We can then represent operations on the basis as elementary row operations on the matrix. More generally, we can represent a general change of basis as left multiplication of X by an invertible $n \times n$ matrix C.

Definition 1.1. For any field $\mathbb{F}$, and any positive integer n, the **vector space** $\mathbb{F}^n$ consists of all n-tuples of elements from $\mathbb{F}$, with the familiar operations of vector addition and scalar multiplication defined by

$$
\mathbf{x} + \mathbf{y} = \begin{bmatrix} x_1 \\ x_2 \\ \vdots \\ x_n \end{bmatrix} + \begin{bmatrix} y_1 \\ y_2 \\ \vdots \\ y_n \end{bmatrix} = \begin{bmatrix} x_1 + y_1 \\ x_2 + y_2 \\ \vdots \\ x_n + y_n \end{bmatrix}, \quad a\mathbf{x} = a \begin{bmatrix} x_1 \\ x_2 \\ \vdots \\ x_n \end{bmatrix} = \begin{bmatrix} ax_1 \\ ax_2 \\ \vdots \\ ax_n \end{bmatrix},
$$

for any $\mathbf{x}, \mathbf{y} \in \mathbb{F}^n$ and any $a \in \mathbb{F}$.

Throughout this book, we will be primarily concerned with the vector space $\mathbb{R}^n$.

Definition 1.2. The **Euclidean space** $\mathbb{R}^n$ consists of all n-tuples of real numbers. We use dot notation for the **scalar product** of vectors $\mathbf{x}, \mathbf{y} \in \mathbb{R}^n$:

$$
\mathbf{x} \cdot \mathbf{y} = \begin{bmatrix} x_1 \\ x_2 \\ \vdots \\ x_n \end{bmatrix} \cdot \begin{bmatrix} y_1 \\ y_2 \\ \vdots \\ y_n \end{bmatrix} = x_1 y_1 + x_2 y_2 + \cdots + x_n y_n = \sum_{i=1}^{n} x_i y_i.
$$

We use single vertical bars for the **length** (or **norm**) of a vector $\mathbf{x} \in \mathbb{R}^n$:

$$
|\mathbf{x}| = \sqrt{\mathbf{x} \cdot \mathbf{x}} = \sqrt{x_1^2 + x_2^2 + \cdots + x_n^2} = \left(\sum_{i=1}^{n} x_i^2 \right)^{1/2}.
$$

We often use the **square-length** instead of the length of a vector $\mathbf{x} \in \mathbb{R}^n$:

$$
|\mathbf{x}|^2 = x_1^2 + x_2^2 + \cdots + x_n^2 = \sum_{i=1}^{n} x_i^2.
$$

We usually do computations for which the input consists of vectors in $\mathbb{Q}^n$ or $\mathbb{Z}^n$: the components are rational numbers or integers. We want to store the intermediate results as exact rational numbers, in order to avoid the issue of rounding error with floating-point arithmetic, and so we use the square-length (which is rational) instead of the length (which is usually irrational).

Definition 1.3. The **angle** θ between nonzero vectors $\mathbf{x}, \mathbf{y} \in \mathbb{R}^n$ is given by

$$\mathbf{x} \cdot \mathbf{y} = |\mathbf{x}|\,|\mathbf{y}|\cos\theta, \qquad \cos\theta = \frac{\mathbf{x} \cdot \mathbf{y}}{|\mathbf{x}|\,|\mathbf{y}|}, \qquad \theta = \arccos\left(\frac{\mathbf{x} \cdot \mathbf{y}}{|\mathbf{x}|\,|\mathbf{y}|}\right).$$

Lemma 1.4. *Two vectors* $\mathbf{x}, \mathbf{y} \in \mathbb{R}^n$ *are orthogonal if and only if* $\mathbf{x} \cdot \mathbf{y} = 0$.

Proof. The cosine is 0 if and only if the angle is an odd multiple of $\pi/2$. $\quad\square$

The angle formulas of Definition 1.3 are closely related to the following famous inequality.

Lemma 1.5. Cauchy-Schwarz inequality. *For any two vectors* $\mathbf{x}, \mathbf{y} \in \mathbb{R}^n$,

$$|\mathbf{x} \cdot \mathbf{y}| \le |\mathbf{x}|\,|\mathbf{y}|.$$

(On the left side, the vertical bars denote the absolute value of the scalar product; on the right side, they denote the lengths of the vectors.)

Given a vector $\mathbf{x} \in \mathbb{R}^n$ and a nonzero vector $\mathbf{y} \in \mathbb{R}^n$, it is often convenient to express $\mathbf{x}$ as a sum of two vectors, $\mathbf{x} = \mathbf{u} + \mathbf{v}$, where $\mathbf{u}$ is parallel to $\mathbf{y}$ (we write $\mathbf{u} \parallel \mathbf{y}$) and $\mathbf{v}$ is orthogonal to $\mathbf{y}$ (we write $\mathbf{v} \perp \mathbf{y}$). If we write $\mathbf{u} = \lambda\mathbf{y}$ where $\lambda \in \mathbb{R}$, then $\mathbf{v} = \mathbf{x} - \mathbf{u} = \mathbf{x} - \lambda\mathbf{y}$ is orthogonal to $\mathbf{y}$, and hence

$$(\mathbf{x} - \lambda\mathbf{y}) \cdot \mathbf{y} = 0.$$

Using the bilinearity of the scalar product we can solve for the scalar λ:

$$\lambda = \frac{\mathbf{x} \cdot \mathbf{y}}{\mathbf{y} \cdot \mathbf{y}} = \frac{\mathbf{x} \cdot \mathbf{y}}{|\mathbf{y}|^2}.$$

It is important for computational reasons to note that if $\mathbf{x}, \mathbf{y} \in \mathbb{Q}^n$ then $\lambda \in \mathbb{Q}$.

Definition 1.6. Given vectors $\mathbf{x}, \mathbf{y} \in \mathbb{R}^n$ with $\mathbf{y} \ne \mathbf{0}$, we write $\mathbf{u}$ and $\mathbf{v}$ for the **components** (or **projections**) of $\mathbf{x}$ **parallel** and **orthogonal** to $\mathbf{y}$:

$$\mathbf{u} = \left(\frac{\mathbf{x} \cdot \mathbf{y}}{\mathbf{y} \cdot \mathbf{y}}\right)\mathbf{y}, \qquad \mathbf{v} = \mathbf{x} - \left(\frac{\mathbf{x} \cdot \mathbf{y}}{\mathbf{y} \cdot \mathbf{y}}\right)\mathbf{y}.$$

Example 1.7. Consider the triangle in $\mathbb{R}^3$ with these points as its vertices:

$$A = (6, 2, -4), \qquad B = (-8, -6, 6), \qquad C = (1, -3, 9).$$

The two sides of the triangle originating at vertex A are

$$\mathbf{x} = \overrightarrow{AB} = \begin{bmatrix} -14 \\ -8 \\ 10 \end{bmatrix}, \qquad \mathbf{y} = \overrightarrow{AC} = \begin{bmatrix} -5 \\ -5 \\ 13 \end{bmatrix}.$$

The scalar product of these vectors is

$$\mathbf{x} \cdot \mathbf{y} = 240.$$

The lengths of these vectors are

$$|\mathbf{x}| = \sqrt{360}, \qquad |\mathbf{y}| = \sqrt{219}.$$

The cosine of the angle θ at vertex A is

$$\cos\theta = \frac{240}{\sqrt{360}\sqrt{219}} \approx 0.8547476863.$$

Therefore

$$\theta \approx 0.5457317946 \text{ radians} \approx 31.26812857 \text{ degrees.}$$

The projection coefficient for $\mathbf{x}$ in the direction of $\mathbf{y}$ is

$$\lambda = \frac{\mathbf{x} \cdot \mathbf{y}}{\mathbf{y} \cdot \mathbf{y}} = \frac{80}{73}.$$

We obtain the decomposition $\mathbf{x} = \mathbf{u} + \mathbf{v}$ where

$$\mathbf{u} = \frac{80}{73}\begin{bmatrix} -5 \\ -5 \\ 13 \end{bmatrix}, \qquad \mathbf{v} = \frac{2}{73}\begin{bmatrix} -311 \\ -92 \\ -155 \end{bmatrix}.$$

We have $\mathbf{u} \parallel \mathbf{y}$ and $\mathbf{v} \perp \mathbf{y}$, and hence $\mathbf{u} \cdot \mathbf{v} = 0$.

Definition 1.8. The vectors $\mathbf{x}_1, \mathbf{x}_2, \ldots, \mathbf{x}_k \in \mathbb{R}^n$ are **linearly dependent** if one of the vectors is a linear combination of the other $k-1$ vectors; equivalently, if there is a non-trivial solution (not all the coefficients are zero) of the equation

$$a_1\mathbf{x}_1 + a_2\mathbf{x}_2 + \cdots + a_k\mathbf{x}_k = \mathbf{0} \qquad (a_1, a_2, \ldots, a_k \in \mathbb{R}).$$

The vectors $\mathbf{x}_1, \mathbf{x}_2, \ldots, \mathbf{x}_k$ are **linearly independent** if this equation has only the trivial solution $a_i = 0$ for $i = 1, 2, \ldots, k$. This implies that $k \leq n$.

The vectors $\mathbf{x}_1, \mathbf{x}_2, \ldots, \mathbf{x}_k \in \mathbf{R}^n$ **span** $\mathbb{R}^n$ if every vector $\mathbf{y} \in \mathbb{R}^n$ is a linear combination of the vectors; equivalently, for every $\mathbf{y} \in \mathbb{R}^n$, the equation

$$a_1\mathbf{x}_1 + a_2\mathbf{x}_2 + \cdots + a_k\mathbf{x}_k = \mathbf{y},$$

has a solution $a_1, a_2, \ldots, a_k \in \mathbb{R}$. This implies that $k \geq n$.

The vectors $\mathbf{x}_1, \mathbf{x}_2, \ldots, \mathbf{x}_k \in \mathbf{R}^n$ form a **basis** of $\mathbb{R}^n$ if they are linearly independent and they span $\mathbb{R}^n$. This implies that $k = n$.

The **standard basis vectors** in $\mathbb{R}^n$ will be denoted $\mathbf{e}_1, \mathbf{e}_2, \ldots, \mathbf{e}_n$; by definition, $\mathbf{e}_i$ has 1 as its i-th component and 0 as its other components.

There are many excellent modern textbooks on elementary linear algebra; we mention in particular those by Anton [11] and Nicholson [112]. At a more advanced level, two standard classical references are Hoffman and Kunze [64] and Jacobson [68]. Computational methods are presented in Golub and van Loan [49] and Trefethen and Bau [137].

1.2   Lattices in $\mathbb{R}^n$

We now introduce the main objects of study in the remainder of this book.

Definition 1.9. Let $n \geq 1$ and let $\mathbf{x}_1$, $\mathbf{x}_2$, ..., $\mathbf{x}_n$ be a basis of $\mathbb{R}^n$. The **lattice** with **dimension** n and **basis** $\mathbf{x}_1$, $\mathbf{x}_2$, ..., $\mathbf{x}_n$ is the set L of all linear combinations of the basis vectors with integral coefficients:

$$L = \mathbb{Z}\mathbf{x}_1 + \mathbb{Z}\mathbf{x}_2 + \cdots + \mathbb{Z}\mathbf{x}_n = \Big\{ \sum_{i=1}^{n} a_i \mathbf{x}_i \mid a_1, a_2, \ldots, a_n \in \mathbb{Z} \Big\}.$$

The basis vectors $\mathbf{x}_1$, $\mathbf{x}_2$, ..., $\mathbf{x}_n$ are said to **generate** or **span** the lattice. For $i = 1, 2, \ldots, n$ we write $\mathbf{x}_i = (x_{i1}, \ldots, x_{in})$ and form the $n \times n$ matrix $X = (x_{ij})$. The **determinant** of the lattice L with basis $\mathbf{x}_1$, $\mathbf{x}_2$, ..., $\mathbf{x}_n$ is

$$\det(L) = |\det(X)|.$$

Note that in this definition we regard the basis vectors as row vectors. We do this so that operations on the basis vectors can be expressed in terms of elementary row operations on the matrix X; equivalently, left multiplication of the matrix X by an integer matrix C with determinant ± 1.

We will prove shortly (Corollary 1.11) that the determinant of the lattice L does not depend on which basis we use. In fact, $\det(L)$ has a natural geometric interpretation: it is the n-dimensional volume of the parallelipiped in $\mathbb{R}^n$ whose edges are the basis vectors $\mathbf{x}_1$, $\mathbf{x}_2$, ..., $\mathbf{x}_n$.

In the trivial case $n = 1$, the lattice L generated by the nonzero real number $\mathbf{x}$ consists of all integral multiples of $\mathbf{x}$. The lattice $L = \mathbb{Z}\mathbf{x}$ has only two bases, namely $\mathbf{x}$ and $-\mathbf{x}$.

If $n \geq 2$, then every lattice has infinitely many different bases. Let $L \subset \mathbb{R}^n$ be the lattice with basis $\mathbf{x}_1, \mathbf{x}_2, \ldots, \mathbf{x}_n$. Let $C = (c_{ij})$ be any $n \times n$ matrix with entries in $\mathbb{Z}$ and $\det(C) = \pm 1$; then C^{-1} also has entries in $\mathbb{Z}$ (see Exercise 1.7). Define vectors $\mathbf{y}_1, \mathbf{y}_2, \ldots, \mathbf{y}_n$ by

$$\mathbf{y}_i = \sum_{j=1}^{n} c_{ij} \mathbf{x}_j \qquad (i = 1, 2, \ldots, n),$$

and let Y be the $n \times n$ matrix with $\mathbf{y}_i$ in row i. We have the matrix equations

$$Y = CX, \qquad X = C^{-1}Y.$$

It follows that any integral linear combination of $\mathbf{x}_1$, $\mathbf{x}_2$, ..., $\mathbf{x}_n$ is also an integral linear combination of $\mathbf{y}_1$, $\mathbf{y}_2$, ..., $\mathbf{y}_n$, and conversely. Hence $\mathbf{y}_1$, $\mathbf{y}_2$, ..., $\mathbf{y}_n$ is another basis for the same lattice L. In fact any two bases for the same lattice are related in this way, as the next lemma shows.

Lemma 1.10. *Let* $\mathbf{x}_1$, $\mathbf{x}_2$, ..., $\mathbf{x}_n$ *and* $\mathbf{y}_1$, $\mathbf{y}_2$, ..., $\mathbf{y}_n$, *be two bases for the same lattice* $L \subset \mathbb{R}^n$. *Let* X *(respectively* Y*) be the* $n \times n$ *matrix with* $\mathbf{x}_i$ *(respectively* $\mathbf{y}_i$*) in row* i *for* $i = 1, 2, \ldots, n$. *Then* $Y = CX$ *for some* $n \times n$ *matrix* C *with integer entries and determinant* ± 1.

Proof. Every $\mathbf{y}_i$ belongs to the lattice with basis $\mathbf{x}_1$, $\mathbf{x}_2$, ..., $\mathbf{x}_n$, and every $\mathbf{x}_i$ belongs to the lattice with basis $\mathbf{y}_1$, $\mathbf{y}_2$, ..., $\mathbf{y}_n$. It follows that

$$\mathbf{x}_i = \sum_{j=1}^{n} b_{ij} \mathbf{y}_j, \qquad \mathbf{y}_i = \sum_{j=1}^{n} c_{ij} \mathbf{x}_j \qquad (i = 1, 2, \ldots, n),$$

where $B = (b_{ij})$ and $C = (c_{ij})$ are $n \times n$ matrices with integer entries. Writing these two equations in matrix form gives $X = BY$ and $Y = CX$, and hence $X = BCX$ and $Y = CBY$. Since both $\mathbf{x}_1$, $\mathbf{x}_2$, ..., $\mathbf{x}_n$ and $\mathbf{y}_1$, $\mathbf{y}_2$, ..., $\mathbf{y}_n$ are bases of $\mathbb{R}^n$, the corresponding matrices X and Y are invertible, and can be canceled from the equations. Therefore $BC = I$ and $CB = I$, and so $\det(B)\det(C) = 1$. Since B and C have integer entries, it follows that either $\det(B) = \det(C) = 1$ or $\det(B) = \det(C) = -1$. □

Corollary 1.11. *The determinant of a lattice does not depend on the basis.*

Proof. Suppose the lattice $L \subset \mathbb{R}^n$ has two bases $\mathbf{x}_1$, $\mathbf{x}_2$, ..., $\mathbf{x}_n$ and $\mathbf{y}_1$, $\mathbf{y}_2$, ..., $\mathbf{y}_n$. Using the notation in the proof of Lemma 1.10, we have

$$|\det(Y)| = |\det(CX)| = |\det(C)\det(X)| = |\pm \det(X)| = |\det(X)|.$$

Since the two bases are arbitrary, this completes the proof. □

Definition 1.12. An $n \times n$ matrix with integer entries and determinant ± 1 will be called **unimodular**.

Definition 1.13. A **unimodular row operation** on a matrix is one of the following elementary row operations:

- multiply any row by -1;

- interchange any two rows;

- add an integral multiple of any row to any other row.

To generate examples of $n \times n$ unimodular matrices, we start with the identity matrix I_n, and then apply any finite sequence of unimodular row operations. The result will be an $n \times n$ unimodular matrix, and in fact any such matrix can be obtained in this way.

If we apply unimodular row operations to the matrix X whose rows contain a basis of the lattice L, then we obtain another basis of the same lattice.

Example 1.14. Start with the 2×2 identity matrix, and apply this sequence of unimodular row operations: add 4 times row 2 to row 1, add 9 times row 1 to row 2, change the sign of row 1, add -4 times row 2 to row 1, change the sign of row 1, change the sign of row 2. We obtain this 2×2 unimodular matrix:

$$C = \begin{bmatrix} 37 & 152 \\ -9 & -37 \end{bmatrix}, \qquad \det(C) = -1.$$

Let L be the lattice in $\mathbb{R}^2$ spanned by the rows of this matrix:

$$X = \begin{bmatrix} 7 & 9 \\ 6 & -5 \end{bmatrix}, \qquad \det(X) = -89.$$

Applying the same sequence of row operations to X gives this matrix Y:

$$Y = CX = \begin{bmatrix} 1171 & -427 \\ -285 & 104 \end{bmatrix}.$$

Writing the the basis vectors as column vectors gives

$$\mathbf{x}_1, \mathbf{x}_2 = \begin{bmatrix} 7 \\ 9 \end{bmatrix}, \begin{bmatrix} 6 \\ -5 \end{bmatrix} \qquad \text{and} \qquad \mathbf{y}_1, \mathbf{y}_2 = \begin{bmatrix} 1171 \\ -427 \end{bmatrix}, \begin{bmatrix} -285 \\ 104 \end{bmatrix}.$$

It is far from obvious that these two bases generate the same lattice in $\mathbb{R}^2$.

We can perform any number of further row operations; a pseudorandom sequence of 100 operations provides this third basis for the same lattice:

$$\mathbf{z}_1, \mathbf{z}_2 = \begin{bmatrix} 91202814184 \\ -26536463447 \end{bmatrix}, \begin{bmatrix} 10682859399 \\ -3108295621 \end{bmatrix}.$$

We can clearly continue this process as long as we want and find bases for the same lattice consisting of arbitrarily long vectors.

The last example shows how easy it is to start with a basis for a lattice consisting of short vectors, and then produce other bases for the same lattice consisting of much longer vectors. Of course, it is much more interesting and important to do exactly the opposite: *Given a basis for a lattice, which in general consists of long vectors, we want to find another "reduced" basis for the same lattice, that is, a basis consisting of short vectors.* This is the problem of lattice basis reduction, the fundamental problem that we will be studying throughout this book.

We now generalize the concept of lattice basis and lattice determinant to any set of m linearly independent vectors in $\mathbb{R}^n$ ($m \le n$).

Definition 1.15. Let $n \ge 1$ and let $\mathbf{x}_1, \mathbf{x}_2, \ldots, \mathbf{x}_m$ ($m \le n$) be a set of m linearly independent vectors in $\mathbb{R}^n$. The m-dimensional **lattice** spanned by these vectors in n-dimensional Euclidean space is defined to be

$$L = \mathbb{Z}\mathbf{x}_1 + \mathbb{Z}\mathbf{x}_2 + \cdots + \mathbb{Z}\mathbf{x}_m = \Big\{ \sum_{i=1}^{m} a_i \mathbf{x}_i \mid a_1, a_2, \ldots, a_m \in \mathbb{Z} \Big\}.$$

For $i = 1, \ldots, m$ we write $\mathbf{x}_i = (x_{i1}, \ldots, x_{in})$ and form the $m \times n$ matrix $X = (x_{ij})$. The **Gram matrix** $\Delta(L)$ of the lattice L is the $m \times m$ matrix in which the (i, j) entry is the scalar product of the i-th and j-th basis vectors:

$$\Delta(L) = (\mathbf{x}_i \cdot \mathbf{x}_j) = XX^t.$$

The determinant of the Gram matrix is always positive (see Exercise 1.11), and we define the **determinant** of the lattice L to be its square root:

$$\det(L) = \sqrt{\det(XX^t)}.$$

If $m = n$ then X is a square matrix, and so

$$\big(\det(L)\big)^2 = \det(XX^t) = \det(X)\det(X^t) = \big(\det(X)\big)^2,$$

which agrees with the previous definition of lattice determinant.

As before, it can easily be shown that the determinant of a lattice does not depend on the choice of basis (see Exercise 1.12). The geometric interpretation is also the same: the determinant is the m-dimensional volume of the parallelipiped in $\mathbb{R}^n$ whose edges are the lattice basis vectors.

Example 1.16. Consider the 3-dimensional lattice L in 5-dimensional Euclidean space spanned by the rows of this matrix:

$$X = \begin{bmatrix} -7 & -7 & 4 & -8 & -8 \\ 1 & 6 & -5 & 8 & -1 \\ -1 & 1 & 4 & -7 & 8 \end{bmatrix}$$

We compute the Gram matrix:

$$\Delta(L) = XX^t = \begin{bmatrix} -7 & -7 & 4 & -8 & -8 \\ 1 & 6 & -5 & 8 & -1 \\ -1 & 1 & 4 & -7 & 8 \end{bmatrix} \begin{bmatrix} -7 & 1 & -1 \\ -7 & 6 & 1 \\ 4 & -5 & 4 \\ -8 & 8 & -7 \\ -8 & -1 & 8 \end{bmatrix}$$

$$= \begin{bmatrix} 242 & -125 & 8 \\ -125 & 127 & -79 \\ 8 & -79 & 131 \end{bmatrix}.$$

The Gram matrix has determinant 618829, and so $\det(L) = \sqrt{618829}$.

In the rest of this section, we consider the problem of extending a linearly independent set of lattice vectors to a basis for the lattice. Our exposition follows Cassels [22], pages 11–14, but we express the results in matrix form as much as possible.

Definition 1.17. Let $L \subset \mathbb{R}^n$ be the lattice with basis $\mathbf{x}_1, \mathbf{x}_2, \ldots, \mathbf{x}_n$. Suppose that $\mathbf{y}_1, \mathbf{y}_2, \ldots, \mathbf{y}_n \in L$ are linearly independent, and let $M \subset \mathbb{R}^n$ be the lattice generated by $\mathbf{y}_1, \mathbf{y}_2, \ldots, \mathbf{y}_n$. We call M a **sublattice** of L and write $M \subseteq L$.

Each basis vector $\mathbf{y}_i$ for the sublattice M belongs to the lattice L, and so

$$\mathbf{y}_i = \sum_{j=1}^{n} c_{ij}\mathbf{x}_j \qquad (i = 1, 2, \ldots, n),$$

where $c_{ij} \in \mathbb{Z}$ for all i, j. As a matrix equation, this says that

$$Y = CX,$$

where $C = (c_{ij})$ is the non-singular $n \times n$ matrix of integer coefficients, and X (respectively Y) is the $n \times n$ matrix containing $\mathbf{x}_i$ (respectively $\mathbf{y}_i$) in row i. Taking the determinant on both sides of this equation gives

$$\det(Y) = \det(C)\det(X), \qquad \det(C) = \frac{\det(Y)}{\det(X)}.$$

Definition 1.18. The **index** ρ of a sublattice M in a lattice L is defined by

$$\rho = |\det(C)| = \frac{|\det(Y)|}{|\det(X)|} = \frac{\det(M)}{\det(L)}.$$

The index is an integer, since the determinant of the sublattice M is an integral multiple of the determinant of the lattice L. (The basis vectors for M span a larger parallelipiped than the basis vectors for L.) It is clear from the above equations that the index depends only on L and M, not on the choice of bases.

Definition 1.19. For any $n \times n$ matrix C, the (i, j) **minor** is the determinant $\det(C_{ij})$ of the $(n-1) \times (n-1)$ matrix C_{ij} obtained by deleting row i and column j, and the (i, j) **cofactor** is $(-1)^{i+j} \det(C_{ij})$. The **adjoint matrix** is the transpose of the matrix of cofactors:

$$\big(\mathrm{adj}(C)\big)_{ij} = (-1)^{i+j}\det(C_{ji}).$$

Lemma 1.20. *The inverse of any non-singular matrix C can be expressed in terms of its adjoint matrix and its determinant:*

$$C^{-1} = \frac{1}{\det(C)}\,\mathrm{adj}(C).$$

Proof. See any textbook on elementary linear algebra. □

Returning to the above discussion of the sublattice M (with matrix Y) of the lattice L (with matrix X), we see that the equation $Y = CX$ implies

$$X = C^{-1}Y = \frac{1}{\det(C)}\,\mathrm{adj}(C)\,Y,$$

and hence

$$\rho\,X = |\det(C)|\,X = \pm\,\mathrm{adj}(C)\,Y.$$

Since the entries of C are integers, so are the entries of $\mathrm{adj}(C)$, and hence every row of the matrix ρX is an integer linear combination of the rows of the matrix Y. We conclude that the lattice ρL, consisting of all multiples by the integer ρ of the vectors in L, is a sublattice of the lattice M.

Lemma 1.21. *If L is a lattice and M is a sublattice of index ρ then*

$$\rho L \subseteq M \subseteq L.$$

We now prove a theorem relating the bases of a lattice L and the bases of a sublattice M. As a corollary we will obtain a necessary and sufficient condition for extending a set of linearly independent lattice vectors to a basis for the lattice.

Theorem 1.22. (Cassels [22], Theorem I, page 11) *Let L be a lattice in $\mathbb{R}^n$ and let M be a sublattice of L. If $\mathbf{x}_1, \mathbf{x}_2, \ldots, \mathbf{x}_n$ is a basis of L, then there exists a basis $\mathbf{y}_1, \mathbf{y}_2, \ldots, \mathbf{y}_n$ of M such that $Y = CX$ where C is a lower-triangular $n \times n$ integer matrix with nonzero entries on the diagonal. That is, we have*

$$\left. \begin{array}{l} \mathbf{y}_1 = c_{11}\mathbf{x}_1 \\ \mathbf{y}_2 = c_{21}\mathbf{x}_1 + c_{22}\mathbf{x}_2 \\ \quad \vdots \\ \mathbf{y}_n = c_{n1}\mathbf{x}_1 + c_{n2}\mathbf{x}_2 + \cdots + c_{nn}\mathbf{x}_n \end{array} \right\} \text{ where } c_{ij} \in \mathbb{Z},\ c_{ii} \neq 0 \text{ for all } i, j.$$

Conversely, if $\mathbf{y}_1, \mathbf{y}_2, \ldots, \mathbf{y}_n$ is any basis of M then there exists a basis $\mathbf{x}_1, \mathbf{x}_2, \ldots, \mathbf{x}_n$ of L satisfying the same conditions.

Proof. Lemma 1.21 shows that $\rho L \subseteq M$, and hence $\rho \mathbf{x}_i \in M$ for all i. It follows that there exist vectors $\mathbf{y}_i \in M$ (not necessarily forming a basis for M) and integers c_{ij} satisfying the conditions of the theorem (in fact, we can take $c_{ii} = \rho$ for all i, and $c_{ij} = 0$ for all $i \neq j$). Thus the set of all n-tuples of vectors $\mathbf{y}_i \in M$ satisfying the conditions of the theorem is non-empty, and so for each i we may take $\mathbf{y}_i \in M$ to be the vector for which the coefficient c_{ii} is positive and as small as possible. We will show that the resulting vectors $\mathbf{y}_1, \mathbf{y}_2, \ldots, \mathbf{y}_n$ form a basis of the sublattice M. Suppose to the contrary that there is a vector $\mathbf{z} \in M$ which is not an integral linear combination of $\mathbf{y}_1, \mathbf{y}_2, \ldots, \mathbf{y}_n$. Writing $\mathbf{z}$ as an integral linear combination of $\mathbf{x}_1, \mathbf{x}_2, \ldots, \mathbf{x}_n$ gives

$$\mathbf{z} = t_1\mathbf{x}_1 + t_2\mathbf{x}_2 + \cdots + t_k\mathbf{x}_k, \quad \text{where } k \leq n \text{ and } t_k \neq 0.$$

We choose $\mathbf{z}$ so that the index k is as small as possible. By assumption $c_{kk} \neq 0$, and so we may perform integer division with remainder of t_k by c_{kk}, obtaining

$$t_k = q c_{kk} + r, \qquad 0 \leq r < c_{kk}.$$

We now consider the vector

$$\begin{aligned} \mathbf{z} - q\mathbf{y}_k &= (t_1\mathbf{x}_1 + t_2\mathbf{x}_2 + \cdots + t_k\mathbf{x}_k) - q(c_{k1}\mathbf{x}_1 + c_{k2}\mathbf{x}_2 + \cdots + c_{kk}\mathbf{x}_k) \\ &= (t_1 - q c_{k1})\mathbf{x}_1 + (t_2 - q c_{k2})\mathbf{x}_2 + \cdots + (t_k - q c_{kk})\mathbf{x}_k. \end{aligned}$$

Since $\mathbf{z}$ and $\mathbf{y}_k$ are in M and q is an integer, we have $\mathbf{z} - q\mathbf{y}_k \in M$. Since $\mathbf{z}$ is not an integral linear combination of $\mathbf{y}_1, \mathbf{y}_2, \ldots, \mathbf{y}_n$ neither is $\mathbf{z} - q\mathbf{y}_k$. But the index k was chosen as small as possible, and so we must have $t_k - qc_{kk} \neq 0$. This implies that the vector $\mathbf{z} - q\mathbf{y}_k \in M$ is an integral linear combination of $\mathbf{x}_1, \mathbf{x}_2, \ldots, \mathbf{x}_k$ whose coefficient of $\mathbf{x}_k$, namely $t_k - qc_{kk} = r$, is nonzero and strictly less than c_{kk}. But this contradicts the choice of $\mathbf{y}_k$. It follows that such a vector $\mathbf{z}$ does not exist, and hence every vector in M must be an integral linear combination of $\mathbf{y}_1, \mathbf{y}_2, \ldots, \mathbf{y}_n$.

For the converse, let $\mathbf{y}_1, \mathbf{y}_2, \ldots, \mathbf{y}_n$ be a basis of M. By Lemma 1.21 we know that $\rho L \subseteq M$, and so we may apply the first part of the proof to the sublattice ρL of the lattice M. We obtain a basis $\rho \mathbf{x}_1, \rho \mathbf{x}_2, \ldots, \rho \mathbf{x}_n$ of ρL such that

$$
\left.
\begin{aligned}
\rho \mathbf{x}_1 &= d_{11}\mathbf{y}_1 \\
\rho \mathbf{x}_2 &= d_{21}\mathbf{y}_1 + d_{22}\mathbf{y}_2 \\
&\vdots \\
\rho \mathbf{x}_n &= d_{n1}\mathbf{y}_1 + d_{n2}\mathbf{y}_2 + \cdots + d_{nn}\mathbf{y}_n
\end{aligned}
\right\} \quad \text{where } d_{ij} \in \mathbb{Z},\ d_{ii} \neq 0 \text{ for all } i, j.
$$

We can write these equations in matrix form as $\rho X = DY$ where $D = (d_{ij})$ is a lower-triangular $n \times n$ integer matrix with nonzero entries on the diagonal. Solving for Y we obtain $Y = \rho D^{-1}X$. It is clear that $\mathbf{x}_1, \mathbf{x}_2, \ldots, \mathbf{x}_n$ form a basis of L, and since $\mathbf{y}_1, \mathbf{y}_2, \ldots, \mathbf{y}_n \in M \subseteq L$, we see that the entries of the matrix ρD^{-1} must be integers, by the uniqueness of the representation of each lattice vector as an (integral) linear combination of basis vectors. $\square$

We note an especially interesting and attractive feature of the last proof: it clearly illustrates the principle that reduction of lattice bases can be naturally regarded as a generalization of integer division with remainder.

We now consider a sequence of corollaries of Theorem 1.22. Recall that E_{ij} is the $n \times n$ matrix in which the (i, j) entry is 1 and the other entries are 0.

Corollary 1.23. *In the first part of Theorem 1.22 we may assume that*

$$c_{ii} > 0 \ (1 \leq i \leq n) \quad \text{and} \quad 0 \leq c_{ij} < c_{jj} \ (1 \leq j < i \leq n).$$

In the second part of Theorem 1.22 we may assume that

$$c_{ii} > 0 \ (1 \leq i \leq n) \quad \text{and} \quad 0 \leq c_{ij} < c_{ii} \ (1 \leq j < i \leq n).$$

Proof. Consider the matrix form of the equations, namely $Y = CX$. If $c_{ii} < 0$ for some i then we left-multiply both sides of the matrix equation by $-E_{ii}$; this corresponds to the unimodular row operation "multiply row i by -1". If $c_{ij} < 0$ or $c_{ij} \geq c_{jj}$ for some i, j then we do integer division with remainder to write $c_{ij} = qc_{jj} + r$ with $0 \leq r < c_{jj}$ (we are now assuming that $c_{jj} > 0$) and then left-multiply both sides of the matrix equation by $I_n - qE_{ij}$; this corresponds to the unimodular row operation "subtract q times row j from row i". We can express the result of all these operations by the matrix equation $UY = UCX$ where U is a unimodular matrix. In fact it is clear that U

is lower-triangular, and hence so is UC. We can therefore replace the basis $\mathbf{y}_1, \mathbf{y}_2, \ldots, \mathbf{y}_n$ of M, consisting of the rows of the matrix Y, by the new basis consisting of the rows of the matrix UY. The second part of the proof is left to the reader (see Exercise 1.16). □

Corollary 1.24. *Let L be an n-dimensional lattice in $\mathbb{R}^n$, and let $\mathbf{y}_1, \mathbf{y}_2, \ldots, \mathbf{y}_m$ ($m \leq n$) be linearly independent vectors in L. There is a basis $\mathbf{x}_1, \mathbf{x}_1, \ldots, \mathbf{x}_n$ of L satisfying the equations*

$$\left.\begin{array}{l} \mathbf{y}_1 = c_{11}\mathbf{x}_1 \\ \mathbf{y}_2 = c_{21}\mathbf{x}_1 + c_{22}\mathbf{x}_2 \\ \quad\vdots \\ \mathbf{y}_m = c_{m1}\mathbf{x}_1 + c_{m2}\mathbf{x}_2 + \cdots + c_{mm}\mathbf{x}_m \end{array}\right\} \;\; where \;\; \left\{\begin{array}{l} c_{ij} \in \mathbb{Z} \; for \; all \; i,j \\ c_{ii} > 0 \; for \; all \; i \\ 0 \leq c_{ij} < c_{ii} \; for \; all \; i,j \end{array}\right.$$

Proof. We can find another $n-m$ vectors $\mathbf{y}_{m+1}, \ldots, \mathbf{y}_n$ in L such that the vectors $\mathbf{y}_1, \mathbf{y}_2, \ldots, \mathbf{y}_n$ are linearly independent. We now apply the second part of Corollary 1.23 to the lattice M with basis $\mathbf{y}_1, \mathbf{y}_2, \ldots, \mathbf{y}_n$. □

Corollary 1.25. *Let L be an n-dimensional lattice in $\mathbb{R}^n$ and let $\mathbf{y}_1, \mathbf{y}_2, \ldots, \mathbf{y}_m$ ($m < n$) be linearly independent vectors in L. These conditions are equivalent:*

(1) There exist another $n-m$ vectors $\mathbf{y}_{m+1}, \ldots, \mathbf{y}_n$ in L such that the vectors $\mathbf{y}_1, \mathbf{y}_2, \ldots, \mathbf{y}_n$ form a basis of L.

(2) Any vector $\mathbf{z} \in L$ which is a (real) linear combination of $\mathbf{y}_1, \mathbf{y}_2, \ldots, \mathbf{y}_m$ is in fact an integral linear combination.

Proof. The implication (1) $\Longrightarrow$ (2) is clear. To prove (2) $\Longrightarrow$ (1), assume that $\mathbf{y}_1, \mathbf{y}_2, \ldots, \mathbf{y}_m$ satisfy condition (2). Since $\mathbf{y}_1, \mathbf{y}_2, \ldots, \mathbf{y}_m$ are linearly independent vectors in L, we may apply Corollary 1.24 to obtain a basis $\mathbf{x}_1, \mathbf{x}_1, \ldots, \mathbf{x}_n$ of L satisfying the given equations. Considering only the first m basis vectors $\mathbf{x}_1, \mathbf{x}_1, \ldots, \mathbf{x}_m$ we have the matrix equation $Y = CX$ where now the matrix C has size $m \times m$. Hence $X = C^{-1}Y$, and now condition (2) implies that the entries of C^{-1} are integers. But C is lower-triangular with diagonal entries $c_{11}, c_{22}, \ldots, c_{mm}$, and hence C^{-1} is lower-triangular with diagonal entries $c_{11}^{-1}, c_{22}^{-1}, \ldots, c_{mm}^{-1}$. Thus for all $i = 1, 2, \ldots, m$ we see that c_{ii} is an integer for which c_{ii}^{-1} is also an integer, and hence $c_{ii} = \pm 1$. Corollary 1.23 now implies that $c_{ii} = 1$ for $1 \leq i \leq m$ and $c_{ij} = 0$ for $1 \leq j < i \leq m$. Thus $C = I_m$, and so $\mathbf{y}_i = \mathbf{x}_i$ for $i = 1, 2, \ldots, m$. To complete the proof we simply set $\mathbf{y}_i = \mathbf{x}_i$ for $i = m+1, \ldots, n$. □

Corollary 1.26. *Let L be an n-dimensional lattice in $\mathbb{R}^n$ with basis $\mathbf{x}_1, \mathbf{x}_2, \ldots, \mathbf{x}_n$. Consider an arbitrary vector $\mathbf{z} \in L$ and write*

$$\mathbf{z} = a_1\mathbf{x}_1 + a_2\mathbf{x}_2 + \cdots + a_n\mathbf{x}_n \qquad (a_1, a_2, \ldots, a_n \in \mathbb{Z}).$$

These conditions are equivalent for any integer $m = 1, 2, \ldots, n$:

(1) There are vectors $\mathbf{y}_{m+1}, \ldots, \mathbf{y}_n \in L$ *such that the following* n *vectors form a basis of* L:

$$\mathbf{x}_1, \mathbf{x}_2, \ldots, \mathbf{x}_{m-1}, \mathbf{z}, \mathbf{y}_{m+1}, \ldots, \mathbf{y}_n.$$

(2) The greatest common divisor of the integers $a_{m+1}, \ldots, a_n$ *is 1.*

Proof. This follows directly from Corollary 1.25 (see Exercise 1.17). □

Up to this point we have been considering "full-rank" sublattices; the dimension of the sublattice M is equal to the dimension of the lattice L. For the next definition and theorem we consider a more general situation.

Definition 1.27. Let L be an n-dimensional lattice in $\mathbb{R}^n$, and let M be an m-dimensional sublattice for some $m < n$: that is, M is the set of all integral linear combinations of m linearly independent vectors in L. We say that M is a **primitive** sublattice if $M = L \cap V$ where V is a subspace of $\mathbb{R}^n$.

Theorem 1.28. (Nguyen [105], Lemma 4, page 28) *The* m-*dimensional sublattice* M *of the* n-*dimensional lattice* $L \subset \mathbb{R}^n$ *is primitive if and only if every basis of* M *can be extended to a basis of* L; *that is, if the vectors* $\mathbf{y}_1, \mathbf{y}_2, \ldots, \mathbf{y}_m$ *form a basis of* M, *then there are vectors* $\mathbf{x}_{m+1}, \ldots, \mathbf{x}_n$ *in* L *such that the vectors* $\mathbf{y}_1, \mathbf{y}_2, \ldots, \mathbf{y}_m, \mathbf{x}_{m+1}, \ldots, \mathbf{x}_n$ *form a basis of* L.

Proof. Exercise 1.18. □

1.3 Geometry of numbers

In this final section, we recall some definitions that will be used in the rest of the book, and state some results without proof.

Definition 1.29. Let L be an m-dimensional lattice in n-dimensional Euclidean space $\mathbb{R}^n$. The **first minimum** of the lattice, denoted $\Lambda_1(L)$, is the length of a shortest nonzero vector $\mathbf{x}_1 \in L$. The **second minimum** of the lattice, denoted $\Lambda_2(L)$, is the smallest real number r such that there exist two linearly independent vectors $\mathbf{x}_1, \mathbf{x}_2 \in L$ such that $|\mathbf{x}_1|, |\mathbf{x}_2| \leq r$. In general, for $i = 1, 2, \ldots, m$, the i-th **successive minimum** of the lattice, denoted $\Lambda_i(L)$, is the smallest real number r such that there exist i linearly independent vectors $\mathbf{x}_1, \mathbf{x}_2, \ldots, \mathbf{x}_i \in L$ such that $|\mathbf{x}_1|, |\mathbf{x}_2|, \ldots, |\mathbf{x}_i| \leq r$. This quantity can be expressed more concisely by the equation

$$\Lambda_i(L) = \min_{\mathbf{x}_1, \ldots, \mathbf{x}_i \in L} \max \left(|\mathbf{x}_1|, \ldots, |\mathbf{x}_i| \right),$$

where the minimum is over all sets of i linearly independent vectors in L.

It is easy to see that the successive minima are weakly increasing:

$$\Lambda_1(L) \le \Lambda_2(L) \le \cdots \le \Lambda_m(L).$$

The best possible basis for an m-dimensional lattice L consists of vectors

$$\mathbf{x}_1, \mathbf{x}_2, \ldots, \mathbf{x}_m \in L \quad \text{such that} \quad |\mathbf{x}_i| = \Lambda_i(L) \quad \text{for} \quad i = 1, 2, \ldots, m.$$

However, such a basis is in general very hard to compute. It is interesting to note that a set of m vectors $\mathbf{x}_1, \mathbf{x}_2, \ldots, \mathbf{x}_m \in L$ which satisfy the conditions $|\mathbf{x}_i| = \Lambda_i(L)$ for $i = 1, 2, \ldots, m$ do not necessarily form a basis of L; for an example with $m = 4$ see Nguyen [105], page 32.

In order to understand better the size of the first minimum $\Lambda_1(L)$, we scale it by the determinant of the lattice. More precisely, we consider

$$\frac{\Lambda_1(L)}{\sqrt[m]{\det(L)}}.$$

Definition 1.30. Hermite's lattice constant, denoted γ_m, is the supremum of the following quantities as L ranges over all m-dimensional lattices:

$$\frac{\Lambda_1(L)^2}{\big(\det(L)\big)^{2/m}}.$$

The quantities γ_m are very difficult to compute, and are known only for $1 \le m \le 8$ and $m = 24$. The following table is from Nguyen [105], page 33:

m	1	2	3	4	5	6	7	8	$\cdots$	24
γ_m	1	$\left(\frac{4}{3}\right)^{1/2}$	$2^{1/3}$	$2^{1/2}$	$8^{1/5}$	$\left(\frac{64}{3}\right)^{1/6}$	$64^{1/7}$	2	$\cdots$	4

Definition 1.31. Let S be an arbitrary subset of n-dimensional Euclidean space $\mathbb{R}^n$. We say that S is **symmetric about the origin** if $\mathbf{x} \in S$ implies $-\mathbf{x} \in S$. We say that S is **convex** if $\mathbf{x}, \mathbf{y} \in S$ implies $\alpha\mathbf{x} + (1-\alpha)\mathbf{y} \in S$ for $0 \le \alpha \le 1$; that is, S contains the line segment joining $\mathbf{x}$ and $\mathbf{y}$.

Theorem 1.32. Minkowski's convex body theorem. *Let L be an n-dimensional lattice in n-dimensional Euclidean space $\mathbb{R}^n$ with determinant $\det(L)$. Let S be a subset of $\mathbb{R}^n$ which is convex and symmetric about the origin; let $\mathrm{vol}(S)$ denote the volume of S. If $\mathrm{vol}(S) > 2^n \det(L)$ then S contains a nonzero vector $\mathbf{x} \in L$.*

Proof. Cassels [22], Theorem II, page 71.   □

1.4 Projects

Project 1.1. Write a computer program that takes as input three points A, B, C in $\mathbb{R}^n$, verifies that the points are the vertices of a triangle (that is, the points are not collinear), and then calculates:

(i) the lengths of the sides of the triangle,

(ii) the angles at the vertices of the triangle,

(iii) for each ordered pair of sides, the components of the first side parallel and orthogonal to the second side.

Test your program on 10 pseudorandom choices of the points A, B, C having coordinates with 1, 2 or 3 digits in the Euclidean space $\mathbb{R}^n$ for $n = 2, 3, \dots, 10$.

Project 1.2. Write a computer program that takes as input an operation count k, a range parameter r, and a basis $\mathbf{x}_1$, $\mathbf{x}_2$, ..., $\mathbf{x}_n$ of $\mathbb{R}^n$ spanning a lattice L, and then applies k unimodular row operations to the corresponding matrix X to obtain another basis of the same lattice. The range parameter is used to limit the scalars: the multiplier m in the third type of row operation ("add an integral multiple of any row to any other row") is a nonzero integer in the range $-r \leq m \leq r$. Test your program for various values of k and r on pseudorandom integral bases of $\mathbb{R}^n$ for $n = 2, 3, \dots, 10$. (You will also need a parameter to limit the components of the pseudorandom basis vectors.)

Project 1.3. Write a computer program that takes as input a basis $\mathbf{x}_1$, $\mathbf{x}_2$, ..., $\mathbf{x}_n$ of the n-dimensional lattice $L \subset \mathbb{R}^n$ together with m vectors $\mathbf{y}_1$, $\mathbf{y}_2$, ..., $\mathbf{y}_m$ in L ($1 \leq m < n$), and determines whether there exist vectors $\mathbf{y}_{m+1}$, ..., $\mathbf{y}_n$ in L such that $\mathbf{y}_1, \mathbf{y}_2, \dots, \mathbf{y}_n$ form a basis of L. Extend your program to find vectors $\mathbf{y}_{m+1}, \dots, \mathbf{y}_n$ satisfying this condition (if they exist).

Project 1.4. Write a survey report on algorithmic aspects of the geometry of numbers and its applications, and give a seminar presentation based on your report. The following survey papers will be useful references: Kannan [72], Vallée [138], and Aardal [1].

1.5 Exercises

Exercise 1.1. Consider the triangle in $\mathbb{R}^2$ with these points as its vertices:

$$A = (-5, -4), \qquad B = (-5, -1), \qquad C = (5, -8).$$

Find the lengths of the sides of this triangle. Calculate the angles at the vertices and verify that their sum is 180 degrees. For each ordered pair of sides, find the components of the first side parallel and orthogonal to the second side.

Exercise 1.2. Same as Exercise 1.1 for these points in $\mathbb{R}^2$:

$$A = (45, -81), \qquad B = (-50, -22), \qquad C = (-16, -9).$$

Exercise 1.3. Same as Exercise 1.1 for these points in $\mathbb{R}^3$:

$$A = (2, -9, 0), \qquad B = (-8, 2, -5), \qquad C = (-9, 7, 7).$$

Exercise 1.4. Same as Exercise 1.1 for these points in $\mathbb{R}^3$:

$$A = (77, 9, 31), \qquad B = (20, -61, -48), \qquad C = (24, 65, 86).$$

Exercise 1.5. Same as Exercise 1.1 for these points in $\mathbb{R}^4$:

$$A = (4, -9, -2, -5), \qquad B = (1, 7, 8, -1), \qquad C = (-6, -8, -2, -2).$$

Exercise 1.6. Same as Exercise 1.1 for these points in $\mathbb{R}^4$:

$$A = (-62, -33, -68, -67), \quad B = (42, 18, -59, 12), \quad C = (52, -13, 82, 72).$$

Exercise 1.7. Let C be an $n \times n$ matrix with integer entries and determinant ± 1. Prove that C^{-1} also has integer entries.

Exercise 1.8. Show that these three bases of $\mathbb{R}^2$ generate the same lattice. For each ordered pair of bases, find a sequence of unimodular row operations which converts from the first basis to the second:

$$\{\, \mathbf{x}_1, \mathbf{x}_2 \,\} = \left\{ \begin{bmatrix} -41 \\ -82 \end{bmatrix}, \begin{bmatrix} 1 \\ -99 \end{bmatrix} \right\},$$

$$\{\, \mathbf{y}_1, \mathbf{y}_2 \,\} = \left\{ \begin{bmatrix} -79 \\ -461 \end{bmatrix}, \begin{bmatrix} -198 \\ -1103 \end{bmatrix} \right\},$$

$$\{\, \mathbf{z}_1, \mathbf{z}_2 \,\} = \left\{ \begin{bmatrix} 26080957 \\ 43756088 \end{bmatrix}, \begin{bmatrix} 3875510 \\ 6501953 \end{bmatrix} \right\}.$$

Exercise 1.9. Same as Exercise 1.8 for these three bases of $\mathbb{R}^3$:

$$\{\, \mathbf{x}_1, \mathbf{x}_2, \mathbf{x}_3 \,\} = \left\{ \begin{bmatrix} 4 \\ -2 \\ 0 \end{bmatrix}, \begin{bmatrix} 3 \\ -3 \\ -3 \end{bmatrix}, \begin{bmatrix} -1 \\ -6 \\ -1 \end{bmatrix} \right\},$$

$$\{\, \mathbf{y}_1, \mathbf{y}_2, \mathbf{y}_3 \,\} = \left\{ \begin{bmatrix} 143 \\ -20 \\ 19 \end{bmatrix}, \begin{bmatrix} -241 \\ -64 \\ -45 \end{bmatrix}, \begin{bmatrix} 110 \\ -5 \\ 16 \end{bmatrix} \right\},$$

$$\{\, \mathbf{z}_1, \mathbf{z}_2, \mathbf{z}_3 \,\} = \left\{ \begin{bmatrix} -26357 \\ 13270 \\ 2307 \end{bmatrix}, \begin{bmatrix} 4836 \\ -2438 \\ -424 \end{bmatrix}, \begin{bmatrix} -105971 \\ 53351 \\ 9275 \end{bmatrix} \right\}.$$

Exercise 1.10. Same as Exercise 1.8 for these three bases of $\mathbb{R}^4$:

$$\left\{ \begin{bmatrix} 5 \\ 0 \\ -5 \\ -1 \end{bmatrix}, \begin{bmatrix} -7 \\ 0 \\ -6 \\ 7 \end{bmatrix}, \begin{bmatrix} 1 \\ -2 \\ -7 \\ 4 \end{bmatrix}, \begin{bmatrix} -1 \\ 7 \\ -3 \\ 1 \end{bmatrix} \right\},$$

$$\left\{ \begin{bmatrix} 82 \\ -371 \\ 271 \\ -129 \end{bmatrix}, \begin{bmatrix} -101 \\ 425 \\ -303 \\ 149 \end{bmatrix}, \begin{bmatrix} -705 \\ 2915 \\ -2090 \\ 1039 \end{bmatrix}, \begin{bmatrix} -2100 \\ 8689 \\ -6240 \\ 3102 \end{bmatrix} \right\},$$

$$\left\{ \begin{bmatrix} 21463771 \\ 1248392 \\ -30241207 \\ -775616 \end{bmatrix}, \begin{bmatrix} 79458521 \\ 4621448 \\ -111952377 \\ -2871329 \end{bmatrix}, \begin{bmatrix} -2726297 \\ -158475 \\ 3841129 \\ 98526 \end{bmatrix}, \begin{bmatrix} 7377273 \\ 428791 \\ -10393946 \\ -266612 \end{bmatrix} \right\}.$$

Exercise 1.11. Let X be any $m \times n$ matrix ($m \le n$) with real entries. Prove that the determinant of the matrix XX^t is always non-negative, and equals 0 if and only if the rows of X are linearly dependent.

Exercise 1.12. Let $\mathbf{x}_1, \mathbf{x}_2, \ldots, \mathbf{x}_m$ be m linearly independent vectors in $\mathbb{R}^n$ spanning the lattice L, and let X be the $m \times n$ matrix with $\mathbf{x}_i$ as row i. Let $\mathbf{y}_1, \mathbf{y}_2, \ldots, \mathbf{y}_m$ be another basis for L with corresponding matrix Y. Prove that there exists a unimodular matrix C such that $Y = CX$ and $X = C^{-1}Y$. Deduce that $\det(XX^t) = \det(YY^t)$, and hence the determinant of L does not depend on the choice of basis.

Exercise 1.13. In each case, find the Gram matrix and the determinant of the m-dimensional lattice L in n-dimensional Euclidean space $\mathbb{R}^n$ spanned by the rows $\mathbf{x}_1, \mathbf{x}_2, \ldots, \mathbf{x}_m$ of the $m \times n$ matrix X:

$$\text{(a)} \quad X = \begin{bmatrix} -5 & -4 & 6 \\ -1 & 1 & -5 \end{bmatrix},$$

$$\text{(b)} \quad X = \begin{bmatrix} 25 & 62 & 58 \\ 53 & 17 & -37 \end{bmatrix},$$

$$\text{(c)} \quad X = \begin{bmatrix} -156 & -142 & 27 \\ 901 & 560 & -733 \end{bmatrix},$$

$$\text{(d)} \quad X = \begin{bmatrix} 5166 & 3296 & -1487 \\ -7461 & 7833 & -5023 \end{bmatrix}.$$

Exercise 1.14. Same as Exercise 2.6 for these matrices:

$$\text{(a)} \quad X = \begin{bmatrix} 7 & 0 & 6 & 3 & 7 \\ 8 & -1 & -2 & -9 & -2 \\ 9 & 6 & 1 & -8 & -6 \end{bmatrix},$$

$$\text{(b)} \quad X = \begin{bmatrix} -59 & -23 & -2 & -31 & 29 \\ 99 & -73 & -83 & 38 & 17 \\ 58 & 30 & -84 & -77 & -63 \end{bmatrix},$$

$$(c) \quad X = \begin{bmatrix} -932 & -95 & -672 & -139 & 784 \\ 989 & -504 & 193 & -489 & 334 \\ -978 & -312 & -712 & -39 & -19 \end{bmatrix}.$$

Exercise 1.15. Same as Exercise 2.6 for these matrices:

$$(a) \quad X = \begin{bmatrix} -9 & -6 & 7 & 4 & 2 & 3 & -8 \\ -6 & 2 & -4 & 9 & -2 & 1 & -8 \\ -8 & -2 & 7 & -8 & 7 & 2 & 5 \\ 7 & -7 & -3 & 6 & -9 & 1 & 9 \end{bmatrix},$$

$$(b) \quad X = \begin{bmatrix} -46 & -42 & 12 & 76 & -51 & -97 & 37 \\ -77 & -84 & 85 & 92 & -34 & 88 & 92 \\ -51 & 65 & 41 & -59 & -4 & 88 & 23 \\ -77 & 54 & -78 & -89 & 0 & -63 & 47 \end{bmatrix},$$

$$(c) \quad X = \begin{bmatrix} 323 & 209 & -629 & 480 & 889 & 91 & -104 \\ -894 & 205 & 691 & 768 & 281 & -242 & 63 \\ -842 & 137 & -399 & 730 & 353 & 586 & 56 \\ -227 & -605 & 130 & 89 & -769 & -409 & -236 \end{bmatrix}.$$

Exercise 1.16. Complete the proof of Corollary 1.23.

Exercise 1.17. Complete the proof of Corollary 1.26.

Exercise 1.18. Prove Theorem 1.28.

Exercise 1.19. (Cassels [22], Lemma 2, page 15). Let $\mathbf{x}_1$, $\mathbf{x}_1$, ..., $\mathbf{x}_n$ be a basis for the n-dimensional lattice $L \subset \mathbb{R}^n$. Consider m lattice vectors,

$$\mathbf{y}_i = \sum_{j=1}^n a_{ij}\mathbf{x}_j, \qquad a_{ij} \in \mathbb{Z}, \qquad i = 1, 2, \ldots, m.$$

Let $A = (a_{ij})$ be the $m \times n$ matrix of coefficients. Prove that the vectors $\mathbf{y}_1$, $\mathbf{y}_2$, ..., $\mathbf{y}_m$ can be extended to a basis of L if and only if the $\binom{n}{m}$ determinants of size $m \times m$ obtained by taking m columns of A have no common factor.

Exercise 1.20. Let C be an $n \times n$ matrix with integer entries, and suppose that only the first m rows of C are known for some $m = 1, 2, \ldots, n-1$. Find necessary and sufficient conditions on the first m rows in order that the remaining $n-m$ rows can be filled in (with integers) to give a unimodular matrix.

Exercise 1.21. Consider the lattice $L \subset \mathbb{R}^2$ with basis $\mathbf{x}_1$, $\mathbf{x}_2$ and the vector $\mathbf{y}_1 \in L$. Determine whether there exists a vector $\mathbf{y}_2 \in L$ such that $\mathbf{y}_1$, $\mathbf{y}_2$ is a basis of L, and find such a vector if it exists:

$$\mathbf{x}_1, \mathbf{x}_2 = \begin{bmatrix} 4 \\ -7 \end{bmatrix}, \begin{bmatrix} -7 \\ -8 \end{bmatrix}; \qquad \mathbf{y}_1 = \begin{bmatrix} -79 \\ -44 \end{bmatrix}.$$

Exercise 1.22. Same as Exercise 1.21 for

$$\mathbf{x}_1, \mathbf{x}_2 = \begin{bmatrix} -72 \\ -32 \end{bmatrix}, \begin{bmatrix} -2 \\ -74 \end{bmatrix}; \qquad \mathbf{y}_1 = \begin{bmatrix} -632 \\ 304 \end{bmatrix}.$$

Exercise 1.23. Consider the lattice $L \subset \mathbb{R}^3$ with basis $\mathbf{x}_1$, $\mathbf{x}_2$, $\mathbf{x}_3$ and the vectors $\mathbf{y}_1, \mathbf{y}_2 \in L$. Determine whether there exists a vector $\mathbf{y}_3 \in L$ such that $\mathbf{y}_1$, $\mathbf{y}_2$, $\mathbf{y}_3$ is a basis of L, and find such a vector if it exists:

$$\mathbf{x}_1, \mathbf{x}_2, \mathbf{x}_3 = \begin{bmatrix} -6 \\ -5 \\ -4 \end{bmatrix}, \begin{bmatrix} 5 \\ -1 \\ 1 \end{bmatrix}, \begin{bmatrix} -8 \\ -5 \\ -3 \end{bmatrix}; \quad \mathbf{y}_1, \mathbf{y}_2 = \begin{bmatrix} 33 \\ 33 \\ 15 \end{bmatrix}, \begin{bmatrix} -54 \\ -16 \\ -15 \end{bmatrix}.$$

Exercise 1.24. Same as Exercise 1.23 for

$$\mathbf{x}_1, \mathbf{x}_2, \mathbf{x}_3 = \begin{bmatrix} 31 \\ 43 \\ 12 \end{bmatrix}, \begin{bmatrix} -50 \\ 25 \\ -2 \end{bmatrix}, \begin{bmatrix} -80 \\ 94 \\ 50 \end{bmatrix}; \quad \mathbf{y}_1, \mathbf{y}_2 = \begin{bmatrix} -795 \\ 267 \\ 74 \end{bmatrix}, \begin{bmatrix} 317 \\ -712 \\ -392 \end{bmatrix}.$$

Exercise 1.25. Consider the lattice $L \subset \mathbb{R}^3$ with basis $\mathbf{x}_1$, $\mathbf{x}_2$, $\mathbf{x}_3$ and the vector $\mathbf{y}_1 \in L$. Determine whether there exist vectors $\mathbf{y}_2, \mathbf{y}_3 \in L$ such that $\mathbf{y}_1$, $\mathbf{y}_2$, $\mathbf{y}_3$ is a basis of L, and find such vectors if they exist:

$$\mathbf{x}_1, \mathbf{x}_2, \mathbf{x}_3 = \begin{bmatrix} 7 \\ 4 \\ -5 \end{bmatrix}, \begin{bmatrix} 8 \\ -9 \\ 5 \end{bmatrix}, \begin{bmatrix} -1 \\ -2 \\ -4 \end{bmatrix}; \qquad \mathbf{y}_1 = \begin{bmatrix} -31 \\ 8 \\ -4 \end{bmatrix}.$$

Exercise 1.26. Same as Exercise 1.25 for

$$\mathbf{x}_1, \mathbf{x}_2, \mathbf{x}_3 = \begin{bmatrix} 18 \\ -62 \\ -67 \end{bmatrix}, \begin{bmatrix} -59 \\ -33 \\ 22 \end{bmatrix}, \begin{bmatrix} 12 \\ -68 \\ 14 \end{bmatrix}; \qquad \mathbf{y}_1 = \begin{bmatrix} 178 \\ 46 \\ -678 \end{bmatrix}.$$

2

Two-Dimensional Lattices

CONTENTS

In this chapter we introduce the simplest case of lattice basis reduction: a lattice in the plane generated by two linearly independent vectors. The algorithm we discuss goes back to Legendre and Gauss in the 19th century. (Scharlau and Opolka [122] provide a very useful history of developments in this area.) This algorithm has a striking resemblance to the classical Euclidean algorithm for computing the greatest common divisor (GCD) of two integers.

2.1 The Euclidean algorithm

We start with a brief review of the Euclidean algorithm, which efficiently computes the GCD (greatest common divisor) of two integers by repeated division with remainder. (One could, at least in principle, compute the GCD by comparing the prime factorizations of the integers, but integer factorization is a hard problem in general.)

Recall that for any two integers a and b with $b \neq 0$, there are unique integers q (quotient) and r (remainder) such that

$$a = qb + r, \qquad 0 \leq r < |b|.$$

The Euclidean algorithm **Euclid**(a, b) for the GCD is given in Figure 2.1.

Example 2.1. We compute the greatest common divisor of $a = 7854$ and $b = 2145$. Set $r_0 = a$ and $r_1 = b$. Dividing r_0 by r_1 gives

$$7854 = 3 \cdot 2145 + 1419 \qquad r_0 = q_2 r_1 + r_2 \qquad q_2 = 3 \qquad r_2 = 1419$$

Now dividing r_1 by r_2 gives

$$2145 = 1 \cdot 1419 + 726 \qquad r_1 = q_3 r_2 + r_3 \qquad q_3 = 1 \qquad r_3 = 726$$

- *Input*: Nonzero integers a and b.

- *Output*: The (positive) greatest common divisor of a and b.

  (1) Set $r_0 \leftarrow \max(|a|, |b|)$ and $r_1 \leftarrow \min(|a|, |b|)$.
  (2) Set $i \leftarrow 1$.
  (3) While $r_i \neq 0$ do:

    (a) Perform integer division with remainder of r_{i-1} by r_i: compute the unique integers q_{i+1} and r_{i+1} such that

    $$r_{i-1} = q_{i+1}r_i + r_{i+1}, \qquad 0 \leq r_{i+1} < r_i.$$

    (b) Set $i \leftarrow i+1$.
  (4) Return r_{i-1}.

FIGURE 2.1
The Euclidean algorithm **Euclid**(a, b) for the greatest common divisor

---

Repeating this process, adding 1 to the subscripts at each step, we obtain

$$
\begin{array}{llll}
1419 = 1 \cdot 726 + 693 & r_2 = q_4 r_3 + r_4 & q_4 = 1 & r_4 = 693 \\
726 = 1 \cdot 693 + 33 & r_3 = q_5 r_4 + r_5 & q_5 = 1 & r_5 = 33 \\
693 = 21 \cdot 33 + 0 & r_4 = q_6 r_5 + r_6 & q_6 = 21 & r_6 = 0
\end{array}
$$

We now have a remainder of 0, so the algorithm terminates. The last nonzero remainder, $r_5 = 33$, is the greatest common divisor of $a = 7854$ and $b = 2145$.

Lemma 2.2. *Let n be the number of iterations of step (3) of the Euclidean algorithm. Writing* log *for the logarithm to the base 2, we have*

$$n < 1 + 2 \log \min(|a|, |b|).$$

Proof. It is clear that

$$r_0 \geq r_1 > r_2 > \cdots > r_n > r_{n+1} = 0,$$

and hence that $q_i \geq 1$ for all i. From this it follows that

$$r_{i-1} = q_{i+1}r_i + r_{i+1} \geq r_i + r_{i+1} > 2r_{i+1},$$

and therefore

$$r_1 \cdots r_{n-2} > 2^{n-2} r_3 \cdots r_n.$$

This implies

$$2^{n-2} < \frac{r_1 r_2}{r_{n-1} r_n}.$$

But $r_1 r_2 < r_1^2$, and $r_{n-1} r_n \geq 2$, so we obtain

$$2^{n-2} < \frac{r_1^2}{2}.$$

Taking logarithms to the base 2 gives

$$n - 2 < -1 + 2 \log r_1,$$

and hence

$$n < 1 + 2 \log r_1.$$

Since $r_1 = \min(|a|, |b|)$, this completes the proof. $\qquad\square$

Remark 2.3. The number of bits (binary digits) required to encode $c \in \mathbb{Z}$ is

$$1 + (\lfloor \log |c| \rfloor + 1) = \lfloor \log |c| \rfloor + 2;$$

the first bit is for the $\pm$ sign and $\lfloor x \rfloor$ is the greatest integer $\leq x$. For example,

$$-4 \longleftrightarrow 1|100, \qquad +7 \longleftrightarrow 0|111.$$

Thus the binary string $c_0 c_1 \cdots c_k$ represents the integer

$$c = (-1)^{c_0} \sum_{i=1}^{k} c_i 2^{k-i}.$$

Hence the number s of bits required to encode the input integers a and b for the Euclidean algorithm satisfies the inequality

$$s = \left(\lfloor \log |a| \rfloor + 2 \right) + \left(\lfloor \log |b| \rfloor + 2 \right) \geq 2 \lfloor \log \min(|a|, |b|) \rfloor + 4.$$

Starting from the inequality in Lemma 2.2 we now obtain

$$\begin{aligned} n &< 1 + 2 \log \min(|a|, |b|) \\ &< 1 + 2 \left(\lfloor \log \min(|a|, |b|) \rfloor + 1 \right) \\ &= 2 \lfloor \log \min(|a|, |b|) \rfloor + 3 \\ &< s. \end{aligned}$$

Hence the number of divisions with remainder performed by the Euclidean algorithm is no greater than the bit size of the input.

The worst case of the Euclidean algorithm occurs when all the quotients q_i are equal to 1; this implies that the remainders r_i decrease as slowly as possible. In this case the equation in step 3(a) of Figure 2.1 becomes

$$r_{i-1} = r_i + r_{i+1}, \qquad 0 \leq r_{i+1} < r_i.$$

This is essentially the recurrence relation for the Fibonacci numbers F_n:

$$F_0 = 0, \qquad F_1 = 1, \qquad F_n = F_{n-1} + F_{n-2} \quad (n \geq 2).$$

In particular, if $r_n = 1$ and $r_{n+1} = 0$ then $a = r_0 = F_{n+1}$ and $b = r_1 = F_n$.

Theorem 2.4. (Knuth [78], page 343) *For $n \geq 1$, let a and b be integers with $a > b > 0$ such that the Euclidean algorithm applied to a and b requires exactly n division steps, and such that a is the smallest integer satisfying these conditions. Then $a = F_{n+2}$ and $b = F_{n+1}$.*

Note that we don't take $a = F_{n+1}$ and $b = F_n$, since when we input consecutive Fibonacci numbers, the Euclidean algorithm terminates one step sooner than expected: the Fibonacci sequence begins 0, 1, 1, 2, 3 and division with remainder gives $3 = 1 \cdot 2 + 1$ but $2 = 2 \cdot 1 + 0$.

There is an exact formula for the Fibonacci numbers (Exercise 2.5):

$$F_n = \frac{1}{\sqrt{5}}\left(\phi_+^n - \phi_-^n\right), \qquad \phi_\pm = \frac{1}{2}\left(1 \pm \sqrt{5}\right).$$

Since $\phi_+ \approx 1.618$ and $\phi_- \approx -0.618$, we get the approximation

$$F_n \approx \frac{1}{\sqrt{5}}\,\phi_+^n, \quad \text{hence} \quad \phi_+^n \approx \sqrt{5}\,F_n.$$

Therefore

$$n \approx \log_{\phi_+}\left(\sqrt{5}F_n\right).$$

It follows that on the inputs $a = F_{n+2}$ and $b = F_{n+1}$ the Euclidean algorithm terminates after n steps where

$$n \approx \log_{\phi_+}(\sqrt{5}\,b) - 1 = \frac{\log(\sqrt{5}\,b)}{\log \phi_+} - 1 = \frac{1}{\log \phi_+}\log b + \left(\frac{\log \sqrt{5}}{\log \phi_+} - 1\right).$$

¿From this we get the estimate

$$n \approx 1.441 \log \min(|a|, |b|) + 0.673, \tag{2.1}$$

which improves the result of Lemma 2.2. This approximate calculation shows that we can get a good estimate on the number of steps required by an algorithm by considering the worst case of the algorithm.

An important variant of the Euclidean algorithm uses symmetric remainders. This means that we modify the inequality on r_{i+1} in step 3(a) of Figure 2.1 as follows:

$$r_{i-1} = q_{i+1}r_i + r_{i+1}, \qquad -\left|\frac{r_i}{2}\right| < r_{i+1} \leq \left|\frac{r_i}{2}\right|.$$

Definition 2.5. We call this modified algorithm the **centered Euclidean algorithm** and denote it by **CEuclid**(a, b). (Other names are the Euclidean algorithm with symmetric remainders, or the Euclidean algorithm with least absolute remainders.)

Example 2.6. We redo Example 2.1 using the centered Euclidean algorithm. For $r_0 = 7854$ and $r_1 = 2145$ we obtain

$$7854 = 4 \cdot 2145 - 726.$$

For $r_1 = 2145$ and $r_2 = -726$ we obtain

$$2145 = (-3) \cdot (-726) - 33.$$

For $r_2 = -726$ and $r_3 = -33$ we obtain

$$-726 = (-22) \cdot (-33) + 0.$$

Algorithm **Euclid** takes 5 steps to compute the gcd, but algorithm **CEuclid** takes only 3 steps.

For the centered Euclidean algorithm, it can be shown that $|q_i| \geq 2$ for $i \geq 3$. In this case the equation in step 3(a) of Figure 2.1 becomes

$$r_{i-1} = 2r_i + r_{i+1}, \qquad -\left|\frac{r_i}{2}\right| < r_{i+1} \leq \left|\frac{r_i}{2}\right|.$$

Hence the worst case corresponds to the recurrence relation

$$G_0 = 0, \qquad G_1 = 1, \qquad G_n = 2G_{n-1} + G_{n-2} \quad (n \geq 2).$$

An exact formula for the numbers G_n is as follows (Exercises 2.6–2.8):

$$G_n = \frac{1}{2\sqrt{2}}(\psi_+^n - \psi_-^n), \qquad \psi_\pm = 1 \pm \sqrt{2}.$$

We can formulate the computation of the greatest common divisor in terms of lattices. Regard $\mathbb{R}$ as the one-dimensional real vector space, and let a and b be nonzero integers which we regard as vectors in $\mathbb{R}$. The lattice $L \subset \mathbb{R}$ generated by a and b is the set of all integral linear combinations of a and b:

$$L = \{\, sa + tb \mid s, t \in \mathbb{Z} \,\}.$$

In elementary number theory, it is shown that $d = \gcd(a, b)$ is the least positive element of L, and that every element of L is an integral multiple of d. In other words, d is a basis of the lattice generated by a and b, and hence d is a shortest vector in this lattice. Therefore, the following problems are equivalent:

- Compute a greatest common divisor of a and b.

- Determine a shortest vector in the lattice generated by a and b.

A very attractive introduction to elementary number theory is Stillwell [134].

2.2 Two-dimensional lattices

Let $\mathbf{x}$ and $\mathbf{y}$ form a basis of $\mathbb{R}^2$. The lattice $L \subset \mathbb{R}^2$ generated by $\mathbf{x}$ and $\mathbf{y}$ is the set of all integral linear combinations of $\mathbf{x}$ and $\mathbf{y}$:

$$L = \{\, a\mathbf{x} + b\mathbf{y} \mid a, b \in \mathbb{Z} \,\}.$$

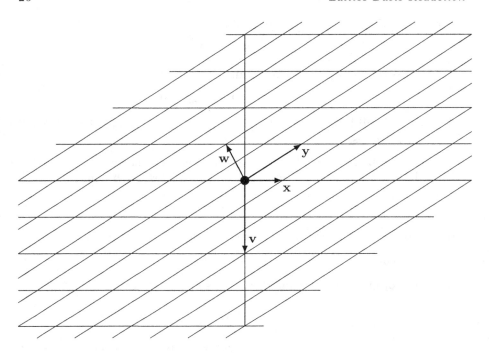

FIGURE 2.2
The two-dimensional lattice L generated by $\mathbf{x} = [2, 0]$ and $\mathbf{y} = [3, 2]$. The vectors $\mathbf{v} = [0, -4]$ and $\mathbf{w} = [-1, 2]$ form another basis of the same lattice.

Figure 2.2 shows a region near the origin of the lattice in the plane $\mathbb{R}^2$ generated by the vectors $\mathbf{x} = [2, 0]$ and $\mathbf{y} = [3, 2]$. (The origin is indicated by a large dot.) If we change basis by unimodular matrix multiplication,

$$\left[\begin{array}{cc} 3 & -2 \\ -2 & 1 \end{array}\right] \left[\begin{array}{cc} 2 & 0 \\ 3 & 2 \end{array}\right] = \left[\begin{array}{cc} 0 & -4 \\ -1 & 2 \end{array}\right],$$

we see that the vectors $\mathbf{v} = [0, -4]$ and $\mathbf{w} = [-1, 2]$ form another basis.

Suppose that L is a lattice in $\mathbb{R}^2$ generated by the vectors $\mathbf{x}$ and $\mathbf{y}$. A basic problem is to find a shortest (nonzero) vector in this lattice; that is, a vector $\mathbf{v}$ for which $|\mathbf{v}| \le |\mathbf{w}|$ for all $\mathbf{w} \in L$, $\mathbf{w} \ne 0$. This is achieved by the algorithm in Figure 2.3, usually attributed to Gauss (but see also the discussion of the earlier contributions of Lagrange in Scharlau and Opolka [122], Chapter 4). In fact this "Gaussian algorithm" finds a minimal basis of a two-dimensional lattice, in the following sense.

Definition 2.7. We say that a basis $\mathbf{x}$, $\mathbf{y}$ of a lattice L in $\mathbb{R}^2$ is **minimal** if $\mathbf{x}$ is a shortest nonzero vector in L and $\mathbf{y}$ is a shortest vector in L which is not a multiple of $\mathbf{x}$.

- *Input*: A basis $\mathbf{x}$, $\mathbf{y}$ of a lattice L in $\mathbb{R}^2$ such that $|\mathbf{x}| \leq |\mathbf{y}|$.

- *Output*: A minimal basis $\mathbf{v}_1$, $\mathbf{v}_2$ of the lattice L.

(1) Set $\mathbf{v}_1 \leftarrow \mathbf{x}$ and $\mathbf{v}_2 \leftarrow \mathbf{y}$. Set finished $\leftarrow$ false.

(2) While not finished do:

    (a) Set $m \leftarrow \left\lceil \dfrac{\mathbf{v}_2 \cdot \mathbf{v}_1}{\mathbf{v}_1 \cdot \mathbf{v}_1} \right\rfloor$.

    (b) Set $\mathbf{v}_2 \leftarrow \mathbf{v}_2 - m\mathbf{v}_1$.

    (c) If $|\mathbf{v}_1| \leq |\mathbf{v}_2|$ then

        (i) set finished $\leftarrow$ true

    else

        (ii) set $\mathbf{u} \leftarrow \mathbf{v}_1$, $\mathbf{v}_1 \leftarrow \mathbf{v}_2$, $\mathbf{v}_2 \leftarrow \mathbf{u}$ (interchange $\mathbf{v}_1$ and $\mathbf{v}_2$).

(3) Return $\mathbf{v}_1$ and $\mathbf{v}_2$.

FIGURE 2.3
The Gaussian algorithm $\mathbf{Gauss}(\mathbf{x}, \mathbf{y})$ for a minimal basis of a lattice in $\mathbb{R}^2$

Definition 2.8. We write $\lceil \mu \rfloor = \lceil \mu - \frac{1}{2} \rceil$ for the **nearest integer** to $\mu \in \mathbb{R}$. Note that the nearest integer to $n + \frac{1}{2}$ ($n \in \mathbb{Z}$) is n, not $n+1$.

In step 2(a) of the Gaussian algorithm we must use the nearest integer m to the orthogonal projection coefficient $\mu = (\mathbf{v}_2 \cdot \mathbf{v}_1)/(\mathbf{v}_1 \cdot \mathbf{v}_1)$ instead of μ itself, because in step 2(b) the vector $\mathbf{v}_2$ must remain in the lattice L.

Example 2.9. Let L be the lattice in $\mathbb{R}^2$ generated by the vectors

$$\mathbf{v}_1 = [-56, 43], \quad |\mathbf{v}_1| \approx 70.60, \qquad \mathbf{v}_2 = [95, -73], \quad |\mathbf{v}_2| \approx 119.8.$$

The first iteration calculates

$$m = \left\lceil \frac{\mathbf{v}_2 \cdot \mathbf{v}_1}{\mathbf{v}_1 \cdot \mathbf{v}_1} \right\rfloor = \left\lceil \frac{-8459}{4985} \right\rfloor \approx \lceil -1.697 \rfloor = -2.$$

We now set

$$\mathbf{v}_2 = \mathbf{v}_2 + 2\mathbf{v}_1 = [-17, 13], \quad |\mathbf{v}_2| \approx 21.40.$$

Since $|\mathbf{v}_2| < |\mathbf{v}_1|$ we interchange the vectors; we now have

$$\mathbf{v}_1 = [-17, 13], \qquad \mathbf{v}_2 = [-56, 43].$$

The second iteration calculates

$$m = \left\lceil \frac{\mathbf{v}_2 \cdot \mathbf{v}_1}{\mathbf{v}_1 \cdot \mathbf{v}_1} \right\rfloor = \left\lceil \frac{1511}{458} \right\rfloor \approx \lceil 3.299 \rfloor = 3.$$

We now set
$$\mathbf{v}_2 = \mathbf{v}_2 - 3\mathbf{v}_1 = [-5, 4], \quad |\mathbf{v}_2| \approx 6.403.$$
Interchanging the vectors gives
$$\mathbf{v}_1 = [-5, 4], \qquad \mathbf{v}_2 = [-17, 13].$$
The third iteration calculates
$$m = \left\lceil \frac{\mathbf{v}_2 \cdot \mathbf{v}_1}{\mathbf{v}_1 \cdot \mathbf{v}_1} \right\rfloor = \left\lceil \frac{137}{41} \right\rfloor \approx \lceil 3.341 \rfloor = 3.$$
We now set
$$\mathbf{v}_2 = \mathbf{v}_2 - 3\mathbf{v}_1 = [-2, 1], \quad |\mathbf{v}_2| \approx 2.236.$$
Interchanging gives
$$\mathbf{v}_1 = [-2, 1], \qquad \mathbf{v}_2 = [-5, 4].$$
The fourth iteration calculates
$$m = \left\lceil \frac{\mathbf{v}_2 \cdot \mathbf{v}_1}{\mathbf{v}_1 \cdot \mathbf{v}_1} \right\rfloor = \left\lceil \frac{14}{5} \right\rfloor = \lceil 2.8 \rfloor = 3.$$
We now set
$$\mathbf{v}_2 = \mathbf{v}_2 - 3\mathbf{v}_1 = [1, 1], \quad |\mathbf{v}_2| \approx 1.414.$$
Interchanging gives
$$\mathbf{v}_1 = [1, 1], \qquad \mathbf{v}_2 = [-2, 1].$$
The fifth iteration calculates
$$m = \left\lceil \frac{\mathbf{v}_2 \cdot \mathbf{v}_1}{\mathbf{v}_1 \cdot \mathbf{v}_1} \right\rfloor = \left\lceil \frac{-1}{2} \right\rfloor = \lceil -0.5 \rfloor = -1.$$
We now set
$$\mathbf{v}_2 = \mathbf{v}_2 + \mathbf{v}_1 = [-1, 2], \quad |\mathbf{v}_2| \approx 2.236.$$
Since $|\mathbf{v}_2| \geq |\mathbf{v}_1|$ we terminate with this minimal basis:
$$\mathbf{v}_1 = [1, 1], \quad |\mathbf{v}_1| \approx 1.414, \qquad \mathbf{v}_2 = [-1, 2], \quad |\mathbf{v}_2| \approx 2.236.$$

It is instructive to compare the Euclidean and Gaussian algorithms. Algorithm **Euclid** repeatedly performs division with remainder, replacing the pair (r_{i-1}, r_i) of integers with the pair (r_i, r_{i+1}) determined by the equation $r_{i+1} = r_{i-1} - q_{i+1} r_i$. Algorithm **Gauss** repeatedly performs orthogonal projection (rounded to the nearest integer), replacing the pair $(\mathbf{v}_1, \mathbf{v}_2)$ of vectors by the pair $(\mathbf{v}_1, \mathbf{v}_2')$ determined by the equation $\mathbf{v}_2' = \mathbf{v}_2 - m\mathbf{v}_1$, and then interchanging the vectors (if necessary). The arithmetic operation of division with remainder in the Euclidean algorithm corresponds to the geometric operation of (rounded) orthogonal projection in the Gaussian algorithm.

Our next goal is to prove that the Gaussian algorithm does indeed produce a minimal basis of the input lattice. The next lemma shows that the new second vector $\mathbf{v}_2'$ is "nearly orthogonal" to the old first vector $\mathbf{v}_1$.

Lemma 2.10. *After the execution of step (2)(b) of Figure 2.3 we have*

$$|\mathbf{v}_2' \cdot \mathbf{v}_1| \le \frac{1}{2}|\mathbf{v}_1|^2,$$

where $\mathbf{v}_2'$ is the new second vector in the basis.

Proof. From the definition of nearest integer, we have

$$m = \left\lceil \frac{\mathbf{v}_2 \cdot \mathbf{v}_1}{\mathbf{v}_1 \cdot \mathbf{v}_1} \right\rfloor \quad \Longrightarrow \quad m - \frac{1}{2} < \frac{\mathbf{v}_2 \cdot \mathbf{v}_1}{\mathbf{v}_1 \cdot \mathbf{v}_1} \le m + \frac{1}{2}.$$

Multiplying this inequality by $\mathbf{v}_1 \cdot \mathbf{v}_1$ gives

$$m(\mathbf{v}_1 \cdot \mathbf{v}_1) - \frac{1}{2}(\mathbf{v}_1 \cdot \mathbf{v}_1) < \mathbf{v}_2 \cdot \mathbf{v}_1 \le m(\mathbf{v}_1 \cdot \mathbf{v}_1) + \frac{1}{2}(\mathbf{v}_1 \cdot \mathbf{v}_1).$$

Subtracting $m(\mathbf{v}_1 \cdot \mathbf{v}_1)$ from all parts gives

$$-\frac{1}{2}(\mathbf{v}_1 \cdot \mathbf{v}_1) < \mathbf{v}_2 \cdot \mathbf{v}_1 - m(\mathbf{v}_1 \cdot \mathbf{v}_1) \le \frac{1}{2}(\mathbf{v}_1 \cdot \mathbf{v}_1).$$

Rewriting the middle term using bilinearity of the scalar product gives

$$-\frac{1}{2}(\mathbf{v}_1 \cdot \mathbf{v}_1) < (\mathbf{v}_2 - m\mathbf{v}_1) \cdot \mathbf{v}_1 \le \frac{1}{2}(\mathbf{v}_1 \cdot \mathbf{v}_1),$$

and therefore

$$|(\mathbf{v}_2 - m\mathbf{v}_1) \cdot \mathbf{v}_1| \le \frac{1}{2}|\mathbf{v}_1|^2,$$

as required. □

Theorem 2.11. *The Gaussian algorithm terminates, and upon termination $\mathbf{v}_1$ is a shortest nonzero vector in the lattice, and $\mathbf{v}_2$ is a shortest vector in the lattice which is not a multiple of $\mathbf{v}_1$.*

Proof. The proof follows Beukers [16]. Regarding $\mathbf{v}_1$ and $\mathbf{v}_2$ as row vectors, we can express step (2)(b) in matrix form as

$$\begin{bmatrix} \mathbf{v}_1' \\ \mathbf{v}_2' \end{bmatrix} = \begin{bmatrix} 1 & 0 \\ -m & 1 \end{bmatrix} \begin{bmatrix} \mathbf{v}_1 \\ \mathbf{v}_2 \end{bmatrix}.$$

Since the matrix has determinant 1, it is clear that step (2) preserves the property that $\mathbf{v}_1$, $\mathbf{v}_2$ is a basis of the lattice L.

The algorithm interchanges $\mathbf{v}_1$ and $\mathbf{v}_2$ in step (2)(c)(ii) when $|\mathbf{v}_2| < |\mathbf{v}_1|$, and so the length of $\mathbf{v}_1$ strictly decreases from one execution of step (2) to the next. For any real number $r > 0$, there are only finitely many elements of the lattice in the disk $\{\mathbf{u} \in \mathbb{R}^2 \mid |\mathbf{u}| \le r\}$. It follows that the algorithm terminates after a finite number of executions of step (2).

Upon termination, step (2)(c) and Lemma 2.10 guarantee that

$$|\mathbf{v}_1| \le |\mathbf{v}_2|, \qquad -\frac{1}{2}|\mathbf{v}_1|^2 \le \mathbf{v}_2 \cdot \mathbf{v}_1 \le \frac{1}{2}|\mathbf{v}_1|^2.$$

Let $\mathbf{u}$ be any nonzero vector in L, so that $\mathbf{u} = a\mathbf{v}_1 + b\mathbf{v}_2$ for some $a, b \in \mathbb{Z}$, not both zero. We have

$$
\begin{aligned}
|\mathbf{u}|^2 &= (a\mathbf{v}_1 + b\mathbf{v}_2) \cdot (a\mathbf{v}_1 + b\mathbf{v}_2) \\
&= a^2|\mathbf{v}_1|^2 + 2ab(\mathbf{v}_1 \cdot \mathbf{v}_2) + b^2|\mathbf{v}_2|^2 && \text{bilinearity} \\
&\ge a^2|\mathbf{v}_1|^2 - |ab||\mathbf{v}_1|^2 + b^2|\mathbf{v}_2|^2 && \text{Lemma 2.10} \\
&\ge a^2|\mathbf{v}_1|^2 - |ab||\mathbf{v}_1|^2 + b^2|\mathbf{v}_1|^2 && |\mathbf{v}_2| \ge |\mathbf{v}_1| \\
&= (a^2 - |ab| + b^2)|\mathbf{v}_1|^2.
\end{aligned}
$$

Since a, b are not both zero, we have $a^2b^2 < (a^2+b^2)^2$ and hence $|ab| < a^2+b^2$. Therefore $|\mathbf{u}|^2 \ge |\mathbf{v}_1|^2$, and so $\mathbf{v}_1$ is a shortest vector in L.

Now suppose that $\mathbf{u} = a\mathbf{v}_1 + b\mathbf{v}_2$ is linearly independent of $\mathbf{v}_1$, or equivalently that $b \ne 0$. We have

$$
\begin{aligned}
|\mathbf{u}|^2 &\ge a^2|\mathbf{v}_1|^2 - |ab||\mathbf{v}_1|^2 + b^2|\mathbf{v}_2|^2 && \text{previous calculation} \\
&= a^2|\mathbf{v}_1|^2 - |ab||\mathbf{v}_1|^2 + \frac{1}{4}b^2|\mathbf{v}_2|^2 + \frac{3}{4}b^2|\mathbf{v}_2|^2 && \text{splitting the last term} \\
&= a^2|\mathbf{v}_1|^2 - |ab||\mathbf{v}_1|^2 + \frac{1}{4}b^2|\mathbf{v}_1|^2 + \frac{3}{4}b^2|\mathbf{v}_2|^2 && |\mathbf{v}_1| \le |\mathbf{v}_2| \\
&= \left(|a| - \frac{1}{2}|b|\right)^2|\mathbf{v}_1|^2 + \frac{3}{4}b^2|\mathbf{v}_2|^2. && \text{completing the square}
\end{aligned}
$$

Hence $|\mathbf{u}|^2 \ge |\mathbf{v}_2|^2$ if $|b| \ne 1$. If $b = \pm 1$ then we have

$$|\mathbf{u}|^2 \ge a^2|\mathbf{v}_1|^2 - |a||\mathbf{v}_1|^2 + |\mathbf{v}_2|^2 = |a|(|a| - 1)|\mathbf{v}_1|^2 + |\mathbf{v}_2|^2.$$

Since $a \in \mathbb{Z}$ we have $|a|(|a| - 1) = 0$ for $|a| \le 1$ and $|a|(|a| - 1) > 0$ for $|a| \ge 2$, and so it follows that $|\mathbf{u}|^2 \ge |\mathbf{v}_2|^2$ in this case also. Therefore $\mathbf{v}_2$ is a shortest vector in L linearly independent of $\mathbf{v}_1$. $\square$

The Gaussian algorithm can be applied to any two linearly independent vectors $\mathbf{v}_1$, $\mathbf{v}_2$ in the Euclidean vector space $\mathbb{R}^n$ for any $n \ge 2$. The vectors $\mathbf{v}_1$, $\mathbf{v}_2$ span a subspace of $\mathbb{R}^n$ linearly isomorphic to $\mathbb{R}^2$, and generate a two-dimensional lattice L in $\mathbb{R}^2$ (as before, this is the set of all integral linear combinations of $\mathbf{v}_1$, $\mathbf{v}_2$).

Example 2.12. Let $n = 3$ and consider the vectors

$$
\begin{aligned}
\mathbf{v}_1 &= [\,-49, \, -70, \, 35\,], && |\mathbf{v}_1| \approx 92.34, \\
\mathbf{v}_2 &= [\,58, \, 89, \, -48\,], && |\mathbf{v}_2| \approx 116.6.
\end{aligned}
$$

The first iteration calculates

$$m = \left\lceil \frac{\mathbf{v}_2 \cdot \mathbf{v}_1}{\mathbf{v}_1 \cdot \mathbf{v}_1} \right\rfloor = \left\lceil \frac{-256}{203} \right\rfloor \approx \lceil -1.261 \rfloor = -1.$$

We now set

$$\mathbf{v}_2 \leftarrow \mathbf{v}_2 + \mathbf{v}_1 = [\, 9,\ 19,\ -13 \,], \qquad |\mathbf{v}_2| \approx 24.72.$$

Since $|\mathbf{v}_2| < |\mathbf{v}_1|$ we interchange the vectors; we now have

$$\mathbf{v}_1 = [\, 9,\ 19,\ -13 \,], \qquad \mathbf{v}_2 = [\, -49,\ -70,\ 35 \,].$$

The second iteration calculates

$$m = \left\lceil \frac{\mathbf{v}_2 \cdot \mathbf{v}_1}{\mathbf{v}_1 \cdot \mathbf{v}_1} \right\rfloor = \left\lceil \frac{-2226}{611} \right\rfloor \approx \lceil -3.643 \rfloor = -4.$$

We now set

$$\mathbf{v}_2 \leftarrow \mathbf{v}_2 + 4\mathbf{v}_1 = [\, -13,\ 6,\ -17 \,], \qquad |\mathbf{v}_2| \approx 22.23.$$

Since $|\mathbf{v}_2| < |\mathbf{v}_1|$ we interchange the vectors; we now have

$$\mathbf{v}_1 = [\, -13,\ 6,\ -17 \,], \qquad \mathbf{v}_2 = [\, 9,\ 19,\ -13 \,].$$

The third iteration calculates

$$m = \left\lceil \frac{\mathbf{v}_2 \cdot \mathbf{v}_1}{\mathbf{v}_1 \cdot \mathbf{v}_1} \right\rfloor = \left\lceil \frac{109}{247} \right\rfloor \approx \lceil 0.4413 \rfloor = 0.$$

Thus $\mathbf{v}_2$ does not change, and the algorithm terminates.

2.3 Vallée's analysis of the Gaussian algorithm

In this section, we consider lattices in $\mathbb{Z}^2$; that is, we assume that the components of the vectors are integers. We follow Vallée [139] and consider a slightly modified version of the Gaussian algorithm. Just as the Euclidean algorithm comes in two variants, the original algorithm and its centered version, so the Gaussian algorithm also comes in two variants, the original algorithm and the centered version displayed in Figure 2.4. The centered Gaussian algorithm permits a more natural and geometric proof that the output consists of a minimal basis.

Definition 2.13. For $\alpha \in \mathbb{R}$ we define $\operatorname{sign}(\alpha) = 1$ if $\alpha \geq 0$ and $\operatorname{sign}(\alpha) = -1$ if $\alpha < 0$. (Note in particular that $\operatorname{sign}(0) = 1$.)

- *Input*: A basis $\mathbf{x}$, $\mathbf{y}$ of a lattice L in $\mathbb{Z}^2$ such that $|\mathbf{x}| \le |\mathbf{y}|$.

- *Output*: A minimal basis $\mathbf{v}_1$, $\mathbf{v}_2$ of the lattice L.

  (1)   Set $\mathbf{v}_1 \leftarrow \mathbf{x}$ and $\mathbf{v}_2 \leftarrow \mathbf{y}$. Set finished $\leftarrow$ false.

  (2)   While not finished do:

   (a)   Set $\mu \leftarrow \dfrac{\mathbf{v}_2 \cdot \mathbf{v}_1}{\mathbf{v}_1 \cdot \mathbf{v}_1}$. Set $m \leftarrow \lceil \mu \rfloor$. Set $\epsilon \leftarrow \mathrm{sign}(\mu - m)$.

   (b)   Set $\mathbf{v}_2 \leftarrow \epsilon(\mathbf{v}_2 - m\mathbf{v}_1)$.

   (c)   If $|\mathbf{v}_1| \le |\mathbf{v}_2|$ then

   (i)   set finished $\leftarrow$ true

   else

   (ii)   set $\mathbf{u} \leftarrow \mathbf{v}_1$, $\mathbf{v}_1 \leftarrow \mathbf{v}_2$, $\mathbf{v}_2 \leftarrow \mathbf{u}$ (interchange $\mathbf{v}_1$ and $\mathbf{v}_2$).

  (3)   Return $\mathbf{v}_1$ and $\mathbf{v}_2$.

FIGURE 2.4
The centered Gaussian algorithm **CGauss(x, y)**

The factor ϵ in step (2)(b) of Figure 2.4 guarantees that the new $\mathbf{v}_2$ makes an acute angle with the old $\mathbf{v}_1$. To see this, let $\mathbf{v}_2' = \epsilon(\mathbf{v}_2 - m\mathbf{v}_1)$ be the new value of $\mathbf{v}_2$. It suffices to show that $\mathbf{v}_2' \cdot \mathbf{v}_1 > 0$. We have

$$\mathbf{v}_2' \cdot \mathbf{v}_1 = \epsilon(\mathbf{v}_2 - m\mathbf{v}_1) \cdot \mathbf{v}_1 = \epsilon(\mathbf{v}_2 \cdot \mathbf{v}_1 - m\mathbf{v}_1 \cdot \mathbf{v}_1)$$
$$= \epsilon\left(\frac{\mathbf{v}_2 \cdot \mathbf{v}_1}{\mathbf{v}_1 \cdot \mathbf{v}_1} - m\right)(\mathbf{v}_1 \cdot \mathbf{v}_1) = \epsilon(\mu - m)|\mathbf{v}_1|^2,$$

and the last quantity is clearly positive by the definition of ϵ.

Theorem 2.14. *The output vectors* $\mathbf{v}_1$, $\mathbf{v}_2$ *of the centered Gaussian algorithm form a minimal basis of the lattice generated by the input vectors* $\mathbf{x}$, $\mathbf{y}$.

Proof. It is clear that $\mathbf{v}_1$ and $\mathbf{v}_2$ satisfy $|\mathbf{v}_1| \le |\mathbf{v}_2|$. Furthermore,

$$0 \le \mathbf{v}_2 \cdot \mathbf{v}_1 \le \frac{1}{2}|\mathbf{v}_1|^2, \qquad (2.2)$$

by the calculation preceding the statement of the Theorem, since

$$-\frac{1}{2} < \mu - m \le \frac{1}{2}, \qquad \epsilon(\mu - m) = |\mu - m|.$$

Hence

$$0 \le \frac{\mathbf{v}_2 \cdot \mathbf{v}_1}{\mathbf{v}_1 \cdot \mathbf{v}_1} \le \frac{1}{2}.$$

Let $\mathbf{w}$ be the component of $\mathbf{v}_2$ orthogonal to $\mathbf{v}_1$. We have

$$\mathbf{v}_2 = \frac{\mathbf{v}_2 \cdot \mathbf{v}_1}{\mathbf{v}_1 \cdot \mathbf{v}_1} \mathbf{v}_1 + \mathbf{w}, \qquad \mathbf{v}_1 \cdot \mathbf{w} = 0,$$

and therefore

$$|\mathbf{v}_2|^2 = \left| \frac{\mathbf{v}_2 \cdot \mathbf{v}_1}{\mathbf{v}_1 \cdot \mathbf{v}_1} \right|^2 |\mathbf{v}_1|^2 + |\mathbf{w}|^2.$$

Dividing all terms by $|\mathbf{v}_2|^2$ gives

$$1 = \left| \frac{\mathbf{v}_2 \cdot \mathbf{v}_1}{\mathbf{v}_1 \cdot \mathbf{v}_1} \right|^2 \frac{|\mathbf{v}_1|^2}{|\mathbf{v}_2|^2} + \frac{|\mathbf{w}|^2}{|\mathbf{v}_2|^2}.$$

The first term on the right side is $\leq \frac{1}{4} \cdot 1$, so we see that

$$\frac{|\mathbf{w}|^2}{|\mathbf{v}_2|^2} \geq \frac{3}{4}, \quad \text{hence} \quad |\mathbf{w}| \geq \frac{\sqrt{3}}{2} |\mathbf{v}_2|. \tag{2.3}$$

Consider the following three subsets of the lattice L generated by $\mathbf{v}_1$ and $\mathbf{v}_2$:

$$V_+ = \{ a\mathbf{v}_1 + \mathbf{v}_2 \mid a \in \mathbb{Z} \},$$
$$V_0 = \{ a\mathbf{v}_1 \mid a \in \mathbb{Z} \},$$
$$V_- = \{ a\mathbf{v}_1 - \mathbf{v}_2 \mid a \in \mathbb{Z} \}.$$

Thus V_0 consists of all integral multiples of $\mathbf{v}_1$, and V_+ (respectively V_-) is the translation of V_0 in the $\mathbf{v}_2$ direction (respectively the $-\mathbf{v}_2$ direction). It follows from inequality (2.3) that all the vectors $\mathbf{z}$ in the lattice L generated by $\mathbf{v}_1$ and $\mathbf{v}_2$, which do not belong to the subset $V_+ \cup V_0 \cup V_-$, satisfy

$$|\mathbf{z}| \geq \sqrt{3} |\mathbf{v}_2| > |\mathbf{v}_2| \geq |\mathbf{v}_1|.$$

(To understand these inequalities, the reader should draw a diagram of the lattice L and the subsets V_+, V_0, V_-.) From these inequalties it is clear that $\mathbf{v}_1$ and $\mathbf{v}_2$ form a minimal basis of the lattice L. $\qquad \square$

Example 2.15. Consider these two vectors in $\mathbb{Z}^2$:

$$\mathbf{v}_1 = [-67, 16], \quad |\mathbf{v}_1| \approx 68.88, \qquad \mathbf{v}_2 = [93, -25], \quad |\mathbf{v}_2| \approx 96.30.$$

The first iteration calculates

$$\mu = \frac{-6631}{4745} \approx -1.397, \qquad m = -1, \qquad \epsilon = -1.$$

Therefore

$$\mathbf{v}_2' = -(\mathbf{v}_2 + \mathbf{v}_1) = [-26, 9], \quad |\mathbf{v}_2'| \approx 27.51.$$

Interchanging the vectors gives

$$\mathbf{v}_1 = [-26, 9], \qquad \mathbf{v}_2 = [-67, 16].$$

The second iteration calculates

$$\mu = \frac{1886}{757} \approx 2.491, \qquad m = 2, \qquad \epsilon = 1.$$

Therefore

$$\mathbf{v}_2' = \mathbf{v}_2 - 2\mathbf{v}_1 = [\,-15,\,-2\,], \quad |\mathbf{v}_2'| \approx 15.13.$$

Interchanging the vectors gives

$$\mathbf{v}_1 = [\,-15,\,-2\,], \qquad \mathbf{v}_2 = [\,-26,\,9\,].$$

The third iteration calculates

$$\mu = \frac{372}{229} \approx 1.624, \qquad m = 2, \qquad \epsilon = -1.$$

Therefore

$$\mathbf{v}_2' = -(\mathbf{v}_2 - 2\mathbf{v}_1) = [\,-4,\,-13\,], \quad |\mathbf{v}_2'| \approx 13.60.$$

Interchanging the vectors gives

$$\mathbf{v}_1 = [\,-4,\,-13\,], \qquad \mathbf{v}_2 = [\,-15,\,-2\,].$$

The fourth iteration calculates

$$\mu = \frac{86}{185} \approx 0.4649, \qquad m = 0, \qquad \epsilon = 1.$$

Therefore $\mathbf{v}_2' = \mathbf{v}_2$ and the algorithm terminates.

In order to analyze the centered Gaussian algorithm, we will consider the parameterized Gaussian algorithm displayed in Figure 2.5. This third version of the algorithm depends on a parameter $t \geq 1$: the termination condition $|\mathbf{v}_1| \leq |\mathbf{v}_2|$ is replaced by $|\mathbf{v}_1| \leq t|\mathbf{v}_2|$. Since the centered Gaussian algorithm is the same as the parameterized Gaussian algorithm for $t = 1$, we are primarily interested in $t > 1$. At first sight, the parameterized algorithm seems to be merely a weakened form of the centered algorithm. But we will see that comparing these two algorithms leads to a very natural analysis of the centered algorithm. Furthermore, the parameterized algorithm is an essential ingredient in the LLL algorithm which we will consider in Chapter 4.

Proposition 2.16. *For the lattice basis* $\mathbf{x}, \mathbf{y} \in \mathbb{Z}^2$, *let* n *be the number of iterations performed by the centered Gaussian algorithm on input* $\mathbf{x}, \mathbf{y}$. *For* $1 < t \leq \sqrt{3}$, *let* $n[t]$ *be the number of iterations performed by the parametrized Gaussian algorithm on input* $\mathbf{x}, \mathbf{y}$. *We have*

$$n[t] \leq n \leq n[t] + 1.$$

- *Input*: A basis $\mathbf{x}$, $\mathbf{y}$ of a lattice L in $\mathbb{Z}^2$ such that $|\mathbf{x}| \leq |\mathbf{y}|$.

- *Output*: A "weakly minimal" basis $\mathbf{v}_1$, $\mathbf{v}_2$ of the lattice L.

  (1) Set $\mathbf{v}_1 \leftarrow x$ and $\mathbf{v}_2 \leftarrow y$. Set finished $\leftarrow$ false.

  (2) While not finished do:

   (a) Set $\mu \leftarrow \dfrac{\mathbf{v}_2 \cdot \mathbf{v}_1}{\mathbf{v}_1 \cdot \mathbf{v}_1}$. Set $m \leftarrow \lceil \mu \rfloor$. Set $\epsilon \leftarrow \operatorname{sign}(\mu - m)$.

   (b) Set $\mathbf{v}_2 \leftarrow \epsilon(\mathbf{v}_2 - m\mathbf{v}_1)$.

   (c) If $|\mathbf{v}_1| \leq t|\mathbf{v}_2|$ then

    (i) set finished $\leftarrow$ true

   else

    (ii) set $\mathbf{u} \leftarrow \mathbf{v}_1$, $\mathbf{v}_1 \leftarrow \mathbf{v}_2$, $\mathbf{v}_2 \leftarrow \mathbf{u}$ (interchange $\mathbf{v}_1$ and $\mathbf{v}_2$).

  (3) Return $\mathbf{v}_1$ and $\mathbf{v}_2$.

FIGURE 2.5
The parameterized Gaussian algorithm $\mathbf{PGauss}[t](x, y)$

Proof. The left-hand inequality is clear, since **CGauss** has a stronger termination condition than **PGauss**[t], and so the centered algorithm terminates no later than the parameterized algorithm. (This holds for all $t \geq 1$.)

For the right-hand inequality, we consider the last iteration of **PGauss**[t], and assume that this is not the last iteration of **CGauss**. It suffices to show that **CGauss** terminates at the next iteration. We have

$$0 \leq \frac{\mathbf{v}_2 \cdot \mathbf{v}_1}{\mathbf{v}_1 \cdot \mathbf{v}_1} \leq \frac{1}{2}, \quad \text{hence} \quad \mathbf{v}_1 \cdot \mathbf{v}_1 - 2\mathbf{v}_2 \cdot \mathbf{v}_1 \geq 0.$$

Since **CGauss** does not terminate at this iteration, we have

$$|\mathbf{v}_2| < |\mathbf{v}_1|.$$

The third edge of the triangle built from $\mathbf{v}_1$ and $\mathbf{v}_2$ is $\mathbf{v}_2 - \mathbf{v}_1$. We have

$$|\mathbf{v}_2 - \mathbf{v}_1|^2 = (\mathbf{v}_2 - \mathbf{v}_1) \cdot (\mathbf{v}_2 - \mathbf{v}_1) = \mathbf{v}_2 \cdot \mathbf{v}_2 - 2\mathbf{v}_2 \cdot \mathbf{v}_1 + \mathbf{v}_1 \cdot \mathbf{v}_1 \geq \mathbf{v}_2 \cdot \mathbf{v}_2 = |\mathbf{v}_2|^2,$$

and so $\mathbf{v}_2$ is the shortest edge of this triangle. The algorithm interchanges the vectors, and so we now have

$$|\mathbf{v}_1| < |\mathbf{v}_2|, \qquad 0 \leq \mathbf{v}_2 \cdot \mathbf{v}_1 \leq \frac{1}{2}|\mathbf{v}_2|^2,$$

where $\mathbf{v}_1$ is the shortest side of the triangle built from $\mathbf{v}_1$ and $\mathbf{v}_2$. Since **PGauss**[t] terminates at this point, we have (using the interchanged vectors)

$$|\mathbf{v}_2| \leq t|\mathbf{v}_1|.$$

Therefore

$$0 \leq \frac{\mathbf{v}_2 \cdot \mathbf{v}_1}{\mathbf{v}_1 \cdot \mathbf{v}_1} = \left(\frac{\mathbf{v}_1 \cdot \mathbf{v}_2}{\mathbf{v}_2 \cdot \mathbf{v}_2} \right) \left(\frac{\mathbf{v}_2 \cdot \mathbf{v}_2}{\mathbf{v}_1 \cdot \mathbf{v}_1} \right) \leq \frac{1}{2} t^2 \leq \frac{3}{2}.$$

The new value of $\mathbf{v}_2$ is therefore either $\mathbf{v}_2' = \mathbf{v}_2$ or $\mathbf{v}_2' = \pm(\mathbf{v}_2 - \mathbf{v}_1)$. Since $\mathbf{v}_1$ is shorter than both of these vectors, the centered algorithm terminates at the next iteration. □

Definition 2.17. We define the **length** $\mathcal{L}$ and the **inertia** $\mathcal{I}$ of the lattice basis $\mathbf{x}$, $\mathbf{y}$ in $\mathbb{Z}^2$ as follows:

$$\mathcal{L}(\mathbf{x}, \mathbf{y}) = \max(|\mathbf{x}|, |\mathbf{y}|), \qquad \mathcal{I}(\mathbf{x}, \mathbf{y}) = |\mathbf{x}|^2 + |\mathbf{y}|^2 \geq \mathcal{L}^2.$$

Lemma 2.18. *For $t > 1$, the number $n[t]$ of iterations of the parameterized Gaussian algorithm on the input basis $\mathbf{x}$, $\mathbf{y}$ of inertia $\mathcal{I}$ satisfies*

$$n[t] \leq \frac{1}{2} \log_t(\mathcal{I}) + 1.$$

Proof. After each iteration, if the algorithm does not terminate, then we have

$$|\mathbf{v}_1| > t|\mathbf{v}_2|, \quad \text{hence} \quad |\mathbf{v}_2| < \frac{1}{t}|\mathbf{v}_1|, \quad \text{hence} \quad |\mathbf{v}_1'| < \frac{1}{t}|\mathbf{v}_1|,$$

where $\mathbf{v}_1'$ is the new value of $\mathbf{v}_1$. Therefore, after n iterations, the shorter vector satisfies

$$|\mathbf{v}_1| < \frac{1}{t^n}|\mathbf{x}|.$$

Since all vectors are in $\mathbb{Z}^2$, the lattice contains no vectors of length < 1, and so the algorithm terminates as soon as

$$\frac{1}{t^n}|\mathbf{x}| \leq 1 \quad \Longleftrightarrow \quad t^n \geq |\mathbf{x}| \quad \Longleftrightarrow \quad n \geq \log_t |\mathbf{x}|.$$

Therefore

$$n[t] \leq \lceil \log_t |\mathbf{x}| \rceil \leq \log_t |\mathbf{x}| + 1 \leq \log_t \mathcal{L} + 1 \leq \frac{1}{2} \log_t \mathcal{I} + 1.$$

This completes the proof. □

Corollary 2.19. *The number of iterations n of the centered Gaussian algorithm on the input basis $\mathbf{x}$, $\mathbf{y}$ of inertia $\mathcal{I}$ satisfies*

$$n \leq \frac{\log \mathcal{I}}{\log 3} + 2.$$

Proof. For $t = \sqrt{3}$, combining Propositions 2.16 and 2.18 gives

$$n \leq n[\sqrt{3}] + 1 \leq \frac{1}{2} \log_{\sqrt{3}}(\mathcal{I}) + 2,$$

and this gives the result. □

2.4 Projects

Project 2.1. Write a report and present a seminar talk on the analysis by Vallée [139] of the "Gauss-acute algorithm". Explain in detail how this is used to determine the worst case configuration of the centered Gaussian algorithm.

Project 2.2. Write a report and present a seminar talk on the analysis of the Euclidean algorithm for the greatest common divisor of two integers. A detailed discussion of this algorithm (over $\mathbb{Z}$ and also over the polynomial ring $\mathbb{F}[x]$ where $\mathbb{F}$ is a field) may be found in von zur Gathen and Gerhard [147], Chapter 3; see also Knuth [78], Section 4.5.3. The first complete analysis of the running time of this algorithm is traditionally assigned to Lamé [86]. The first complete proof of precise upper bounds for the number of steps taken by both versions of the algorithm (original and centered) was given by Dupré [40]. A historical discussion of research on the Euclidean algorithm may be found in Shallit [128]. The first complete analysis of the average number of steps taken by several variants of the algorithm was given by Vallée [140].

Project 2.3. Another interesting topic for a written report and/or a seminar presentation: Daudé et al. [36] show that "the average-case complexity of the Gaussian algorithm, measured in the number of iterations performed, is asymptotically constant, and thus essentially independent of the size of the input vectors". The proofs involve continued fractions, continuant polynomials, complex analysis, and functional operators. This paper establishes that the average-case complexity of the algorithm is asymptotic to the constant $1.3511315744\ldots$, which can be expressed in terms of the tetralogarithm and the Riemann zeta function.

Project 2.4. Another interesting topic for a written report and/or a seminar presentation: B. Vallée and A. Vera [141] establish a connection between the Gaussian algorithm and Eisenstein series from the theory of modular forms. They study three main parameters related to the output basis: the first minimum, the Hermite constant, and the projection of a second minimum orthogonal to the first minimum. They obtain sharp estimates for the distribution of these quantities by relating them to the hyperbolic geometry of the complex plane. They also provide a dynamical analysis of parameters which describe the execution of the algorithm in a more precise way than the number of iterations, in particular the additive costs and the bit complexity.

Project 2.5. Another interesting topic for a written report and/or a seminar presentation: Kaib and Schnorr [69] generalize the Gaussian algorithm, which uses the Euclidean norm on $\mathbb{R}^2$, to the case of an arbitrary norm on $\mathbb{R}^2$, and extend Vallée's analysis [139] of the algorithm to this more general case.

Project 2.6. Another interesting topic for a written report and/or a seminar presentation: P. Q. Nguyen and D. Stehlé [106] study a "greedy" algorithm for lattice basis reduction in dimensions ≥ 3 with respect to the Euclidean norm; their algorithm is a natural generalization of the Gaussian algorithm. For dimensions 3 and 4, the output of their algorithm is optimal, and the bit-complexity is quadratic without fast integer arithmetic.

2.5 Exercises

Exercise 2.1. Use the original Euclidean algorithm to compute the greatest common divisor of

$$a = 10946, \qquad b = 3840.$$

Exercise 2.2. Repeat Exercise 2.1 using the centered Euclidean algorithm.

Exercise 2.3. Use the original Euclidean algorithm to compute the greatest common divisor of

$$a = F_{21} = 10395, \qquad b = F_{20} = 6765.$$

Exercise 2.4. Repeat Exercise 2.3 using the centered Euclidean algorithm.

Exercise 2.5. (a) Explain why $|q_i| \geq 1$ for all i in the original Euclidean algorithm. (b) Prove the exact formula for the Fibonacci numbers. Hint: Express the Fibonacci recurrence as the matrix equation,

$$\begin{bmatrix} F_{n+1} \\ F_n \end{bmatrix} = \begin{bmatrix} 1 & 1 \\ 1 & 0 \end{bmatrix} \begin{bmatrix} F_n \\ F_{n-1} \end{bmatrix}.$$

From this it follows that

$$\begin{bmatrix} F_{n+1} \\ F_n \end{bmatrix} = A^n \begin{bmatrix} 1 \\ 0 \end{bmatrix}, \qquad A = \begin{bmatrix} 1 & 1 \\ 1 & 0 \end{bmatrix}.$$

Diagonalize the matrix A: find a diagonal matrix D and an invertible matrix C such that $A = C^{-1}DC$. Then use the fact that $A^n = C^{-1}D^nC$.

Exercise 2.6. (a) Explain why $|q_i| \geq 2$ for all $i \geq 2$ in the centered Euclidean algorithm. (b) Prove an exact formula for the numbers G_n defined by the recurrence relation

$$G_0 = 0, \qquad G_1 = 1, \qquad G_n = 2\,G_{n-1} + G_{n-2} \quad (n \geq 2).$$

Hint: See the hint for Exercise 2.5.

Exercise 2.7. Lemma 2.2 gives an upper bound for the number of divisions in the original Euclidean algorithm. Prove an analogous upper bound for the centered Euclidean algorithm.

Exercise 2.8. Equation (2.1) gives a better upper bound for the number of divisions in the original Euclidean algorithm. Use the results of Exercise 2.6 to prove an analogous better upper bound for the centered Euclidean algorithm.

Exercise 2.9. Prove that the following three bases of $\mathbb{R}^2$ all generate the same two-dimensional lattice:

$$\begin{array}{lll}
\text{basis 1:} & \mathbf{x} = [\,0,\, 2\,] & \mathbf{y} = [\,5,\, 1\,] \\
\text{basis 2:} & \mathbf{x} = [\,85,\, -31\,] & \mathbf{y} = [\,-60,\, 22\,] \\
\text{basis 3:} & \mathbf{x} = [\,-230,\, 84\,] & \mathbf{y} = [\,545,\, -199\,]
\end{array}$$

Exercise 2.10. Apply the original Gaussian algorithm to these vectors in $\mathbb{R}^2$:

$$\mathbf{x} = [\,69,\, 11\,], \qquad \mathbf{y} = [\,88,\, 14\,].$$

Exercise 2.11. Apply the original Gaussian algorithm to these vectors in $\mathbb{R}^3$:

$$\mathbf{x} = [\,1,\, -38,\, 86\,], \qquad \mathbf{y} = [\,0,\, -27,\, 61\,].$$

Exercise 2.12. Apply the original Gaussian algorithm to these vectors in $\mathbb{R}^4$:

$$\mathbf{x} = [\,40,\, -82,\, -74,\, -5\,], \qquad \mathbf{y} = [\,29,\, -58,\, -45,\, -12\,].$$

Exercise 2.13. Apply the centered Gaussian algorithm to these vectors:

$$\mathbf{x} = [\,60,\, -33\,], \qquad \mathbf{y} = [\,97,\, -53\,].$$

Exercise 2.14. Apply the centered Gaussian algorithm to these vectors:

$$\mathbf{x} = [\,16,\, -26,\, 31\,], \qquad \mathbf{y} = [\,-42,\, 67,\, -80\,].$$

Exercise 2.15. Apply the centered Gaussian algorithm to these vectors:

$$\mathbf{x} = [\,-98,\, -40,\, 59,\, -92\,], \qquad \mathbf{y} = [\,73,\, 22,\, -35,\, 69\,].$$

Exercise 2.16. Redo Exercise 2.13 using the parameterized Gaussian algorithm with $t = \sqrt{3}$.

Exercise 2.17. Redo Exercise 2.14 using the parameterized Gaussian algorithm with $t = \sqrt{3}$.

Exercise 2.18. Redo Exercise 2.15 using the parameterized Gaussian algorithm with $t = \sqrt{3}$.

Exercise 2.19. Suppose that $\mathbf{x}$ and $\mathbf{y}$ are in $\mathbb{R}^2$, but not necessarily in $\mathbb{Z}^2$. Let ℓ be the length of the shortest (nonzero) vector in the lattice generated by $\mathbf{x}$ and $\mathbf{y}$. Prove that for $t > 1$, the number $n[t]$ of iterations of the parameterized Gaussian algorithm satisfies (see Definition 2.17)

$$n[t] \leq \frac{1}{2} \log_t \left(\frac{\mathcal{I}}{\ell} \right) + 1.$$

Exercise 2.20. Prove that for fixed $t > 1$, the number of iterations of the parameterized Gaussian algorithm is bounded by a polynomial in the size of the input.

3

Gram-Schmidt Orthogonalization

CONTENTS

In this chapter we review the classical Gram-Schmidt algorithm for converting an arbitrary basis of $\mathbb{R}^n$ into an orthogonal basis. This is a standard topic in elementary linear algebra, but we develop this material with a view to its application to the LLL algorithm.

3.1 The Gram-Schmidt theorem

Definition 3.1. Let $\mathbf{x}_1, \ldots, \mathbf{x}_n$ be a basis of $\mathbb{R}^n$. The **Gram-Schmidt orthogonalization (GSO)** of $\mathbf{x}_1, \ldots, \mathbf{x}_n$ is the following basis $\mathbf{x}_1^*, \ldots, \mathbf{x}_n^*$:

$$\mathbf{x}_1^* = \mathbf{x}_1,$$

$$\mathbf{x}_i^* = \mathbf{x}_i - \sum_{j=1}^{i-1} \mu_{ij} \mathbf{x}_j^* \quad (2 \leq i \leq n), \qquad \mu_{ij} = \frac{\mathbf{x}_i \cdot \mathbf{x}_j^*}{\mathbf{x}_j^* \cdot \mathbf{x}_j^*} \quad (1 \leq j < i \leq n).$$

We do not normalize the vectors. We note that if the vectors $\mathbf{x}_1, \ldots, \mathbf{x}_n$ are in $\mathbb{Q}^n$, then so are the vectors $\mathbf{x}_1^*, \ldots, \mathbf{x}_n^*$.

It is important to note that the Gram-Schmidt basis vectors $\mathbf{x}_1^*, \ldots, \mathbf{x}_n^*$ are usually not in the lattice generated by $\mathbf{x}_1, \ldots, \mathbf{x}_n$: in general $\mathbf{x}_1^*, \ldots, \mathbf{x}_n^*$ are not integral linear combinations of $\mathbf{x}_1, \ldots, \mathbf{x}_n$.

Remark 3.2. If we set $\mu_{ii} = 1$ for $1 \leq i \leq n$ then we have

$$\mathbf{x}_i = \sum_{j=1}^{i} \mu_{ij} \mathbf{x}_j^*.$$

We write $\mathbf{x}_i = (x_{i1}, \ldots, x_{in})$ and form the matrix $X = (x_{ij})$ in which row i is the vector $\mathbf{x}_i$, and similarly $X^* = (x_{ij}^*)$. If we set $\mu_{ij} = 0$ for $1 \le i < j \le n$ then Definition 3.1 can be written as the matrix equation

$$X = MX^*, \qquad M = (\mu_{ij}).$$

The matrix M is lower triangular with $\mu_{ii} = 1$ for all i, so it is invertible, and hence we also have $X^* = M^{-1}X$.

Example 3.3. A basis of $\mathbb{R}^3$ consists of the rows $\mathbf{x}_1, \mathbf{x}_2, \mathbf{x}_3$ of the matrix

$$X = \begin{bmatrix} 3 & -1 & 5 \\ -5 & 2 & -1 \\ -3 & 9 & 2 \end{bmatrix}.$$

Gram-Schmidt orthogonalization gives the orthogonal rows of this matrix

$$X^* = \begin{bmatrix} 3 & -1 & 5 \\ \frac{-109}{35} & \frac{48}{35} & \frac{15}{7} \\ \frac{1521}{566} & \frac{1859}{283} & \frac{-169}{566} \end{bmatrix}.$$

The corresponding matrix equation $X = MX^*$ is

$$\begin{bmatrix} 3 & -1 & 5 \\ -5 & 2 & -1 \\ -3 & 9 & 2 \end{bmatrix} = \begin{bmatrix} 1 & 0 & 0 \\ \frac{-22}{35} & 1 & 0 \\ \frac{-8}{35} & \frac{909}{566} & 1 \end{bmatrix} \begin{bmatrix} 3 & -1 & 5 \\ \frac{-109}{35} & \frac{48}{35} & \frac{15}{7} \\ \frac{1521}{566} & \frac{1859}{283} & \frac{-169}{566} \end{bmatrix}.$$

Inverting this gives the equation $X^* = M^{-1}X$:

$$\begin{bmatrix} 3 & -1 & 5 \\ \frac{-109}{35} & \frac{48}{35} & \frac{15}{7} \\ \frac{1521}{566} & \frac{1859}{283} & \frac{-169}{566} \end{bmatrix} = \begin{bmatrix} 1 & 0 & 0 \\ \frac{22}{35} & 1 & 0 \\ \frac{-221}{283} & \frac{-909}{566} & 1 \end{bmatrix} \begin{bmatrix} 3 & -1 & 5 \\ -5 & 2 & -1 \\ -3 & 9 & 2 \end{bmatrix}.$$

Theorem 3.4. Gram-Schmidt Theorem. *Let* $\mathbf{x}_1, \ldots, \mathbf{x}_n$ *be a basis of* $\mathbb{R}^n$ *and let* $\mathbf{x}_1^*, \ldots, \mathbf{x}_n^*$ *be its Gram-Schmidt orthogonalization. Let X (respectively X^*) be the $n \times n$ matrix in which row i is the vector $\mathbf{x}_i$ (respectively $\mathbf{x}_i^*$) for $1 \le i \le n$. We have:*

(a) $\mathbf{x}_i^* \cdot \mathbf{x}_j^* = 0$ *for* $1 \le i < j \le n$.

(b) $\mathrm{span}(\mathbf{x}_1^*, \ldots, \mathbf{x}_k^*) = \mathrm{span}(\mathbf{x}_1, \ldots, \mathbf{x}_k)$ *for* $1 \le k \le n$.

(c) *For* $1 \le k \le n$, *the vector* $\mathbf{x}_k^*$ *is the projection of* $\mathbf{x}_k$ *onto the orthogonal complement of* $\mathrm{span}(\mathbf{x}_1, \ldots, \mathbf{x}_{k-1})$.

(d) $|\mathbf{x}_k^*| \le |\mathbf{x}_k|$ *for* $1 \le k \le n$.

(e) $\det(X^*) = \det(X)$.

Proof. (a) Induction on j. For $j = 1$ there is nothing to prove. Assume that the claim holds for some $j \geq 1$. For $1 \leq i < j+1$ we have

$$
\mathbf{x}_i^* \cdot \mathbf{x}_{j+1}^* = \mathbf{x}_i^* \cdot \left(\mathbf{x}_{j+1} - \sum_{k=1}^{j} \mu_{j+1,k} \mathbf{x}_k^* \right) \qquad \text{definition of } \mathbf{x}_{j+1}^*
$$

$$
= \mathbf{x}_i^* \cdot \mathbf{x}_{j+1} - \sum_{k=1}^{j} \mu_{j+1,k} (\mathbf{x}_i^* \cdot \mathbf{x}_k^*) \qquad \text{bilinearity of scalar product}
$$

$$
= \mathbf{x}_i^* \cdot \mathbf{x}_{j+1} - \mu_{j+1,i} (\mathbf{x}_i^* \cdot \mathbf{x}_i^*) \qquad \text{inductive hypothesis}
$$

$$
= \mathbf{x}_i^* \cdot \mathbf{x}_{j+1} - \frac{\mathbf{x}_{j+1} \cdot \mathbf{x}_i^*}{\mathbf{x}_i^* \cdot \mathbf{x}_i^*} (\mathbf{x}_i^* \cdot \mathbf{x}_i^*) \qquad \text{definition of } \mu_{j+1,i}
$$

$$
= 0 \qquad \text{symmetry of scalar product}
$$

(b) By Remark 3.2 we have $\mathbf{x}_i \in \mathrm{span}(\mathbf{x}_1^*, \ldots, \mathbf{x}_k^*)$ for $1 \leq i \leq k$, and hence

$$
\mathrm{span}(\mathbf{x}_1, \ldots, \mathbf{x}_k) \subseteq \mathrm{span}(\mathbf{x}_1^*, \ldots, \mathbf{x}_k^*) \,.
$$

For the reverse inclusion we use induction on k. For $k = 1$ we have $\mathbf{x}_1^* = \mathbf{x}_1$ and so the claim is obvious. Assume that the claim holds for some $k \geq 1$. Using Definition 3.1 again we get

$$
\mathbf{x}_{k+1}^* = \mathbf{x}_{k+1} - \sum_{j=1}^{k} \mu_{k+1,j} \mathbf{x}_j^* = \mathbf{x}_{k+1} + \mathbf{y}, \qquad \mathbf{y} \in \mathrm{span}(\mathbf{x}_1^*, \ldots, \mathbf{x}_k^*) \,.
$$

The inductive hypothesis gives $\mathrm{span}(\mathbf{x}_1^*, \ldots, \mathbf{x}_k^*) \subseteq \mathrm{span}(\mathbf{x}_1, \ldots, \mathbf{x}_k)$, and so the last equation implies $\mathbf{x}_{k+1}^* \in \mathrm{span}(\mathbf{x}_1, \ldots, \mathbf{x}_{k+1})$. Therefore

$$
\mathrm{span}(\mathbf{x}_1^*, \ldots, \mathbf{x}_{k+1}^*) \subseteq \mathrm{span}(\mathbf{x}_1, \ldots, \mathbf{x}_{k+1}) \,.
$$

(c) To simplify notation we write $U = \mathrm{span}(\mathbf{x}_1, \ldots, \mathbf{x}_{k-1})$; then $U^\perp$ is the subspace of $\mathbb{R}^n$ consisting of all vectors $\mathbf{y}$ such that $\mathbf{y} \cdot \mathbf{x} = 0$ for every vector $\mathbf{x} \in U$. There is a unique decomposition $\mathbf{x}_k = \mathbf{x}_k' + \mathbf{y}$ where $\mathbf{x}_k' \in U^\perp$ and $\mathbf{y} \in U$; here $\mathbf{x}_k'$ is the projection of $\mathbf{x}_k$ onto the orthogonal complement of U. Remark 3.2 gives

$$
\mathbf{x}_k = \mathbf{x}_k^* + \sum_{j=1}^{k-1} \mu_{kj} \mathbf{x}_j^* \,.
$$

By part (b) we have $U = \mathrm{span}(\mathbf{x}_1^*, \ldots, \mathbf{x}_{k-1}^*)$, and therefore $\mathbf{x}_k^* = \mathbf{x}_k'$.

(d) Using part (a) we see that

$$
\mathbf{x}_k = \mathbf{x}_k^* + \sum_{j=1}^{k-1} \mu_{kj} \mathbf{x}_j^* \quad \text{implies} \quad |\mathbf{x}_k|^2 = |\mathbf{x}_k^*|^2 + \sum_{j=1}^{k-1} \mu_{kj}^2 |\mathbf{x}_j^*|^2 \,.
$$

Since every term in the sum is nonnegative, this proves the claim.

(e) By Remark 3.2 we have $X = MX^*$ where $M = (\mu_{ij})$ is a lower triangular matrix with $\mu_{ii} = 1$ for $1 \leq i \leq n$. Hence $\det(M) = 1$ and therefore $\det(X) = \det(M) \det(X^*) = \det(X^*)$. $\qquad \square$

A direct consequence of Theorem 3.4 is the following famous inequality for the determinant of a matrix.

Corollary 3.5. Hadamard's Inequality. *Let $X = (x_{ij})$ be an $n \times n$ matrix over $\mathbb{R}$, and let $B = \max_{i,j} |x_{ij}|$ be the maximum of the absolute values of its entries. Then*

$$|\det(X)| \leq n^{n/2} B^n.$$

Proof. Let $\mathbf{x}_i = (x_{i1}, \ldots, x_{in})$ for $1 \leq i \leq n$ be the row vectors of X. If the rows are linearly dependent, then $\det(X) = 0$ and the result is clear. If the row vectors are linearly independent, then let X^* be the matrix whose rows are the Gram-Schmidt orthogonal basis vectors $\mathbf{x}_1^*, \mathbf{x}_2^*, \ldots, \mathbf{x}_n^*$. By Theorem 3.4(e),

$$|\det(X)| = |\det(X^*)|.$$

Since $|\det(X^*)|$ is the volume of the n-dimensional parallelepiped spanned by the orthogonal vectors $\mathbf{x}_1^*, \mathbf{x}_2^*, \ldots, \mathbf{x}_n^*$, Theorem 3.4(a) implies

$$|\det(X^*)| = |\mathbf{x}_1^*||\mathbf{x}_2^*| \cdots |\mathbf{x}_n^*|.$$

Theorem 3.4(d) now gives

$$|\det(X)| \leq |\mathbf{x}_1||\mathbf{x}_2| \cdots |\mathbf{x}_n|.$$

For $1 \leq i \leq n$ we have

$$|\mathbf{x}_i| = \sqrt{x_{i1}^2 + \cdots + x_{in}^2} \leq \left(nB^2\right)^{1/2} = n^{1/2}B.$$

Therefore $|\det(X)| \leq \left(n^{1/2}B\right)^n$, which completes the proof. $\qquad\square$

Let the rows of the matrix $X = (x_{ij})$ be a basis $\mathbf{x}_1, \mathbf{x}_2, \ldots, \mathbf{x}_n$ of $\mathbb{R}^n$. Let the rows $\mathbf{x}_1^*, \mathbf{x}_2^*, \ldots, \mathbf{x}_n^*$ of the matrix $X^* = (x_{ij}^*)$ be the Gram-Schmidt orthogonalization. The matrix X^* is orthogonal in the sense that $X^*(X^*)^t$ is a diagonal matrix; its diagonal entries are $|\mathbf{x}_1^*|^2, \ldots, |\mathbf{x}_n^*|^2$. We obtain an orthonormal basis $\overline{\mathbf{x}}_1^*, \ldots, \overline{\mathbf{x}}_n^*$ by setting

$$\overline{\mathbf{x}}_i^* = \frac{\mathbf{x}_i^*}{|\mathbf{x}_i^*|} \quad (1 \leq i \leq n).$$

Let $\overline{X}^* = (\overline{x}_{ij}^*)$ be the matrix in which row i contains the vector $\overline{\mathbf{x}}_i^*$. Let D be the diagonal matrix with $d_{ii} = |\mathbf{x}_i^*|$ for $1 \leq i \leq n$. Since $\mathbf{x}_i^*$ is nonzero for all i, we know that D is invertible. We have $\overline{X}^* = D^{-1}X^*$, and $\overline{X}^*$ is orthonormal in the sense that $\overline{X}^*\left(\overline{X}^*\right)^t = I_n$. From $X = MX^*$ we obtain

$$X = \left(MD\right)\left(D^{-1}X^*\right).$$

This is the LQ decomposition of the matrix X, since we have a lower triangular matrix $L = MD$ on the left and an orthonormal matrix $Q = D^{-1}X^*$ on the right. If we transpose the matrices in the last equation, we get

$$X^t = (D^{-1}X^*)^t (MD)^t.$$

This is the QR decomposition of X^t, with an orthonormal matrix $Q = (DX^*)^t$ on the left and an upper triangular matrix $R = (MD^{-1})^t$ on the right.

Example 3.6. Continuing from Example 3.3, we see that the norms of the orthogonal basis vectors $\mathbf{x}_1^*$, $\mathbf{x}_2^*$, $\mathbf{x}_3^*$, are the diagonal entries of D:

$$D = \begin{bmatrix} |\mathbf{x}_1^*| & 0 & 0 \\ 0 & |\mathbf{x}_2^*| & 0 \\ 0 & 0 & |\mathbf{x}_3^*| \end{bmatrix} = \begin{bmatrix} \sqrt{35} & 0 & 0 \\ 0 & \sqrt{\frac{566}{35}} & 0 \\ 0 & 0 & \frac{169}{\sqrt{566}} \end{bmatrix}$$

Floating point arithmetic gives an approximate LQ decomposition of X:

$$X = MD \approx \begin{bmatrix} 5.916079783 & 0 & 0 \\ -3.718678721 & 4.021371480 & 0 \\ -1.352246808 & 6.458351016 & 7.103599844 \end{bmatrix}$$

$$Q = D^{-1}X^* \approx \begin{bmatrix} 0.5070925528 & -0.1690308509 & 0.8451542550 \\ -0.7744337302 & 0.3410350371 & 0.5328672455 \\ 0.3782982166 & 0.9247289739 & -0.0420331352 \end{bmatrix}$$

Considering Hadamard's inequality, we have $\det(X) = -169$ and

$$|\mathbf{x}_1^*|\,|\mathbf{x}_2^*|\,|\mathbf{x}_3^*| = \sqrt{35}\sqrt{\frac{566}{35}}\frac{169}{\sqrt{566}} = 169 = |\det(X)|\,,$$

$$|\mathbf{x}_1|\,|\mathbf{x}_2|\,|\mathbf{x}_3| = \sqrt{35}\sqrt{30}\sqrt{94} \approx 314.1655614\,.$$

Since $n = 3$ and $B = 9$ for the matrix X, we get

$$n^{n/2}B^n = 3^{3/2}9^3 \approx 3787.995117\,.$$

The inequalities in the proof of Corollary 3.5 can be summarized as

$$|\det(X)| \le |\mathbf{x}_1|\,|\mathbf{x}_2|\,|\mathbf{x}_3| \le n^{n/2}B^n \qquad \text{or} \qquad 169 \le 314.166 \le 3787.995\,.$$

In this example the upper bound on $|\det(X)|$ is very weak.

We reformulate the definition of the Gram matrix of a basis of $\mathbb{R}^n$.

Definition 3.7. Let $\mathbf{x}_1, \ldots, \mathbf{x}_n$ be a basis of $\mathbb{R}^n$, and let X be the $n \times n$ matrix with $\mathbf{x}_i$ in row i for $1 \le i \le n$. For $1 \le k \le n$, let X_k be the $k \times n$ matrix consisting of the first k rows of F. The **k-th Gram matrix** of this basis is the $k \times k$ symmetric matrix

$$G_k = X_k X_k^t.$$

The k-**th Gram determinant** of this basis is

$$d_k = \det(G_k).$$

For convenience we set $d_0 = 1$. If $\mathbf{x}_i \in \mathbb{Z}^n$ for all i then $d_k \in \mathbb{Z}$ for $0 \leq k \leq n$. This fact will be important later in our analysis of the LLL algorithm.

Proposition 3.8. *Let $\mathbf{x}_1, \ldots, \mathbf{x}_n$ be a basis of $\mathbb{R}^n$, and let $\mathbf{x}_1^*, \ldots, \mathbf{x}_n^*$ be its Gram-Schmidt orthogonalization. For $1 \leq k \leq n$ the k-th Gram determinant of the basis is the product of the square-lengths of the GSO vectors:*

$$d_k = \prod_{i=1}^{k} |\mathbf{x}_i^*|^2 \,.$$

Proof. By Remark 3.2 we can express the Gram-Schmidt orthogonalization as the matrix equation $X = MX^*$. Let M_k be the upper left $k \times k$ submatrix of M, and let X_k^* be the $k \times n$ matrix consisting of the first k rows of X^*. We have the factorization $X_k = M_k X_k^*$ where $\det(M_k) = 1$. Therefore

$$
\begin{aligned}
d_k &= \det\left(X_k X_k^t\right) && \text{definition of } d_k \\
&= \det\left((M_k X_k^*)(M_k X_k^*)^t\right) && \text{factorization of } X_k \\
&= \det\left(M_k \left(X_k^*(X_k^*)^t\right) M_k^t\right) && \text{apply transpose and associativity} \\
&= \det(M_k) \det\left(X_k^*(X_k^*)^t\right) \det(M_k^t) && \det(ABC) = \det(A)\det(B)\det(C) \\
&= \det\left(X_k^*(X_k^*)^t\right) && \det(M_k) = 1.
\end{aligned}
$$

Since the rows $\mathbf{x}_1^*, \mathbf{x}_2^*, \ldots, \mathbf{x}_k^*$ of the matrix X_k^* are orthogonal by Theorem 3.4(a), we see that $X_k^*(X_k^*)^t$ is a diagonal matrix with diagonal entries $|\mathbf{x}_1^*|^2$, $|\mathbf{x}_2^*|^2, \ldots, |\mathbf{x}_k^*|^2$, and this completes the proof. $\qquad\square$

Example 3.9. Continuing from Example 3.6, for $k = 1$ we get

$$X_1 = \begin{bmatrix} 3 & -1 & 5 \end{bmatrix}, \qquad X_1 X_1^t = \begin{bmatrix} 35 \end{bmatrix}, \qquad d_1 = 35 = |\mathbf{x}_1^*|^2.$$

For $k = 2$ we get

$$X_2 = \begin{bmatrix} 3 & -1 & 5 \\ -5 & 2 & -1 \end{bmatrix}, \qquad X_2 X_2^t = \begin{bmatrix} 35 & -22 \\ -22 & 30 \end{bmatrix},$$

and so $d_2 = 566 = |\mathbf{x}_1^*|^2 |\mathbf{x}_2^*|^2$. For $k = 3$ we get

$$X_3 = \begin{bmatrix} 3 & -1 & 5 \\ -5 & 2 & -1 \\ -3 & 9 & 2 \end{bmatrix}, \qquad X_3 X_3^t = \begin{bmatrix} 35 & -22 & -8 \\ -22 & 30 & 31 \\ -8 & 31 & 94 \end{bmatrix},$$

and so $d_3 = 28561 = |\mathbf{x}_1^*|^2 |\mathbf{x}_2^*|^2 |\mathbf{x}_3^*|^2$.

3.2 Complexity of the Gram-Schmidt process

The next lemma is a familiar result from elementary linear algebra.

Lemma 3.10. Cramer's Rule. *Let A be an $n \times n$ matrix over $\mathbb{R}$ with $\det(A) \neq 0$, let $\mathbf{y} \in \mathbb{R}^n$ be a column vector, and let $\mathbf{x} = [x_1, \ldots, x_n]^t$ be the unique solution of the linear system $A\mathbf{x} = \mathbf{y}$. For $1 \leq i \leq n$ we have*

$$x_i = \frac{\det(A_i)}{\det(A)},$$

where A_i is the matrix obtained from A by replacing column i by $\mathbf{y}$.

We can now give bounds on the denominators of the rational numbers that appear in the Gram-Schmidt orthogonalization of vectors with integral components. This will be important in our analysis of the LLL algorithm.

Proposition 3.11. *Let $\mathbf{x}_1, \ldots, \mathbf{x}_n$ be a basis of $\mathbb{R}^n$ with $\mathbf{x}_i \in \mathbb{Z}^n$ for $1 \leq i \leq n$. Let $\mathbf{x}_1^*, \ldots, \mathbf{x}_n^* \in \mathbb{Q}^n$ be its Gram-Schmidt orthogonalization with coefficients $\mu_{ij} \in \mathbb{Q}$. For $1 \leq k \leq n$, let d_k be the corresponding Gram determinant. Then:*

(a) The vector $d_{k-1}\mathbf{x}_k^*$ has integral components for $1 \leq k \leq n$.

(b) The quantity $d_j\mu_{kj}$ is an integer for $1 \leq j \leq k$.

(c) We have $|\mu_{kj}| \leq d_{j-1}^{1/2}\,|\mathbf{x}_k|$ for $1 \leq j \leq k$.

Proof. (a) We can express the Gram-Schmidt orthogonalization by the matrix equation $X = MX^*$, or equivalently, $X^* = M^{-1}X$, where M^{-1} (like M) is a lower triangular matrix with every diagonal entry equal to 1. Therefore

$$\mathbf{x}_k^* = \mathbf{x}_k - \sum_{j=1}^{k-1} \lambda_{kj}\mathbf{x}_j, \qquad \lambda_{kj} \in \mathbb{Q}. \tag{3.1}$$

Here, unlike in Definition 3.1, all the vectors on the right side are "unstarred"; the λ_{kj} are the "inverse Gram-Schmidt coefficients". Consider the vector $\mathbf{x}_i$ for $1 \leq i \leq k-1$. By Theorem 3.4(a,b) we have $\mathbf{x}_i \cdot \mathbf{x}_k^* = 0$. Taking the scalar product of both sides of equation (3.1) with $\mathbf{x}_i$ we get

$$0 = \mathbf{x}_i \cdot \mathbf{x}_k - \sum_{j=1}^{k-1} \lambda_{kj}(\mathbf{x}_i \cdot \mathbf{x}_j).$$

Rearranging this equation gives

$$\sum_{j=1}^{k-1} (\mathbf{x}_i \cdot \mathbf{x}_j)\lambda_{kj} = \mathbf{x}_i \cdot \mathbf{x}_k.$$

For fixed k, there are $k-1$ choices for i, and so we have a linear system of $k-1$ equations in $k-1$ variables $\lambda_{k1}, \lambda_{k2}, \ldots, \lambda_{k,k-1}$. The coefficient matrix of this system is the Gram matrix G_{k-1}, and the determinant of G_{k-1} is d_{k-1}. Cramer's Rule implies that $d_{k-1}\lambda_{kj} \in \mathbb{Z}$ for $1 \leq j \leq k-1$. We now have

$$d_{k-1}\mathbf{x}_k^* = d_{k-1}\left(\mathbf{x}_k - \sum_{j=1}^{k-1} \lambda_{kj}\mathbf{x}_j\right) = d_{k-1}\mathbf{x}_k - \sum_{j=1}^{k-1}\left(d_{k-1}\lambda_{kj}\right)\mathbf{x}_j.$$

Thus for $1 \leq k \leq n$ the vector $d_{k-1}\mathbf{x}_k^*$ is a linear combination with integral coefficients of vectors with integral components.

(b) From Proposition 3.8 we see that $|\mathbf{x}_j^*|^2 = d_j/d_{j-1}$, and so using the formula for μ_{kj} from Definition 3.1 we have

$$d_j\mu_{kj} = d_j\frac{\mathbf{x}_k \cdot \mathbf{x}_j^*}{\mathbf{x}_j^* \cdot \mathbf{x}_j^*} = d_j\frac{\mathbf{x}_k \cdot \mathbf{x}_j^*}{|\mathbf{x}_j^*|^2} = d_{j-1}\left(\mathbf{x}_k \cdot \mathbf{x}_j^*\right) = \mathbf{x}_k \cdot \left(d_{j-1}\mathbf{x}_j^*\right).$$

By assumption $\mathbf{x}_k \in \mathbb{Z}^n$, and by part (a), $d_{j-1}\mathbf{x}_j^* \in \mathbb{Z}^n$; hence $d_j\mu_{kj} \in \mathbb{Z}$.

(c) Using the formula for μ_{kj} from Definition 3.1 and the Cauchy-Schwarz inequality, we get

$$|\mu_{kj}| = \left|\frac{\mathbf{x}_k \cdot \mathbf{x}_j^*}{\mathbf{x}_j^* \cdot \mathbf{x}_j^*}\right| = \frac{|\mathbf{x}_k \cdot \mathbf{x}_j^*|}{|\mathbf{x}_j^*|^2} \leq \frac{|\mathbf{x}_k||\mathbf{x}_j^*|}{|\mathbf{x}_j^*|^2} = \frac{|\mathbf{x}_k|}{|\mathbf{x}_j^*|}.$$

Since $\mathbf{x}_1, \ldots, \mathbf{x}_j \in \mathbb{Z}^n$, we see by Definition 3.7 that $d_j \in \mathbb{Z}$, and since $d_j > 0$ we get $d_j \geq 1$. Using this and Proposition 3.8 we get

$$|\mathbf{x}_j^*|^2 = \frac{d_j}{d_{j-1}} \geq \frac{1}{d_{j-1}}, \quad \text{hence} \quad |\mathbf{x}_j^*| \geq \frac{1}{d_{j-1}^{1/2}}.$$

This completes the proof.   □

Example 3.12. Continuing from Example 3.9, part (a) of Proposition 3.11 says that if we multiply row k of X^* by d_{k-1}, then we clear the denominators in the orthogonal basis vectors:

$$\begin{aligned}
d_0 &= 1 &\implies& d_0f_1^* = [3, -1, 5], \\
d_1 &= 35 &\implies& d_1f_2^* = [-109, 48, 75], \\
d_2 &= 566 &\implies& d_2f_3^* = [1521, 3718, -169].
\end{aligned}$$

Part (b) of Proposition 3.11 says that if we multiply column j of the matrix M by d_j, then we clear denominators in the μ_{kj}:

$$d_1\begin{bmatrix} \mu_{11} \\ \mu_{21} \\ \mu_{31} \end{bmatrix} = 35\begin{bmatrix} 1 \\ \frac{-22}{35} \\ \frac{-8}{35} \end{bmatrix} = \begin{bmatrix} 35 \\ -22 \\ -8 \end{bmatrix},$$

$$d_2 \begin{bmatrix} \mu_{12} \\ \mu_{22} \\ \mu_{32} \end{bmatrix} = 566 \begin{bmatrix} 0 \\ 1 \\ \frac{909}{566} \end{bmatrix} = \begin{bmatrix} 0 \\ 566 \\ 909 \end{bmatrix}.$$

This is trivial for $k = 3$.

3.3 Further results on the Gram-Schmidt process

An important part of the LLL algorithm is the repeated exchange of consecutive basis vectors. The next proposition shows how these exchanges affect the Gram-Schmidt orthogonalization of the basis.

Proposition 3.13. *Let* $\mathbf{x}_1, \mathbf{x}_2, \ldots, \mathbf{x}_n$ *be a basis of* $\mathbb{R}^n$*, and let* $\mathbf{x}_1^*, \mathbf{x}_2^*, \ldots, \mathbf{x}_n^*$ *be its Gram-Schmidt orthogonalization. Let* j *be in the range* $1 \le j \le n-1$*, and let* $\widehat{\mathbf{x}}_1, \widehat{\mathbf{x}}_2, \ldots, \widehat{\mathbf{x}}_n$ *be the new basis obtained by exchanging* $\mathbf{x}_j$ *and* $\mathbf{x}_{j+1}$*:*

$$\widehat{\mathbf{x}}_j = \mathbf{x}_{j+1}, \qquad \widehat{\mathbf{x}}_{j+1} = \mathbf{x}_j, \qquad \widehat{\mathbf{x}}_i = \mathbf{x}_i \quad (i \ne j, j+1).$$

Let $\widehat{\mathbf{x}}_1^*, \widehat{\mathbf{x}}_2^*, \ldots, \widehat{\mathbf{x}}_n^*$ *be the Gram-Schmidt orthogonalization of the new basis. Then* $\widehat{\mathbf{x}}_i^* = \mathbf{x}_i^*$ *for* $i \ne j, j+1$ *but*

$$\widehat{\mathbf{x}}_j^* = \mathbf{x}_{j+1}^* + \mu_{j+1,j}\mathbf{x}_j^*, \qquad \widehat{\mathbf{x}}_{j+1}^* = \frac{|\mathbf{x}_{j+1}^*|^2}{|\widehat{\mathbf{x}}_j^*|^2}\mathbf{x}_j^* - \mu_{j+1,j}\frac{|\mathbf{x}_j^*|^2}{|\widehat{\mathbf{x}}_j^*|^2}\mathbf{x}_{j+1}^*.$$

Proof. Since $\widehat{\mathbf{x}}_i = \mathbf{x}_i$ for $i \ne j, j+1$ it is clear that $\widehat{\mathbf{x}}_i^* = \mathbf{x}_i^*$ for $i < j$, and Theorem 3.4(c) implies the same result for $i > j+1$. For $\widehat{\mathbf{x}}_j^*$ we have

$$\widehat{\mathbf{x}}_j^* = \widehat{\mathbf{x}}_j - \sum_{i=1}^{j-1} \frac{\widehat{\mathbf{x}}_j \cdot \widehat{\mathbf{x}}_i^*}{\widehat{\mathbf{x}}_i^* \cdot \widehat{\mathbf{x}}_i^*}\widehat{\mathbf{x}}_i^* \qquad \text{Definition 3.1}$$

$$= \mathbf{x}_{j+1} - \sum_{i=1}^{j-1} \frac{\mathbf{x}_{j+1} \cdot \mathbf{x}_i^*}{\mathbf{x}_i^* \cdot \mathbf{x}_i^*}\mathbf{x}_i^* \qquad \text{definition of } \widehat{\mathbf{x}}_1, \ldots, \widehat{\mathbf{x}}_n$$

$$= \mathbf{x}_{j+1} - \sum_{i=1}^{j-1} \mu_{j+1,i}\mathbf{x}_i^* \qquad \text{Definition 3.1}$$

$$= \mathbf{x}_{j+1} - \sum_{i=1}^{j} \mu_{j+1,i}\mathbf{x}_i^* + \mu_{j+1,j}\mathbf{x}_j^* \qquad \text{add and subtract term}$$

$$= \mathbf{x}_{j+1}^* + \mu_{j+1,j}\mathbf{x}_j^* \qquad \text{Definition 3.1}$$

Since $\mathbf{x}_j^*$ and $\mathbf{x}_{j+1}^*$ are orthogonal, it follows that

$$|\widehat{\mathbf{x}}_j^*|^2 = |\mathbf{x}_{j+1}^*|^2 + \mu_{j+1,j}^2|\mathbf{x}_j^*|^2. \tag{3.2}$$

We also have

$$
\begin{aligned}
\mathbf{x}_j \cdot \widehat{\mathbf{x}}_j^* &= \mathbf{x}_j \cdot \left(\mathbf{x}_{j+1}^* + \mu_{j+1,j} \mathbf{x}_j^* \right) \\
&= \mathbf{x}_j \cdot \mathbf{x}_{j+1}^* + \mu_{j+1,j}\, \mathbf{x}_j \cdot \mathbf{x}_j^* && \text{bilinearity of scalar product} \\
&= \mu_{j+1,j} \left(\sum_{k=1}^{j} \mu_{kj} \mathbf{x}_k^* \right) \cdot \mathbf{x}_j^* && \mathbf{x}_j \cdot \mathbf{x}_{j+1}^* = 0 \text{ and Remark 3.2} \\
&= \mu_{j+1,j} \sum_{k=1}^{j} \mu_{kj} \left(\mathbf{x}_k^* \cdot \mathbf{x}_j^* \right) && \text{bilinearity of scalar product} \\
&= \mu_{j+1,j}\, \mathbf{x}_j^* \cdot \mathbf{x}_j^* && \mathbf{x}_k \cdot \mathbf{x}_j^* = 0 \text{ for } k < j, \text{ and } \mu_{jj} = 1
\end{aligned}
$$

Therefore

$$
\mathbf{x}_j \cdot \widehat{\mathbf{x}}_j^* = \mu_{j+1,j} |\mathbf{x}_j^*|^2. \tag{3.3}
$$

For $\widehat{\mathbf{x}}_{j+1}^*$ we have

$$
\begin{aligned}
\widehat{\mathbf{x}}_{j+1}^* &= \widehat{\mathbf{x}}_{j+1} - \sum_{i=1}^{j} \frac{\widehat{\mathbf{x}}_{j+1} \cdot \widehat{\mathbf{x}}_i^*}{\widehat{\mathbf{x}}_i^* \cdot \widehat{\mathbf{x}}_i^*} \widehat{\mathbf{x}}_i^* && \text{Definition 3.1} \\
&= \mathbf{x}_j - \sum_{i=1}^{j} \frac{\mathbf{x}_j \cdot \widehat{\mathbf{x}}_i^*}{\widehat{\mathbf{x}}_i^* \cdot \widehat{\mathbf{x}}_i^*} \widehat{\mathbf{x}}_i^* && \text{definition of } \widehat{\mathbf{x}}_{j+1} \\
&= \mathbf{x}_j - \sum_{i=1}^{j-1} \frac{\mathbf{x}_j \cdot \widehat{\mathbf{x}}_i^*}{\widehat{\mathbf{x}}_i^* \cdot \widehat{\mathbf{x}}_i^*} \widehat{\mathbf{x}}_i^* - \frac{\mathbf{x}_j \cdot \widehat{\mathbf{x}}_j^*}{\widehat{\mathbf{x}}_j^* \cdot \widehat{\mathbf{x}}_j^*} \widehat{\mathbf{x}}_j^* && \text{separate last term} \\
&= \mathbf{x}_j - \sum_{i=1}^{j-1} \frac{\mathbf{x}_j \cdot \mathbf{x}_i^*}{\mathbf{x}_i^* \cdot \mathbf{x}_i^*} \mathbf{x}_i^* - \frac{\mathbf{x}_j \cdot \widehat{\mathbf{x}}_j^*}{\widehat{\mathbf{x}}_j^* \cdot \widehat{\mathbf{x}}_j^*} \widehat{\mathbf{x}}_j^* && \text{since } \widehat{\mathbf{x}}_i^* = \mathbf{x}_i^* \text{ for } i < j \\
&= \mathbf{x}_j^* - \frac{\mathbf{x}_j \cdot \widehat{\mathbf{x}}_j^*}{\widehat{\mathbf{x}}_j^* \cdot \widehat{\mathbf{x}}_j^*} \widehat{\mathbf{x}}_j^* && \text{definition of } \mathbf{x}_j^* \\
&= \mathbf{x}_j^* - \frac{\mathbf{x}_j \cdot \widehat{\mathbf{x}}_j^*}{\widehat{\mathbf{x}}_j^* \cdot \widehat{\mathbf{x}}_j^*} \left(\mathbf{x}_{j+1}^* + \mu_{j+1,j} \mathbf{x}_j^* \right) && \text{by the result for } \widehat{\mathbf{x}}_j^* \\
&= \mathbf{x}_j^* - \frac{\mu_{j+1,j} |\mathbf{x}_j^*|^2}{|\widehat{\mathbf{x}}_j^*|^2} \left(\mathbf{x}_{j+1}^* + \mu_{j+1,j} \mathbf{x}_j^* \right) && \text{equation (3.3)} \\
&= \left(1 - \frac{\mu_{j+1,j}^2 |\mathbf{x}_j^*|^2}{|\widehat{\mathbf{x}}_j^*|^2} \right) \mathbf{x}_j^* - \frac{\mu_{j+1,j} |\mathbf{x}_j^*|^2}{|\widehat{\mathbf{x}}_j^*|^2} \mathbf{x}_{j+1}^* && \text{rearranging} \\
&= \frac{|\mathbf{x}_{j+1}^*|^2}{|\widehat{\mathbf{x}}_j^*|^2} \mathbf{x}_j^* - \mu_{j+1,j} \frac{|\mathbf{x}_j^*|^2}{|\widehat{\mathbf{x}}_j^*|^2} \mathbf{x}_{j+1}^* && \text{equation (3.2)}
\end{aligned}
$$

This completes the proof.   □

Note that we can use equation (3.2) to write $\widehat{\mathbf{x}}_{j+1}^*$ entirely in terms of "unhatted" vectors:

$$\widehat{\mathbf{x}}^*_{j+1} = \frac{|\mathbf{x}^*_{j+1}|^2}{|\mathbf{x}^*_{j+1}|^2 + \mu^2_{j+1,j}|\mathbf{x}^*_j|^2}\mathbf{x}^*_j - \mu_{j+1,j}\frac{|\mathbf{x}^*_j|^2}{|\mathbf{x}^*_{j+1}|^2 + \mu^2_{j+1,j}|\mathbf{x}^*_j|^2}\mathbf{x}^*_{j+1}.$$

The next result connects Gram-Schmidt orthogonalization with short vectors in lattices: it gives a lower bound for the length of a nonzero lattice vector in terms of the Gram-Schmidt orthogonalization of the lattice basis.

Proposition 3.14. *Let $\mathbf{x}_1$, $\mathbf{x}_2$, ..., $\mathbf{x}_n$ be a basis of $\mathbb{R}^n$, and let $\mathbf{x}^*_1$, $\mathbf{x}^*_2$, ..., $\mathbf{x}^*_n$ be its Gram-Schmidt orthogonalization. Let L be the lattice generated by $\mathbf{x}_1$, $\mathbf{x}_2$, ..., $\mathbf{x}_n$. For any nonzero $\mathbf{y} \in L$ we have*

$$|\mathbf{y}| \geq \min\{|\mathbf{x}^*_1|, |\mathbf{x}^*_2|, \ldots, |\mathbf{x}^*_n|\}.$$

That is, any nonzero lattice vector is at least as long as the shortest vector in the Gram-Schmidt orthogonalization.

Proof. Let $\mathbf{y}$ be any nonzero element of L:

$$\mathbf{y} = \sum_{i=1}^n r_i\mathbf{x}_i,$$

where $r_i \in \mathbb{Z}$ for $1 \leq i \leq n$. Since $\mathbf{y} \neq 0$ we have $r_i \neq 0$ for some i; let k be the largest index with $r_k \neq 0$. Using Definition 3.1, we can express $\mathbf{x}_1$, $\mathbf{x}_2$, ..., $\mathbf{x}_n$ in terms of $\mathbf{x}^*_1$, $\mathbf{x}^*_2$, ..., $\mathbf{x}^*_n$:

$$\mathbf{y} = \sum_{i=1}^k r_i \sum_{j=1}^i \mu_{ij}\mathbf{x}^*_j = \sum_{i=1}^k \sum_{j=1}^i r_i\mu_{ij}\mathbf{x}^*_j.$$

Reversing the order of summation, and using $\mu_{kk} = 1$, we obtain

$$\mathbf{y} = \sum_{j=1}^k \left(\sum_{i=j}^k r_i\mu_{ij}\right)\mathbf{x}^*_j = r_k\mathbf{x}^*_k + \sum_{j=1}^{k-1}\nu_j\mathbf{x}^*_j,$$

for some $\nu_1, \ldots, \nu_{k-1} \in \mathbb{R}$. Since $\mathbf{x}^*_1$, $\mathbf{x}^*_2$, ..., $\mathbf{x}^*_n$ are orthogonal, we obtain

$$|\mathbf{y}|^2 = r_k^2|\mathbf{x}^*_k|^2 + \sum_{j=1}^{k-1}\nu_j^2|\mathbf{x}^*_j|^2.$$

Since r_k is a nonzero integer, we have $r_k^2 \geq 1$, and so

$$|\mathbf{y}|^2 \geq |\mathbf{x}^*_k|^2 + \sum_{j=1}^{k-1}\nu_j^2|\mathbf{x}^*_j|^2.$$

All the terms in the sum are non-negative, and hence

$$|\mathbf{y}|^2 \geq |\mathbf{x}^*_k|^2 \geq \min\{|\mathbf{x}^*_1|^2, \ldots, |\mathbf{x}^*_n|^2\}.$$

Taking square roots completes the proof. □

- *Input*: A basis $\mathbf{x}_1, \ldots, \mathbf{x}_n$ of $\mathbb{R}^n$.

- *Output*: An orthogonal basis $\mathbf{x}_1^*, \ldots, \mathbf{x}_n^*$ of $\mathbb{R}^n$ satisfying Theorem 3.4.

    (1) For $i = 1, 2, \ldots, n$ do:
        (a) Set $\mathbf{x}_i^* \leftarrow \mathbf{x}_i$.
        (b) For $j = 1, 2, \ldots, i-1$ do:
            (i) Set $\mu_{ij} \leftarrow \dfrac{\mathbf{x}_i \cdot \mathbf{x}_j^*}{\mathbf{x}_j^* \cdot \mathbf{x}_j^*}$.
            (ii) Set $\mathbf{x}_i^* \leftarrow \mathbf{x}_i^* - \mu_{ij}\mathbf{x}_j^*$.

FIGURE 3.1
The Gram-Schmidt algorithm to compute an orthogonal basis

In Figure 3.1 we reformulate Gram-Schmidt orthogonalization (Definition 3.1) in the algorithmic form that we will need in the next chapter. (Note that step (1)(b) does nothing when $i = 1$.)

3.4   Projects

Project 3.1. The Hadamard inequality, combined with the Chinese remainder theorem, is a useful tool in modular computations of large integer determinants. (See for example, von zur Gathen and Gerhard [147], Chapter 5.) Write a report and present a seminar talk on this topic.

Project 3.2. Study the behavior of the Gram-Schmidt algorithm when floating point arithmetic is used instead of exact rational arithmetic. Useful references are Golub and van Loan [49] and Trefethen and Bau [137].

Project 3.3. Study the application of Gram-Schmidt and other orthogonalization methods to least squares solutions of overdetermined linear systems. Useful references are Golub and van Loan [49] and Trefethen and Bau [137].

3.5 Exercises

Exercise 3.1. Consider this matrix:

$$X = \begin{bmatrix} 4 & 5 & 1 \\ 4 & 8 & 2 \\ 6 & 2 & 6 \end{bmatrix}$$

Let $\mathbf{x}_1, \mathbf{x}_2, \mathbf{x}_3$ be the rows of X.

    (a) Compute the Gram-Schmidt orthogonalization $\mathbf{x}_1^*, \mathbf{x}_2^*, \mathbf{x}_3^*$.

    (b) Express the GSO in matrix form: $X = MX^*$ where $M = (\mu_{ij})$.

    (c) Verify that $|\mathbf{x}_i^*| \leq |\mathbf{x}_i|$ for $i = 1, \ldots, n$.

    (d) Verify Hadamard's inequality for X.

Exercise 3.2. Same as Exercise 3.1, but for $n = 4$ and the matrix

$$X = \begin{bmatrix} 8 & 6 & 2 & 6 \\ 2 & 4 & 8 & 4 \\ 1 & -1 & -6 & -8 \\ 2 & 6 & -9 & 6 \end{bmatrix}$$

Exercise 3.3. Same as Exercise 3.1, but for $n = 5$ and the matrix

$$X = \begin{bmatrix} -6 & -5 & -8 & -4 & -1 \\ 5 & -4 & -6 & 9 & 4 \\ -8 & 1 & 6 & 2 & -7 \\ -5 & -3 & 6 & 0 & -7 \\ -1 & 9 & 2 & 0 & -8 \end{bmatrix}$$

Exercise 3.4. Let $\mathbf{x}_1, \mathbf{x}_2, \ldots, \mathbf{x}_n$ be a basis of $\mathbb{R}^n$, and let X be the $n \times n$ matrix with $\mathbf{x}_i$ as row i for $1 \leq i \leq n$. Let $X = MX^*$ be a factorization of X satisfying these two conditions:

    (a) $M = (\mu_{ij})$ is lower triangular and $\mu_{ii} = 1$ for $1 \leq i \leq n$,

    (b) $X^* (X^*)^t$ is a diagonal matrix.

Prove that $X = MX^*$ is the Gram-Schmidt orthogonalization of X. In other words, the GSO is uniquely determined by conditions (a) and (b).

Exercise 3.5. Find the LQ decomposition of the matrix X in Exercise 3.1.

Exercise 3.6. Find the LQ decomposition of the matrix X in Exercise 3.2.

Exercise 3.7. Find the LQ decomposition of the matrix X in Exercise 3.3.

Exercise 3.8. Find the Gram matrices and Gram determinants for the matrix X in Exercise 3.1.

Exercise 3.9. Find the Gram matrices and Gram determinants for the matrix X in Exercise 3.2.

Exercise 3.10. Find the Gram matrices and Gram determinants for the matrix X in Exercise 3.3.

Exercise 3.11. Prove Cramer's Rule.

Exercise 3.12. Verify the claims of Proposition 3.11 for the matrix X of Exercise 3.1.

Exercise 3.13. Verify the claims of Proposition 3.11 for the matrix X of Exercise 3.2.

Exercise 3.14. Verify the claims of Proposition 3.11 for the matrix X of Exercise 3.3.

Exercise 3.15. For the matrix X of Exercise 3.1, transpose rows 2 and 3, and recompute the GSO. Verify the claims of Proposition 3.14 in this case.

Exercise 3.16. For the matrix X of Exercise 3.2, transpose rows 3 and 4, and recompute the GSO. Verify the claims of Proposition 3.14 in this case.

Exercise 3.17. For the matrix X of Exercise 3.3, transpose rows 4 and 5, and recompute the GSO. Verify the claims of Proposition 3.14 in this case.

Exercise 3.18. Find a lower bound for the length of the shortest nonzero vector in the set of all integral linear combinations of the rows of the matrix X from Exercise 3.1.

Exercise 3.19. Find a lower bound for the length of the shortest nonzero vector in the set of all integral linear combinations of the rows of the matrix X from Exercise 3.2.

Exercise 3.20. Find a lower bound for the length of the shortest nonzero vector in the set of all integral linear combinations of the rows of the matrix X from Exercise 3.3.

Exercise 3.21. Let $\mathbf{x}_1, \mathbf{x}_2, \ldots, \mathbf{x}_n$ be a basis of $\mathbb{R}^n$, and let L be the set of all integral linear combinations of $\mathbf{x}_1, \mathbf{x}_2, \ldots, \mathbf{x}_n$. Prove that for any vector $\mathbf{z} \in \mathbb{R}^n$ there is a vector $\mathbf{y} \in L$ for which

$$|\mathbf{z} - \mathbf{y}|^2 \le \frac{1}{4}\Big(|\mathbf{x}_1|^2 + |\mathbf{x}_2|^2 + \cdots + |\mathbf{x}_n|^2\Big).$$

Exercise 3.22. Proposition 3.13 shows what happens to the Gram-Schmidt basis vectors after an exchange of two consecutive vectors in the original basis. Derive the corresponding result for the Gram-Schmidt coefficients μ_{ij}.

Exercise 3.23. What happens if the input vectors to the Gram-Schmidt process are linearly dependent?

4

The LLL Algorithm

CONTENTS

This chapter provides a detailed exposition of the first section of the famous paper *Factoring polynomials with rational coefficients* [88] by A. K. Lenstra, H. W. Lenstra Jr. and L. Lovász. This paper introduced the most important algorithm for lattice basis reduction. This algorithm is called the LLL algorithm after the initials of its authors. For a historical paper on the origins of the LLL algorithm, see Smeets [132].

4.1 Reduced lattice bases

We start with an example.

Example 4.1. The rows $\mathbf{x}_1, \mathbf{x}_2, \mathbf{x}_3, \mathbf{x}_4$ of the following 4×4 matrix X form a basis of a lattice L in $\mathbb{R}^4$:

$$X = \begin{bmatrix} -2 & 7 & 7 & -5 \\ 3 & -2 & 6 & -1 \\ 2 & -8 & -9 & -7 \\ 8 & -9 & 6 & -4 \end{bmatrix}, \qquad \det(F) = 632.$$

We can find another basis of the same lattice using a unimodular matrix C:

$$C = \begin{bmatrix} -13071 & -5406 & -9282 & -2303 \\ -20726 & -8571 & -14772 & -3651 \\ -2867 & -1186 & -2043 & -505 \\ -14338 & -5936 & -10216 & -2525 \end{bmatrix}, \qquad \det(C) = -1.$$

The rows $\mathbf{y}_1, \mathbf{y}_2, \mathbf{y}_3, \mathbf{y}_4$ of the matrix $Y = CX$ form another basis of L:

$$Y = CX = \begin{bmatrix} -27064 & 14298 & -54213 & 144947 \\ -43013 & 23095 & -85466 & 230209 \\ -5950 & 3192 & -11828 & 31842 \\ -29764 & 15959 & -59188 & 159238 \end{bmatrix}, \qquad \det(G) = -632.$$

Suppose we are given a "bad" basis (such as $\mathbf{y}_1, \mathbf{y}_2, \mathbf{y}_3, \mathbf{y}_4$) of a lattice L. How do we find a "good" basis (such as $\mathbf{x}_1, \mathbf{x}_2, \mathbf{x}_3, \mathbf{x}_4$) of L, assuming of course that we do not know the change of basis matrix C? This is the fundamental problem of lattice basis reduction, that we can solve using the LLL algorithm.

Remark 4.2. The matrix C in Example 4.1 was obtained by starting with the identity matrix I_4 and applying 50 pseudorandom unimodular row operations:

(i) choose i, and multiply row i by -1,

(ii) choose an unordered set $\{i, j\}$, and interchange rows i and j,

(iii) choose an ordered pair (i, j) and an integer m with $|m| \leq 9$, and add m times row i to row j.

Definition 4.3. The **reduction parameter** is a real number α such that

$$\frac{1}{4} < \alpha < 1.$$

The **standard value** of the parameter is

$$\alpha = \frac{3}{4}.$$

Let $\mathbf{x}_1, \mathbf{x}_2, \ldots, \mathbf{x}_n$ be an ordered basis of the lattice L in $\mathbb{R}^n$, and let $\mathbf{x}_1^*, \mathbf{x}_2^*, \ldots, \mathbf{x}_n^*$ be its Gram-Schmidt orthogonalization. We write $X = MX^*$ where X (respectively X^*) is the matrix with $\mathbf{x}_i$ (respectively $\mathbf{x}_i^*$) as row i, and $M = (\mu_{ij})$ is the matrix of GSO coefficients. The basis $\mathbf{x}_1, \mathbf{x}_2, \ldots, \mathbf{x}_n$ is called α-**reduced** (or **LLL-reduced with parameter** α) if it satisfies

(1) $|\mu_{ij}| \leq \frac{1}{2}$ for $1 \leq j < i \leq n$,

(2) $|\mathbf{x}_i^* + \mu_{i,i-1}\mathbf{x}_{i-1}^*|^2 \geq \alpha|\mathbf{x}_{i-1}^*|^2$ for $2 \leq i \leq n$.

Condition (2) is called the **exchange condition**. Since $\mathbf{x}_1^*, \mathbf{x}_2^*, \ldots, \mathbf{x}_n^*$ are orthogonal, condition (2) can be written as

$(2')$ $|\mathbf{x}_i^*|^2 \geq (\alpha - \mu_{i,i-1}^2)|\mathbf{x}_{i-1}^*|^2$ for $2 \leq i \leq n$.

Condition (1) says that each basis vector $\mathbf{x}_i$ is "almost orthogonal" to the span of the previous vectors, since by Theorem 3.4 we have

$$\mathrm{span}(\mathbf{x}_1, \ldots, \mathbf{x}_{i-1}) = \mathrm{span}(\mathbf{x}_1^*, \ldots, \mathbf{x}_{i-1}^*).$$

Conditions (2) and (2′) say that exchanging $\mathbf{x}_{i-1}$ and $\mathbf{x}_i$ and then recomputing the GSO can produce a new shorter vector

$$\widehat{\mathbf{x}}_{i-1}^* = \mathbf{x}_i^* + \mu_{i,i-1}\mathbf{x}_{i-1}^*,$$

but not "too much" shorter; this uses Proposition 3.13.

We will prove that, for any lattice L in $\mathbb{R}^n$, any basis $\mathbf{x}_1, \mathbf{x}_2, \ldots, \mathbf{x}_n$ of L, and any $\alpha \in (\frac{1}{4}, 1)$, the LLL algorithm produces an α-reduced basis of L in a number of steps which is bounded by a polynomial in the size of the input. We may also consider $\alpha = 1$, the "limiting case" of Definition 4.3, but for this value of the reduction parameter we cannot prove that the LLL algorithm terminates in polynomial time.

Definition 4.4. We define the **auxiliary parameter** β as follows:

$$\beta = \frac{4}{4\alpha - 1} \quad \text{so that} \quad \beta > \frac{4}{3} \quad \text{and} \quad \frac{1}{\beta} = \alpha - \frac{1}{4}.$$

For the standard value $\alpha = \frac{3}{4}$ we obtain $\beta = 2$.

Example 4.5. Let L be the lattice of Example 4.1; the rows $\mathbf{x}_1, \mathbf{x}_2, \mathbf{x}_3, \mathbf{x}_4$ of the matrix X form a basis of L. The Gram-Schmidt orthogonalization can be written as the rows $\mathbf{x}_1^*, \mathbf{x}_2^*, \mathbf{x}_3^*, \mathbf{x}_4^*$ of the matrix X^*. The matrices X and X^* are related by the equation $X = MX^*$ where $M = (\mu_{ij})$:

$$
X = \begin{bmatrix} -2 & 7 & 7 & -5 \\ 3 & -2 & 6 & -1 \\ 2 & -8 & -9 & -7 \\ 8 & -9 & 6 & -4 \end{bmatrix}
$$

$$
= \begin{bmatrix} 1 & 0 & 0 & 0 \\ \frac{27}{127} & 1 & 0 & 0 \\ -\frac{88}{127} & -\frac{799}{5621} & 1 & 0 \\ -\frac{17}{127} & \frac{10873}{5621} & \frac{350695}{765183} & 1 \end{bmatrix}
\begin{bmatrix} -2 & 7 & 7 & -5 \\ \frac{435}{127} & -\frac{443}{127} & \frac{573}{127} & \frac{8}{127} \\ \frac{6189}{5621} & -\frac{20491}{5621} & -\frac{19720}{5621} & -\frac{58771}{5621} \\ \frac{153576}{255061} & \frac{271760}{765183} & -\frac{139672}{765183} & \frac{632}{765183} \end{bmatrix}
$$

$$= MX^*.$$

¿From the matrix M it is clear that the basis $\mathbf{x}_1, \mathbf{x}_2, \mathbf{x}_3, \mathbf{x}_4$ is not α-reduced for any α, since even the first condition in Definition 4.3 does not hold:

$$|\mu_{31}|, \; |\mu_{42}| > \frac{1}{2}.$$

Now consider the matrix C with entries in $\mathbb{Z}$ and $\det(C) = -1$, and the new basis $\mathbf{y}_1$, $\mathbf{y}_2$, $\mathbf{y}_3$, $\mathbf{y}_4$ for the lattice L given by the rows of the matrix $Y = CX$:

$$
Y = CX = \begin{bmatrix} 1 & -8 & -2 & 4 \\ 1 & -6 & -1 & 3 \\ 0 & 4 & 1 & -2 \\ 0 & 1 & 0 & 0 \end{bmatrix} \begin{bmatrix} -2 & 7 & 7 & -5 \\ 3 & -2 & 6 & -1 \\ 2 & -8 & -9 & -7 \\ 8 & -9 & 6 & -4 \end{bmatrix}
$$

$$
= \begin{bmatrix} 2 & 3 & 1 & 1 \\ 2 & 0 & -2 & -4 \\ -2 & 2 & 3 & -3 \\ 3 & -2 & 6 & -1 \end{bmatrix}.
$$

The Gram-Schmidt orthogonalization of $\mathbf{y}_1$, $\mathbf{y}_2$, $\mathbf{y}_3$, $\mathbf{y}_4$ can be written as the rows $\mathbf{y}_1^*$, $\mathbf{y}_2^*$, $\mathbf{y}_3^*$, $\mathbf{y}_4^*$ of the matrix Y^* which satisfies the matrix equation $Y = \widehat{M}Y^*$ where $\widehat{M} = (\widehat{\mu}_{ij})$:

$$
Y = \widehat{M}Y^* = \begin{bmatrix} 1 & 0 & 0 & 0 \\ -\frac{2}{15} & 1 & 0 & 0 \\ \frac{2}{15} & \frac{17}{178} & 1 & 0 \\ \frac{1}{3} & -\frac{5}{89} & \frac{931}{2271} & 1 \end{bmatrix} \begin{bmatrix} 2 & 3 & 1 & 1 \\ \frac{34}{15} & \frac{2}{5} & -\frac{28}{15} & -\frac{58}{15} \\ -\frac{221}{89} & \frac{139}{89} & \frac{271}{89} & -\frac{246}{89} \\ \frac{7900}{2271} & -\frac{8216}{2271} & \frac{9796}{2271} & -\frac{316}{757} \end{bmatrix}.
$$

From this equation we can verify that the basis $\mathbf{y}_1$, $\mathbf{y}_2$, $\mathbf{y}_3$, $\mathbf{y}_4$ is α-reduced for all $\alpha < 1$. The first condition of Definition 4.3 clearly holds since $\widehat{\mu}_{ij} \leq \frac{1}{2}$ for all $1 \leq j < i \leq 4$. For the second condition, it suffices to consider the limiting value $\alpha = 1$, since it is easy to see that if a lattice basis is α-reduced then it is α'-reduced for any $\alpha' < \alpha$. We now calculate as follows:

$$|\mathbf{y}_1^*|^2 = 15,$$

$$|\mathbf{y}_2^*|^2 = \tfrac{356}{15} \approx 27.73, \qquad \widehat{\mu}_{21}^2 = \tfrac{4}{225}, \qquad \left(1 - \widehat{\mu}_{21}^2\right)|\mathbf{y}_1^*|^2 = \tfrac{221}{15} \approx 14.73,$$

$$|\mathbf{y}_3^*|^2 = \tfrac{2271}{89} \approx 25.52, \qquad \widehat{\mu}_{32}^2 = \tfrac{289}{31684}, \qquad \left(1 - \widehat{\mu}_{32}^2\right)|\mathbf{y}_2^*|^2 = \tfrac{2093}{89} \approx 23.52,$$

$$|\mathbf{y}_4^*|^2 = \tfrac{99856}{2271} \approx 43.97, \qquad \widehat{\mu}_{43}^2 = \tfrac{866761}{5157441}, \qquad \left(1 - \widehat{\mu}_{43}^2\right)|\mathbf{y}_3^*|^2 = \tfrac{4290680}{202119} \approx 21.23.$$

From this it is clear that for $i = 2, 3, 4$ we have

$$|\mathbf{y}_i^*|^2 \geq \left(1 - \widehat{\mu}_{i,i-1}^2\right)|\mathbf{y}_{i-1}^*|^2.$$

Thus the basis $\mathbf{y}_1$, $\mathbf{y}_2$, $\mathbf{y}_3$, $\mathbf{y}_4$ is α-reduced for any α with $\frac{1}{4} < \alpha < 1$.

Part (c) of the next result (Proposition 4.6) gives an upper bound for the length $|\mathbf{x}_1|$ of the first vector in an α-reduced lattice basis in terms of the determinant of the lattice. The subsequent result (Theorem 4.7) gives an upper bound for $|\mathbf{x}_1|$ in terms of the length of a shortest (nonzero) vector in the lattice.

Proposition 4.6. *If $\mathbf{x}_1$, $\mathbf{x}_2$, $\ldots$, $\mathbf{x}_n$ is an α-reduced basis of the lattice L in $\mathbb{R}^n$, and $\mathbf{x}_1^*$, $\mathbf{x}_2^*$, $\ldots$, $\mathbf{x}_n^*$ is its Gram-Schmidt orthogonalization, then*

(a) $|\mathbf{x}_j|^2 \leq \beta^{i-j}|\mathbf{x}_i^*|^2$ for $1 \leq j \leq i \leq n$,

(b) $\det(L) \leq |\mathbf{x}_1|\,|\mathbf{x}_2| \cdots |\mathbf{x}_n| \leq \beta^{n(n-1)/4}\det(L)$,

(c) $|\mathbf{x}_1| \leq \beta^{(n-1)/4}\big(\det(L)\big)^{1/n}$,

where β is the auxiliary parameter from Definition 4.4.

Proof. The two conditions in Definition 4.3 imply that for $1 < i \leq n$ we have

$$|\mathbf{x}_i^*|^2 \geq \big(\alpha - \mu_{i,i-1}^2\big)|\mathbf{x}_{i-1}^*|^2 \geq \big(\alpha - \tfrac{1}{4}\big)|\mathbf{x}_{i-1}^*|^2 = \frac{1}{\beta}|\mathbf{x}_{i-1}^*|^2.$$

Therefore $|\mathbf{x}_{i-1}^*|^2 \leq \beta|\mathbf{x}_i^*|^2$, and so an easy induction gives

$$|\mathbf{x}_j^*|^2 \leq \beta^{i-j}|\mathbf{x}_i^*|^2 \qquad (1 \leq j \leq i \leq n). \tag{4.1}$$

The definition of $\mathbf{x}_i^*$ in the GSO can be rewritten as

$$\mathbf{x}_i = \mathbf{x}_i^* + \sum_{j=1}^{i-1} \mu_{ij}\mathbf{x}_j^*,$$

and since $\mathbf{x}_1^*, \ldots, \mathbf{x}_n^*$ are orthogonal, we get

$$|\mathbf{x}_i|^2 = |\mathbf{x}_i^*|^2 + \sum_{j=1}^{i-1} \mu_{ij}^2|\mathbf{x}_j^*|^2.$$

Definition 4.3 and equation (4.1) now give

$$|\mathbf{x}_i|^2 \leq |\mathbf{x}_i^*|^2 + \sum_{j=1}^{i-1} \frac{1}{4}\beta^{i-j}|\mathbf{x}_i^*|^2 = \left(1 + \frac{1}{4}\sum_{j=1}^{i-1}\beta^{i-j}\right)|\mathbf{x}_i^*|^2.$$

Using the summation formula for a geometric sequence, we obtain

$$|\mathbf{x}_i|^2 \leq \left(1 + \frac{1}{4}\cdot\frac{\beta^i-\beta}{\beta-1}\right)|\mathbf{x}_i^*|^2.$$

We show by induction on i that

$$1 + \frac{1}{4}\cdot\frac{\beta^i-\beta}{\beta-1} \leq \beta^{i-1}.$$

The basis $i = 1$ is trivial. For the inductive step it suffices to show that

$$1 + \frac{1}{4}\cdot\frac{\beta^{i+1}-\beta}{\beta-1} \leq \beta\left(1 + \frac{\beta^i-\beta}{4(\beta-1)}\right).$$

Since $\beta > \tfrac{4}{3}$, multiplying by $4(\beta-1)$ gives an equivalent inequality; simplification then gives

$$(\beta - 1)(3\beta - 4) \geq 0,$$

which is clear since $\beta > \frac{4}{3}$. We now have

$$|\mathbf{x}_i|^2 \le \beta^{i-1}|\mathbf{x}_i^*|^2. \tag{4.2}$$

Using this and equation (4.1) gives

$$|\mathbf{x}_j|^2 \le \beta^{j-1}|\mathbf{x}_j^*|^2 \le \beta^{i-1}|\mathbf{x}_i^*|^2 \qquad (1 \le j \le i \le n),$$

which proves (a). From Theorem 3.4 we know that

$$\det(L) = |\mathbf{x}_1^*|\,|\mathbf{x}_2^*|\,\cdots\,|\mathbf{x}_n^*| \le |\mathbf{x}_1|\,|\mathbf{x}_2|\,\cdots\,|\mathbf{x}_n|,$$

which proves the left inequality in (b). Equation (4.2) implies

$$|\mathbf{x}_1|^2\,|\mathbf{x}_2|^2\,\cdots\,|\mathbf{x}_n|^2 \le \beta^{0+1+2+\cdots+(n-1)}\,|\mathbf{x}_1^*|^2\,|\mathbf{x}_2^*|^2\,\cdots\,|\mathbf{x}_n^*|^2,$$

and therefore

$$|\mathbf{x}_1|\,|\mathbf{x}_2|\,\cdots\,|\mathbf{x}_n| \le \beta^{n(n-1)/4}|\mathbf{x}_1^*|\,|\mathbf{x}_2^*|\,\cdots\,|\mathbf{x}_n^*| = \beta^{n(n-1)/4}\det(L),$$

which proves the right inequality in (b). Setting $j = 1$ in (a) gives

$$|\mathbf{x}_1|^2 \le \beta^{i-1}|\mathbf{x}_i^*|^2 \qquad (1 \le i \le n),$$

and taking the product over $i = 1, 2, \ldots, n$ gives

$$|\mathbf{x}_1|^{2n} \le \beta^{0+1+2+\cdots+(n-1)}\,|\mathbf{x}_1^*|^2\,|\mathbf{x}_2^*|^2\,\cdots\,|\mathbf{x}_n^*|^2 = \beta^{n(n-1)/2}\big(\det(L)\big)^2.$$

Now taking $2n$-th roots proves (c). $\qquad\square$

The upper bound for $|\mathbf{x}_1|$ in the next result is exponential, but it depends only on the reduction parameter α and the dimension n, so it applies uniformly to all lattices of dimension n.

Theorem 4.7. LLL Theorem. *If $\mathbf{x}_1, \mathbf{x}_2, \ldots, \mathbf{x}_n$ is an α-reduced basis of the lattice L in $\mathbb{R}^n$, and $\mathbf{y} \in L$ is any nonzero lattice vector, then*

$$|\mathbf{x}_1| \le \beta^{(n-1)/2}|\mathbf{y}|.$$

In particular, the first vector in the α-reduced basis is no longer than $\beta^{(n-1)/2}$ times the shortest nonzero vector in L.

Proof. Let $\mathbf{x}_1^*, \mathbf{x}_2^*, \ldots, \mathbf{x}_n^*$ be the Gram-Schmidt orthogonalization of $\mathbf{x}_1, \mathbf{x}_2, \ldots, \mathbf{x}_n$. From the definition of an α-reduced basis, for $2 \le i \le n$ we have

$$|\mathbf{x}_i^*|^2 \ge \big(\alpha - \mu_{i,i-1}^2\big)|\mathbf{x}_{i-1}^*|^2 \ge \big(\alpha - \tfrac{1}{4}\big)|\mathbf{x}_{i-1}^*|^2 = \frac{1}{\beta}|\mathbf{x}_{i-1}^*|^2.$$

Since $\mathbf{x}_1^* = \mathbf{x}_1$ this gives

$$|\mathbf{x}_1|^2 = |\mathbf{x}_1^*|^2 \le \beta|\mathbf{x}_2^*|^2 \le \beta^2|\mathbf{x}_3^*|^2 \le \cdots \le \beta^{n-1}|\mathbf{x}_n^*|^2,$$

and hence for $1 \leq i \leq n$ we have

$$|\mathbf{x}_i^*|^2 \geq \beta^{-(i-1)}|\mathbf{x}_1|^2.$$

Proposition 3.14 now shows that for any nonzero vector $g \in L$ we have

$$|\mathbf{y}| \geq \min\{|\mathbf{x}_1^*|, \ldots, |\mathbf{x}_n^*|\} \geq \beta^{-(n-1)/2}|\mathbf{x}_1|^2,$$

and this completes the proof. $\qquad\square$

The previous result is the case $m = 1$ of the next result, which gives upper bounds for the lengths of all the vectors in an α-reduced basis.

Theorem 4.8. *If* $\mathbf{x}_1$, $\mathbf{x}_2$, $\ldots$, $\mathbf{x}_n$ *is an α-reduced basis of the lattice L in $\mathbb{R}^n$, and* $\mathbf{y}_1$, $\mathbf{y}_2$, $\ldots$, $\mathbf{y}_m \in L$ *are any m linearly independent lattice vectors, then for $1 \leq j \leq m$ we have*

$$|\mathbf{x}_j| \leq \beta^{(n-1)/2} \max\{|\mathbf{y}_1|, \ldots, |\mathbf{y}_m|\}.$$

Proof. We start by writing $\mathbf{y}_1$, $\mathbf{y}_2$, $\ldots$, $\mathbf{y}_m$ as integral linear combinations of $\mathbf{x}_1$, $\mathbf{x}_2$, $\ldots$, $\mathbf{x}_n$:

$$\mathbf{y}_j = \sum_{i=1}^{n} r_{ij}\mathbf{x}_i \quad (r_{ij} \in \mathbb{Z}, \ 1 \leq i \leq n, \ 1 \leq j \leq m).$$

Let $i(j)$ denote the largest index i for which $r_{ij} \neq 0$; we think of $i(j)$ as a function i of a variable j. The definition of the GSO gives

$$\mathbf{y}_j = \sum_{i=1}^{i(j)} r_{ij}\mathbf{x}_i = \sum_{i=1}^{i(j)} r_{ij} \sum_{k=1}^{i} \mu_{ik}\mathbf{x}_k^* = \sum_{i=1}^{i(j)} \sum_{k=1}^{i} r_{ij}\mu_{ik}\mathbf{x}_k^* \quad (1 \leq j \leq m).$$

Considering the term in $\mathbf{x}_{i(j)}^*$, and observing that $r_{i,i(j)} \in \mathbb{Z}$ with $r_{i,i(j)} \neq 0$, and $\mu_{i(j),i(j)} = 1$, we get

$$|\mathbf{y}_j|^2 \geq |\mathbf{x}_{i(j)}^*|^2 \quad (1 \leq j \leq m). \tag{4.3}$$

Since $\{\mathbf{y}_1, \mathbf{y}_2, \ldots, \mathbf{y}_m\}$ is an unordered set, we may assume without loss of generality that

$$i(1) \leq i(2) \leq \cdots \leq i(m).$$

We claim that $j \leq i(j)$ for $1 \leq j \leq m$: if $i(j) < j$ for some j then $\mathbf{y}_1$, $\mathbf{y}_2$, $\ldots$, $\mathbf{y}_j$ all belong to the linear span of $\mathbf{x}_1$, $\mathbf{x}_2$, $\ldots$, $\mathbf{x}_{i(j)}$, contradicting the linear independence of $\mathbf{y}_1$, $\mathbf{y}_2$, $\ldots$, $\mathbf{y}_j$. Hence we may take $i = i(j)$ in part (a) of Proposition 4.6, and then use equation (4.3) to obtain

$$|\mathbf{x}_j|^2 \leq \beta^{i(j)-1}|\mathbf{x}_{i(j)}^*|^2 \leq \beta^{n-1}|\mathbf{x}_{i(j)}^*|^2 \leq \beta^{n-1}|\mathbf{y}_j|^2$$
$$\leq \beta^{n-1} \max\{|\mathbf{y}_1|^2, \ldots, |\mathbf{y}_m|^2\},$$

for $1 \leq j \leq m$. This completes the proof. $\qquad\square$

Recall Definition 1.29 of the successive minima of a lattice. Let $\mathbf{x}_1, \mathbf{x}_2, \ldots,$ $\mathbf{x}_n$ be an α-reduced basis of the lattice L in $\mathbb{R}^n$, and let $\mathbf{x}_1^*, \mathbf{x}_2^*, \ldots, \mathbf{x}_n^*$ be its GSO. Part (a) of Proposition 4.6 states

$$|\mathbf{x}_j|^2 \leq \beta^{i-1} |\mathbf{x}_i^*|^2 \qquad (1 \leq j \leq i \leq n),$$

and hence

$$\beta^{1-i} |\mathbf{x}_j|^2 \leq |\mathbf{x}_i^*|^2 \leq |\mathbf{x}_i|^2 \qquad (1 \leq j \leq i \leq n),$$

using Theorem 3.4. Hence

$$\beta^{1-i} \max \left\{ |\mathbf{x}_1|^2, |\mathbf{x}_2|^2, \ldots, |\mathbf{x}_i|^2 \right\} \leq |\mathbf{x}_i|^2 \qquad (1 \leq i \leq n).$$

Since $\mathbf{x}_1, \mathbf{x}_2, \ldots, \mathbf{x}_i$ are linearly independent vectors in L, Theorem 4.8 gives

$$|\mathbf{x}_i|^2 \leq \beta^{n-1} \max \left\{ |\mathbf{x}_1|^2, |\mathbf{x}_2|^2, \ldots, |\mathbf{x}_i|^2 \right\} \qquad (1 \leq i \leq n).$$

Combining the last two inequalities shows that for $1 \leq i \leq n$ we have

$$\beta^{1-i} \max \left\{ |\mathbf{x}_1|^2, \ldots, |\mathbf{x}_i|^2 \right\} \leq |\mathbf{x}_i|^2 \leq \beta^{n-1} \max \left\{ |\mathbf{x}_1|^2, \ldots, |\mathbf{x}_i|^2 \right\}. \qquad (4.4)$$

Now suppose that $\mathbf{y}_1, \mathbf{y}_2, \ldots, \mathbf{y}_i$ are linearly independent lattice vectors which achieve the i-th minimum of the lattice; clearly

$$\max \left\{ |\mathbf{y}_1|^2, |\mathbf{y}_2|^2, \ldots, |\mathbf{y}_i|^2 \right\} \leq \max \left\{ |\mathbf{x}_1|^2, |\mathbf{x}_2|^2, \ldots, |\mathbf{x}_i|^2 \right\}.$$

Using this for the leftmost term in the inequality (4.4), and the general case of Theorem 4.8 for the rightmost term, we obtain

$$\beta^{(1-i)/2} \max\{|\mathbf{y}_1|, \ldots, |\mathbf{y}_i|\} \leq |\mathbf{x}_i| \leq \beta^{(n-1)/2} \max\{|\mathbf{y}_1|, \ldots, |\mathbf{y}_i|\}.$$

This can also be written as

$$\beta^{(1-n)/2} |\mathbf{x}_i| \leq \max\{|\mathbf{y}_1|, \ldots, |\mathbf{y}_i|\} \leq \beta^{(i-1)/2} |\mathbf{x}_i|.$$

This shows that $|\mathbf{x}_i|$ can be regarded as an approximation to the i-th sucessive minimum of the lattice.

4.2 The original LLL algorithm

The original LLL algorithm for lattice basis reduction, rewritten in a more "structured" form (without "goto" statements), appears in Figure 4.1. The input consists of a basis $\mathbf{x}_1, \mathbf{x}_2, \ldots, \mathbf{x}_n$ of the lattice $L \subset \mathbb{R}^n$, and a reduction parameter $\alpha \in \mathbb{R}$ in the range $\frac{1}{4} < \alpha < 1$. The output consists of an α-reduced basis $\mathbf{y}_1, \mathbf{y}_2, \ldots, \mathbf{y}_n$ of the lattice L.

Step (1) of the main loop simply makes a copy $\mathbf{y}_1, \mathbf{y}_2, \ldots, \mathbf{y}_n$ of the input

- *Input*: A basis $\mathbf{x}_1, \mathbf{x}_2, \ldots, \mathbf{x}_n$ of the lattice $L \subset \mathbb{R}^n$, and a reduction parameter $\alpha \in \mathbb{R}$ in the range $\frac{1}{4} < \alpha < 1$.

- *Output*: An α-reduced basis $\mathbf{y}_1, \mathbf{y}_2, \ldots, \mathbf{y}_n$ of the lattice L.

- *Procedure* $\mathtt{reduce}(k, \ell)$:

  If $|\mu_{k\ell}| > \frac{1}{2}$ then
  - (1) Set $\mathbf{y}_k \leftarrow \mathbf{y}_k - \lceil \mu_{k\ell} \rceil \mathbf{y}_\ell$.
  - (2) For $j = 1, 2, \ldots, \ell-1$ do: Set $\mu_{kj} \leftarrow \mu_{kj} - \lceil \mu_{k\ell} \rceil \mu_{\ell j}$.
  - (3) Set $\mu_{k\ell} \leftarrow \mu_{k\ell} - \lceil \mu_{k\ell} \rceil$.

- *Procedure* $\mathtt{exchange}(k)$:
  - (1) Set $\mathbf{z} \leftarrow \mathbf{y}_{k-1}$, $\mathbf{y}_{k-1} \leftarrow \mathbf{y}_k$, $\mathbf{y}_k \leftarrow \mathbf{z}$ *(exchange $\mathbf{y}_{k-1}$ and $\mathbf{y}_k$)*.
  - (2) Set $\nu \leftarrow \mu_{k,k-1}$. Set $\delta \leftarrow \gamma_k^* + \nu^2 \gamma_{k-1}^*$.
  - (3) Set $\mu_{k,k-1} \leftarrow \nu \gamma_{k-1}^*/\delta$. Set $\gamma_k^* \leftarrow \gamma_k^* \gamma_{k-1}^*/\delta$. Set $\gamma_{k-1}^* \leftarrow \delta$.
  - (4) For $j = 1, 2, \ldots, k-2$ do: Set $t \leftarrow \mu_{k-1,j}$, $\mu_{k-1,j} \leftarrow \mu_{kj}$, $\mu_{kj} \leftarrow t$ *(exchange $\mu_{k-1,j}$ and μ_{kj})*.
  - (5) For $i = k+1, \ldots, n$ do:
    - (a) Set $\xi \leftarrow \mu_{ik}$. Set $\mu_{ik} \leftarrow \mu_{i,k-1} - \nu \mu_{ik}$.
    - (b) Set $\mu_{i,k-1} \leftarrow \mu_{k,k-1} \mu_{ik} + \xi$.

- *Main loop*:
  - (1) For $i = 1, 2, \ldots, n$ do: Set $\mathbf{y}_i \leftarrow \mathbf{x}_i$.
  - (2) For $i = 1, 2, \ldots, n$ do:
    - (a) Set $\mathbf{y}_i^* \leftarrow \mathbf{y}_i$.
    - (b) For $j = 1, 2, \ldots, i-1$ do:
      Set $\mu_{ij} \leftarrow (\mathbf{y}_i \cdot \mathbf{y}_j^*)/\gamma_j^*$ and $\mathbf{y}_i^* \leftarrow \mathbf{y}_i^* - \mu_{ij} \mathbf{y}_j^*$.
    - (c) Set $\gamma_i^* \leftarrow \mathbf{y}_i^* \cdot \mathbf{y}_i^*$.
  - (3) Set $k \leftarrow 2$.
  - (4) While $k \leq n$ do:
    - (a) Call $\mathtt{reduce}(k, k-1)$.
    - (b) If $\gamma_k^* \geq (\alpha - \mu_{k,k-1}^2)\gamma_{k-1}^*$ then
      - (i) For $\ell = k-2, \ldots, 2, 1$ do: Call $\mathtt{reduce}(k, \ell)$.
      - (ii) Set $k \leftarrow k+1$.

      else
      - (iii) Call $\mathtt{exchange}(k)$.
      - (iv) If $k > 2$ then set $k \leftarrow k-1$.

FIGURE 4.1
The original LLL algorithm for lattice basis reduction

vectors $\mathbf{x}_1, \mathbf{x}_2, \ldots, \mathbf{x}_n$. Step (2) computes the Gram-Schmidt orthogonalization of the vectors $\mathbf{y}_1, \mathbf{y}_2, \ldots, \mathbf{y}_n$.

Step (3) initializes the index k of the vector $\mathbf{y}_k$ currently being processed. Step (4) performs the basis reduction; it repeatedly calls two procedures which reduce and exchange the vectors $\mathbf{y}_1, \mathbf{y}_2, \ldots, \mathbf{y}_n$.

Procedure `reduce`(k, ℓ) makes $\mathbf{y}_k$ almost orthogonal to $\mathbf{y}_\ell$. If $|\mu_{k\ell}| \leq \frac{1}{2}$ then the procedure does nothing; otherwise, it reduces $\mathbf{y}_k$ by subtracting the integral multiple $\lceil \mu_{k\ell} \rfloor$ of $\mathbf{y}_\ell$. Since $\lceil \mu_{k\ell} \rfloor$ is the nearest integer to the Gram-Schmidt coefficient $\mu_{k\ell}$, this is the best possible reduction we can perform, given that $\mathbf{y}_k$ must remain in the lattice generated by $\mathbf{x}_1, \mathbf{x}_2, \ldots, \mathbf{x}_n$. The procedure then updates the GSO basis and coefficients.

Procedure `exchange`(k) interchanges the vectors $\mathbf{y}_{k-1}$ and $\mathbf{y}_k$, and then updates the GSO basis and coefficients according to Proposition 3.13.

The vectors $\mathbf{y}_1, \mathbf{y}_2, \ldots, \mathbf{y}_n$ are modified continually throughout the algorithm, but in such a way that they always form a basis for the lattice L. Step 4(a) of the main loop reduces $\mathbf{y}_k$ using $\mathbf{y}_{k-1}$; this is done before testing the reduction condition in Step 4(b) since that condition depends only on these two vectors. In Step 4(b), if the reduction condition is satisfied, $\mathbf{y}_k$ is completely reduced using $\mathbf{y}_{k-2}, \ldots, \mathbf{y}_2, \mathbf{y}_1$, and then the index k is increased; otherwise, an exchange is performed, and the index k is decreased (if it does not already have the minimum value $k = 2$).

It is not immediately clear that the LLL algorithm terminates. In the rest of this chapter, we prove termination, verify that on termination the output is an α-reduced basis, and then analyze the complexity of the algorithm to demonstrate that the running time is bounded by a polynomial in the size of the input. But first an example.

Example 4.9. We start with the vectors $\mathbf{x}_1, \mathbf{x}_2, \mathbf{x}_3, \mathbf{x}_4$ from Example 4.5 which form a basis of the lattice L in $\mathbb{R}^4$. We will trace the LLL algorithm using the limiting value $\alpha = 1$ of the reduction parameter.

Step (1) makes a copy $\mathbf{y}_1, \mathbf{y}_2, \mathbf{y}_3, \mathbf{y}_4$ of the lattice basis vectors, and Step (2) computes its Gram-Schmidt orthogonalization $\mathbf{y}_1^*, \mathbf{y}_2^*, \mathbf{y}_3^*, \mathbf{y}_4^*$. Throughout this example we will write the current state of the algorithm as a triple

$$Y, \quad M, \quad \gamma^*,$$

where Y is the matrix with $\mathbf{y}_1, \mathbf{y}_2, \mathbf{y}_3, \mathbf{y}_4$ as its rows, $M = (\mu_{ij})$ is the matrix of Gram-Schmidt coefficients, and γ^* is the column vector of squared-lengths of the GSO vectors $\mathbf{y}_1^*, \mathbf{y}_2^*, \mathbf{y}_3^*, \mathbf{y}_4^*$.

The initial state (iteration 0) is

$$
\begin{bmatrix}
-2 & 7 & 7 & -5 \\
3 & -2 & 6 & -1 \\
2 & -8 & -9 & -7 \\
8 & -9 & 6 & -4
\end{bmatrix},
\quad
\begin{bmatrix}
1 & 0 & 0 & 0 \\
\frac{27}{127} & 1 & 0 & 0 \\
-\frac{88}{127} & -\frac{799}{5621} & 1 & 0 \\
-\frac{17}{127} & \frac{10873}{5621} & \frac{350695}{765183} & 1
\end{bmatrix},
\quad
\begin{bmatrix}
127 \\
\frac{5621}{127} \\
\frac{765183}{5621} \\
\frac{399424}{765183}
\end{bmatrix}.
$$

Step (3) sets $k = 2$.

Step (4)(a) calls `reduce(2,1)`; since $|\mu_{21}| = 27/127 < 1/2$ the procedure performs no action. Step (4)(b) tests the exchange condition; we have

$$\gamma_2^* = \frac{5621}{127} < \frac{15400}{127} = (1 - \mu_{21}^2)\gamma_1^*.$$

Hence Step (4)(b)(iii) calls `exchange(2)`; the current state (iteration 1) is

$$
\begin{bmatrix}
3 & -2 & 6 & -1 \\
-2 & 7 & 7 & -5 \\
2 & -8 & -9 & -7 \\
8 & -9 & 6 & -4
\end{bmatrix},
\begin{bmatrix}
1 & 0 & 0 & 0 \\
\frac{27}{50} & 1 & 0 & 0 \\
-\frac{1}{2} & -\frac{3725}{5621} & 1 & 0 \\
\frac{41}{25} & -\frac{3064}{5621} & \frac{350695}{765183} & 1
\end{bmatrix},
\begin{bmatrix}
50 \\
\frac{5621}{50} \\
\frac{765183}{5621} \\
\frac{399424}{765183}
\end{bmatrix}.
$$

Since $k = 2$ we do not decrement k in Step (4)(b)(iv); we return to the top of the loop with $k = 2$.

Step (4)(a) calls `reduce(2,1)`; since $|\mu_{21}| = 27/50 > 1/2$ and $\lceil\mu_{21}\rfloor = 1$, the procedure reduces $\mathbf{y}_2$ by subtracting $\mathbf{y}_1$. The current state (iteration 2) is

$$
\begin{bmatrix}
3 & -2 & 6 & -1 \\
-5 & 9 & 1 & -4 \\
2 & -8 & -9 & -7 \\
8 & -9 & 6 & -4
\end{bmatrix},
\begin{bmatrix}
1 & 0 & 0 & 0 \\
-\frac{23}{50} & 1 & 0 & 0 \\
-\frac{1}{2} & -\frac{3725}{5621} & 1 & 0 \\
\frac{41}{25} & -\frac{3064}{5621} & \frac{350695}{765183} & 1
\end{bmatrix},
\begin{bmatrix}
50 \\
\frac{5621}{50} \\
\frac{765183}{5621} \\
\frac{399424}{765183}
\end{bmatrix}.
$$

Step (4)(b) tests the exchange condition; we have

$$\gamma_2^* = \frac{5621}{50} \geq \frac{1971}{50} = (1 - \mu_{21}^2)\gamma_1^*.$$

Since $k = 2$, Step (4)(b)(i) does nothing, and Step (4)(b)(ii) increments k; we now return to the top of the loop with $k = 3$.

Step (4)(a) calls `reduce(3,2)`; since $|\mu_{32}| = 3725/5621 > 1/2$ and $\lceil\mu_{32}\rfloor = -1$, the procedure reduces $\mathbf{y}_3$ by adding $\mathbf{y}_2$. The current state (iteration 3) is

$$
\begin{bmatrix}
3 & -2 & 6 & -1 \\
-5 & 9 & 1 & -4 \\
-3 & 1 & -8 & -11 \\
8 & -9 & 6 & -4
\end{bmatrix},
\begin{bmatrix}
1 & 0 & 0 & 0 \\
-\frac{23}{50} & 1 & 0 & 0 \\
-\frac{24}{25} & \frac{1896}{5621} & 1 & 0 \\
\frac{41}{25} & -\frac{3064}{5621} & \frac{350695}{765183} & 1
\end{bmatrix},
\begin{bmatrix}
50 \\
\frac{5621}{50} \\
\frac{765183}{5621} \\
\frac{399424}{765183}
\end{bmatrix}.
$$

Step (4)(b) tests the exchange condition; we have

$$\gamma_3^* = \frac{765183}{5621} \geq \frac{1120033}{11242} = (1 - \mu_{32}^2)\gamma_2^*.$$

Step (4)(b)(i) calls `reduce(3, 1)`; since $|\mu_{31}| = 24/25 > 1/2$ and $\lceil\mu_{31}\rfloor = -1$,

iteration 5	exchange	$k = 4$		
iteration 6	reduce	$k = 3$	$\ell = 2$	$\lceil \mu_{k\ell} \rfloor = -1$
iteration 7	exchange	$k = 3$		
iteration 8	reduce	$k = 2$	$\ell = 1$	$\lceil \mu_{k\ell} \rfloor = 1$
iteration 9	reduce	$k = 3$	$\ell = 2$	$\lceil \mu_{k\ell} \rfloor = 1$
iteration 10	reduce	$k = 3$	$\ell = 1$	$\lceil \mu_{k\ell} \rfloor = -1$
iteration 11	reduce	$k = 4$	$\ell = 3$	$\lceil \mu_{k\ell} \rfloor = -1$
iteration 12	exchange	$k = 4$		
iteration 13	reduce	$k = 3$	$\ell = 2$	$\lceil \mu_{k\ell} \rfloor = 1$
iteration 14	exchange	$k = 3$		
iteration 15	exchange	$k = 2$		
iteration 16	exchange	$k = 3$		
iteration 17	reduce	$k = 2$	$\ell = 1$	$\lceil \mu_{k\ell} \rfloor = 1$
iteration 18	exchange	$k = 2$		
iteration 19	exchange	$k = 4$		
iteration 20	reduce	$k = 3$	$\ell = 2$	$\lceil \mu_{k\ell} \rfloor = 1$
iteration 21	exchange	$k = 3$		
iteration 22	reduce	$k = 2$	$\ell = 1$	$\lceil \mu_{k\ell} \rfloor = -1$
iteration 23	exchange	$k = 2$		

FIGURE 4.2
Reduce and exchange operations for Example 4.9

---

the procedure reduces $\mathbf{y}_3$ by adding $\mathbf{y}_1$. The current state (iteration 4) is

$$
\begin{bmatrix} 3 & -2 & 6 & -1 \\ -5 & 9 & 1 & -4 \\ 0 & -1 & -2 & -12 \\ 8 & -9 & 6 & -4 \end{bmatrix}, \quad
\begin{bmatrix} 1 & 0 & 0 & 0 \\ -\frac{23}{50} & 1 & 0 & 0 \\ \frac{1}{25} & \frac{1896}{5621} & 1 & 0 \\ \frac{41}{25} & -\frac{3064}{5621} & \frac{350695}{765183} & 1 \end{bmatrix}, \quad
\begin{bmatrix} 50 \\ \frac{5621}{50} \\ \frac{765183}{5621} \\ \frac{399424}{765183} \end{bmatrix}.
$$

Step $(4)(b)(ii)$ increments k, and we return to the top of the loop with $k = 4$.

After another 19 iterations, the algorithm terminates. Figure 4.2 gives a summary of the remaining reduce and exchange operations performed.

The final state of the algorithm is

$$
\begin{bmatrix} 2 & 3 & 1 & 1 \\ 2 & 0 & -2 & -4 \\ -2 & 2 & 3 & -3 \\ 3 & -2 & 6 & -1 \end{bmatrix}, \quad
\begin{bmatrix} 1 & 0 & 0 & 0 \\ -\frac{2}{15} & 1 & 0 & 0 \\ \frac{2}{15} & \frac{17}{178} & 1 & 0 \\ \frac{1}{3} & -\frac{5}{89} & \frac{931}{2271} & 1 \end{bmatrix}, \quad
\begin{bmatrix} 15 \\ \frac{356}{15} \\ \frac{2271}{89} \\ \frac{99856}{2271} \end{bmatrix}.
$$

This is how the matrix Y in Example 4.5 was obtained. The squared lengths of the original basis vectors $\mathbf{x}_1$, $\mathbf{x}_2$, $\mathbf{x}_3$, $\mathbf{x}_4$ are 127, 50, 198, 197; the squared lengths of the reduced basis vectors $\mathbf{y}_1$, $\mathbf{y}_2$, $\mathbf{y}_3$, $\mathbf{y}_4$ are 15, 24, 26, 50. Even in this small example of four vectors with single-digit components, the LLL algorithm has substantially improved the basis vectors: the longest reduced vector has the same length as the shortest original vector.

Remark 4.10. We can easily modify the LLL algorithm so that it also computes the matrix C for which $Y = CX$. In Step (1) we initialize C to the $n \times n$ identity matrix I_n. In Step (1) of procedure `reduce`, we apply the same elementary row operation to C as to Y: adding $-\lceil \mu_{k\ell} \rfloor$ times row ℓ to row k. In Step (1) of procedure `exchange`, we apply the same elementary row operation to C as to Y: interchanging rows $k-1$ and k.

Maple code for the LLL algorithm is presented in two parts (but one procedure) in Figures 4.3 and 4.4. This code uses procedures from the Maple package `LinearAlgebra`. It takes two arguments: the original lattice basis as the rows of the matrix Y, and the reduction parameter α. The code accepts a non-square matrix Y: it can apply the LLL algorithm to an m-dimensional lattice in n-dimensional space. The reduce and exchange procedures are included in the main procedure. In other respects, including variable names, the structure of the code is very similar to the algorithm of Figure 4.1.

4.3 Analysis of the LLL algorithm

In order to study the complexity of the LLL algorithm, we must first determine how the Gram-Schmidt basis and coefficients change during the reduction in Step (4)(b)(i) and during the exchange in Step (4)(b)(iii).

Lemma 4.11. Reduction Lemma. *Consider one call to* `reduce` *in Step (4)(b)(i) with given k and ℓ, and write $\nu = \lceil \mu_{k\ell} \rfloor$. Let $Y = MY^*$ and $Z = NZ^*$ be the matrix equations for the GSO before and after the call to* `reduce`(k, ℓ)*. Let $E = I_n - \nu E_{k\ell}$ be the elementary matrix which represents subtracting ν times row ℓ from row k (thus $E_{ii} = 1$ for $1 \leq i \leq n$, $E_{k\ell} = -\nu$, $E_{ij} = 0$ otherwise). We have*

$$Z = EY, \qquad N = EM, \qquad Z^* = Y^*.$$

In particular, the Gram-Schmidt orthogonal basis does not change. Before the call to `reduce`(k, ℓ) *we have*

$$|\mu_{kj}| \leq \tfrac{1}{2} \qquad (\ell < j < k).$$

After the execution of the loop on ℓ in Step (4)(b)(i) we have

$$|\mu_{kj}| \leq \tfrac{1}{2} \qquad (1 \leq j < k).$$

```
with( LinearAlgebra ):

LLL := proc( X, alpha )
local Y, Ystar, gstar, delta, mu, nu, xi,
      i, j, k, l, m, n, r, t:
Y := copy( X ):
m := RowDimension( Y ):
n := ColumnDimension( Y ):
Ystar := Matrix( m, n ):
mu := Matrix( m, m ):
gstar := Vector( m ):
for i to m do
  for t to n do Ystar[i,t] := Y[i,t] od:
  for j to i-1 do
    mu[i,j] :=
      DotProduct( Row(Y,i), Row(Ystar,j) ) / gstar[j]:
    Ystar := RowOperation( Ystar,[i,j], -mu[i,j] )
  od:
  gstar[i] := DotProduct( Row(Ystar,i), Row(Ystar,i) )
od:
```

FIGURE 4.3
Maple code for the LLL algorithm: part 1

```
k := 2:
while k <= m do
  if abs(mu[k,k-1]) > 1/2 then
    r := ceil( mu[k,k-1]-1/2 ):
    Y := RowOperation( Y, [k,k-1], -r ):
    for j to k-2 do mu[k,j] := mu[k,j] - r*mu[k-1,j] od:
    mu[k,k-1] := mu[k,k-1] - r
  fi:
  if gstar[k] >= (alpha-mu[k,k-1]^2) * gstar[k-1] then
    for l from k-2 to 1 by -1 do
      if abs(mu[k,l]) > 1/2 then
        r := ceil( mu[k,l]-1/2 ):
        Y := RowOperation( Y, [k,l], -r ):
        for j to l-1 do mu[k,j] := mu[k,j] - r*mu[l,j] od:
        mu[k,l] := mu[k,l] - r
      fi
    od:
    k := k + 1
  else
    nu := mu[k,k-1]:
    delta := gstar[k] + nu^2*gstar[k-1]:
    mu[k,k-1] := nu*gstar[k-1]/delta:
    gstar[k] := gstar[k-1]*gstar[k]/delta:
    gstar[k-1] := delta:
    Y := RowOperation( Y, [k-1,k] ):
    for j to k-2 do
      t := mu[k-1,j]: mu[k-1,j] := mu[k,j]: mu[k,j] := t
    od:
    for i from k+1 to m do
      xi := mu[i,k]:
      mu[i,k] := mu[i,k-1] - nu*mu[i,k]:
      mu[i,k-1] := mu[k,k-1]*mu[i,k] + xi
    od:
    if k > 2 then k := k - 1 fi
  fi
od:
RETURN( Y )
end:
```

FIGURE 4.4
Maple code for the LLL algorithm: part 2

Proof. Since Z is obtained from Y by subtracting ν times row ℓ from row k, we have $Z = EY$. Since $\ell < k$, the span of the vectors $\mathbf{y}_1, \mathbf{y}_2, \ldots, \mathbf{y}_i$ remains the same for all i, and hence the orthogonal basis $\mathbf{y}_1^*, \mathbf{y}_2^*, \ldots, \mathbf{y}_n^*$ does not change. Therefore $Z^* = Y^*$, and this gives

$$(EM)Y^* = E(MY^*) = EY = Z = NZ^* = NY^*,$$

and since Y^* is invertible we get $EM = N$. At the beginning of the loop in Step (4)(a)(i) we have $\ell = k-1$, so there are no values of j in the range $\ell < j < k$; hence the condition $|\mu_{kj}| \le \frac{1}{2}$ for $\ell < j < k$ is trivially satisfied. Suppose the condition holds before the call to `reduce(k, ℓ)`. The operation $Z = EY$ changes only row k of Y, and so for $\ell < j < k$ the old coefficient

$$\mu_{kj} = \frac{\mathbf{y}_k \cdot \mathbf{y}_j^*}{\mathbf{y}_j^* \cdot \mathbf{y}_j^*}$$

becomes the new coefficient

$$\mu_{kj}' = \frac{(\mathbf{y}_k - \nu\mathbf{y}_\ell) \cdot \mathbf{y}_j^*}{\mathbf{y}_j^* \cdot \mathbf{y}_j^*} = \frac{\mathbf{y}_k \cdot \mathbf{y}_j^*}{\mathbf{y}_j^* \cdot \mathbf{y}_j^*} - \nu\frac{\mathbf{y}_\ell \cdot \mathbf{y}_j^*}{\mathbf{y}_j^* \cdot \mathbf{y}_j^*} = \mu_{kj} - \nu\mu_{\ell j} = \mu_{kj}.$$

since $j > \ell$ implies $\mu_{\ell j} = 0$. As for $\mu_{k\ell}$, it becomes

$$\mu_{k\ell} - \nu\mu_{\ell\ell} = \mu_{k\ell} - \lceil \mu_{k\ell} \rfloor,$$

since $\mu_{\ell\ell} = 1$, and by definition of $\lceil \mu_{k\ell} \rfloor$ we have

$$\left| \mu_{k\ell} - \lceil \mu_{k\ell} \rfloor \right| \le \tfrac{1}{2}.$$

This completes the proof. □

The next lemma shows that the product of the squared lengths of the orthogonal basis vectors decreases by a factor of at least the reduction parameter α during every execution of Step (4)(b)(iii). This will be the key fact in the proof that the LLL algorithm terminates, and also explains why we need to assume that α is strictly less than 1.

Lemma 4.12. Exchange Lemma. *Consider the call to* `exchange` *with a given value of k in Step (4)(b)(iii). Let $Y = MY^*$ and $Z = NZ^*$ be the matrix equations for the GSO before and after the exchange. We have*

$$\mathbf{z}_i^* = \mathbf{y}_i^* \quad (i \ne k-1, k), \qquad |\mathbf{z}_{k-1}^*|^2 < \alpha|\mathbf{y}_{k-1}^*|^2, \qquad |\mathbf{z}_k^*| \le |\mathbf{y}_{k-1}^*|.$$

Proof. For all i except $i = k-1, k$ we have $\mathbf{y}_i = \mathbf{z}_i$ and the span of $\mathbf{y}_1, \mathbf{y}_2, \ldots, \mathbf{y}_{i-1}$ equals the span of $\mathbf{z}_1, \mathbf{z}_2, \ldots, \mathbf{z}_{i-1}$. Since $\mathbf{y}_i^*$ (respectively $\mathbf{z}_i^*$) is the projection of $\mathbf{y}_i$ (respectively $\mathbf{z}_i$) onto the orthogonal complement of the span of $\mathbf{y}_1, \ldots, \mathbf{y}_{i-1}$ (respectively $\mathbf{z}_1, \ldots, \mathbf{z}_{i-1}$), the first claim follows. Since

$\mathbf{z}_k = \mathbf{y}_{k-1}$ and $\mathbf{z}_{k-1} = \mathbf{y}_k$, the vector $\mathbf{z}_{k-1}^*$ is the component of $\mathbf{y}_k$ orthogonal to the span of $\mathbf{y}_1, \mathbf{y}_2, \ldots, \mathbf{y}_{k-2}$. By definition of the coefficients μ_{ij} we have

$$\mathbf{y}_k = \mathbf{y}_k^* + \sum_{\ell=1}^{k-1} \mu_{k\ell} \mathbf{y}_\ell^*, \quad \text{hence} \quad \mathbf{z}_{k-1}^* = \mathbf{y}_k^* + \mu_{k,k-1} \mathbf{y}_{k-1}^*.$$

Since $\mathbf{y}_k^*$ and $\mathbf{y}_{k-1}^*$ are orthogonal, we get

$$|\mathbf{z}_{k-1}^*|^2 = |\mathbf{y}_k^*|^2 + \mu_{k,k-1}^2 |\mathbf{y}_{k-1}^*|^2.$$

Since by assumption the exchange condition $\gamma_k^* \geq (\alpha - \mu_{k,k-1}^2)\gamma_{k-1}^*$ in Step (4)(b) is false, we have

$$|\mathbf{y}_k^*|^2 < (\alpha - \mu_{k,k-1}^2)|\mathbf{y}_{k-1}^*|^2,$$

and therefore

$$|\mathbf{z}_{k-1}^*|^2 < \alpha |\mathbf{y}_{k-1}^*|^2.$$

This proves the second claim. As for $\mathbf{z}_k^*$, it is the component of $\mathbf{y}_{k-1}$ orthogonal to the span of $\mathbf{y}_1, \ldots, \mathbf{y}_{k-2}, \mathbf{y}_k$ (note that $\mathbf{y}_{k-1}$ is omitted). We write U for the span of $\mathbf{y}_1, \mathbf{y}_2, \ldots, \mathbf{y}_{k-2}$. We have

$$\mathbf{y}_{k-1} = \mathbf{y}_{k-1}^* + \sum_{\ell=1}^{k-2} \mu_{k-1,\ell} \mathbf{y}_\ell^* = \mathbf{y}_{k-1}^* + \mathbf{u}, \quad \text{where} \quad \mathbf{u} = \sum_{\ell=1}^{k-2} \mu_{k-1,\ell} \mathbf{y}_\ell^*.$$

Therefore $\mathbf{z}_k^*$ is the component of $\mathbf{y}_{k-1}^* + \mathbf{u}$ orthogonal to the subspace $U + \mathbb{R}\mathbf{y}_k$. Using Theorem 3.4, we get

$$\mathbf{u} \in \text{span}(\mathbf{y}_1^*, \ldots, \mathbf{y}_{k-2}^*) = \text{span}(\mathbf{y}_1, \ldots, \mathbf{y}_{k-2}) = U \subset U + \mathbb{R}\mathbf{y}_k.$$

Hence $\mathbf{z}_k^*$ is the component of $\mathbf{y}_{k-1}^*$ orthogonal to the subspace $U + \mathbb{R}\mathbf{y}_k$, and this completes the proof. $\square$

Lemma 4.13. *At the start of each iteration of Step (4)(b), the following conditions hold:*

$$|\mu_{ij}| \leq \tfrac{1}{2} \ (1 \leq j < i < k), \qquad |\mathbf{y}_i^* + \mu_{i,i-1}\mathbf{y}_{i-1}^*|^2 \geq \alpha |\mathbf{y}_{i-1}^*|^2 \ (2 \leq i < k).$$

If the algorithm terminates, then the output $\mathbf{y}_1, \mathbf{y}_2, \ldots, \mathbf{y}_n$ *is a reduced basis of the lattice.*

Proof. Exercise 4.13. $\square$

Our next goal is to find an upper bound for the number of iterations of the loop in Step 4 of the LLL algorithm. To simplify the argument, we assume that the original basis $\mathbf{x}_1, \mathbf{x}_2, \ldots, \mathbf{x}_n$ consists of vectors with integral components; that is, $\mathbf{x}_i \in \mathbb{Z}^n$ for $i = 1, 2, \ldots, n$. (Remark 4.24 below explains how this assumption may be removed.)

Recall the definition of the Gram determinant (Definition 3.7): for any $k = 1, 2, \ldots, n$ we consider the first k basis vectors $\mathbf{y}_1, \mathbf{y}_2, \ldots, \mathbf{y}_n$ and let G_k be the $k \times n$ matrix with $\mathbf{y}_i$ as row i for $1 \leq i \leq k$. Since we are now assuming that $\mathbf{y}_i \in \mathbb{Z}^n$ for all i, the Gram matrix $G_k G_k^t$ is a $k \times k$ matrix with integer entries, and the Gram determinant $d_k = \det(G_k G_k^t)$ is an integer. By Proposition 3.8 we have

$$d_k = \prod_{\ell=1}^{k} |g_\ell^*|^2. \tag{4.5}$$

Lemma 4.14. *During the calls to* reduce(k, ℓ) *in the LLL algorithm, the Gram determinants d_i do not change. During the calls to* exchange(k), *the Gram determinants d_i do not change for $i \neq k-1$, but d_{k-1} changes to a new value $d'_{k-1} \leq \alpha d_{k-1}$, where α is the reduction parameter.*

Proof. Lemma 4.11 shows that the orthogonal basis $\mathbf{y}_1^*, \mathbf{y}_2^*, \ldots, \mathbf{y}_n^*$ does not change during calls to reduce, so the first claim follows from equation (4.5).

For $i < k-1$, the call to exchange(k) has no effect on the Gam matrix G_i, and hence no effect on d_i. For $i > k-1$, the call to exchange(k) transposes two rows of G_i and two columns of G_i^t, multiplying $\det(G_i G_i^t)$ by $(-1)^2$, so again there is no effect on d_i. For $i = k-1$ we write $\mathbf{y}_\ell^*$ and $\mathbf{z}_\ell^*$ for the vectors before and after the call to exchange(k). We have

$$d'_{k-1} = \prod_{\ell=1}^{k-1} |\mathbf{z}_\ell^*|^2 \qquad \text{equation (4.5)}$$

$$= |\mathbf{z}_{k-1}^*|^2 \prod_{\ell=1}^{k-2} |\mathbf{z}_\ell^*|^2 \qquad \text{separate last factor}$$

$$\leq \alpha |\mathbf{y}_{k-1}^*|^2 \prod_{\ell=1}^{k-2} |\mathbf{z}_\ell^*|^2 \qquad \text{Lemma 4.12}$$

$$= \alpha |\mathbf{y}_{k-1}^*|^2 \prod_{\ell=1}^{k-2} |\mathbf{y}_\ell^*|^2 \qquad \mathbf{z}_\ell^* = \mathbf{y}_\ell^* \text{ for } 1 \leq \ell \leq k-2$$

$$= \alpha \prod_{\ell=1}^{k-1} |\mathbf{y}_\ell^*|^2 \qquad \text{include last factor}$$

$$= \alpha d_{k-1} \qquad \text{equation (4.5)}$$

This completes the proof. $\qquad \square$

Definition 4.15. The **loop invariant** is the quantity

$$D = \prod_{k=1}^{n-1} d_k.$$

We write D_0 for the value of D at the start of the algorithm. The assumption that the original basis vectors $\mathbf{x}_1$, $\mathbf{x}_2$, ..., $\mathbf{x}_n$ are in $\mathbb{Z}^n$ implies that D is a positive integer throughout the algorithm. (The term "invariant" is perhaps unfortunate; our goal is to prove that this quantity strictly decreases during the execution of the LLL algorithm.)

Lemma 4.16. *We have*

$$D_0 \leq B^{n(n-1)} \text{ where } B = \max\left(|\mathbf{x}_1|, |\mathbf{x}_2|, \ldots, |\mathbf{x}_n|\right).$$

Proof. We have

$$D_0 = \prod_{k=1}^{n-1} d_k \qquad\qquad \text{Definition 4.15}$$

$$= \prod_{k=1}^{n-1} \prod_{\ell=1}^{k} |\mathbf{x}_\ell^*|^2 \qquad\qquad \text{equation (4.5)}$$

$$= \left(|\mathbf{x}_1^*|^2\right)\left(|\mathbf{x}_1^*|^2|\mathbf{x}_2^*|^2\right) \cdots \left(|\mathbf{x}_1^*|^2|\mathbf{x}_2^*|^2 \cdots |\mathbf{x}_{n-1}^*|^2\right) \qquad\qquad \text{expand products}$$

$$= \prod_{k=1}^{n-1} |\mathbf{x}_k^*|^{2(n-k)} \qquad\qquad \text{collect factors}$$

$$\leq \prod_{k=1}^{n-1} |\mathbf{x}_k|^{2(n-k)} \qquad\qquad \text{Theorem 3.4}$$

$$\leq \prod_{k=1}^{n-1} B^{2(n-k)} \qquad\qquad \text{definition of } B$$

$$= \prod_{k=1}^{n-1} B^{2k} \qquad\qquad \text{reverse order}$$

$$= B^{n(n-1)} \qquad\qquad \sum_{k=1}^{n-1} 2k = n(n-1)$$

This completes the proof. $\qquad\qquad\qquad\qquad\qquad\qquad\qquad\qquad\qquad\qquad\qquad\qquad \square$

Definition 4.17. We write E for the total number of calls to `exchange` performed throughout the LLL algorithm.

Lemma 4.18. *We have*

$$E \leq -\frac{\log B}{\log \alpha}\, n(n-1).$$

(A ratio of logarithms to the same base does not depend on the base.)

Proof. Lemma 4.14 implies that D decreases to at most αD after each call to `exchange`. Since D is a positive integer throughout the algorithm, we have

$$1 \leq \alpha^E D_0, \quad \text{equivalently,} \quad \alpha^{-E} \leq D_0.$$

Lemma 4.16 implies

$$\alpha^{-E} \le B^{n(n-1)}, \quad \text{equivalently,} \quad -E \log \alpha \le n(n-1) \log B.$$

We divide by $-\log \alpha$ (which is > 0 since $\alpha < 1$) to complete the proof. $\qquad \square$

From the last lemma, it follows that the LLL algorithm terminates: the last call to **exchange** is the last time the index k decreases; every subsequent pass through the loop will call **reduce** and increase k. Since $2 \le k \le n$ throughout the algorithm, there will be at most $n-1$ more passes through the loop.

Theorem 4.19. Termination Theorem. *The total number of passes through the loop in Step (4) of the LLL algorithm is at most*

$$-\frac{2 \log B}{\log \alpha} n(n-1) + (n-1).$$

Proof. By definition, E is the number of times the algorithm passes through Steps (4)(b)(iii) and (4)(b)(iv). We write E' for the number of times the algorithm passes through Steps (4)(b)(i) and (4)(b)(ii). Thus $E + E'$ is the total number of passes through the loop in Step (4). Every time E increases by 1, the index k decreases by 1; every time E' increases by 1, the index k increases by 1. It follows that the integer $k + E - E'$ remains constant throughout the algorithm. At the start, we have $k = 2$ and $E = E' = 0$, so $k + E - E' = 2$. At the end, we have $k = n + 1$, so $n + 1 + E - E' = 2$, and hence $E' - E = n - 1$. This gives $E' + E = 2E + n - 1$, and now Lemma 4.18 completes the proof. $\qquad \square$

Example 4.20. If $n = 4$ and the components of the original lattice basis vectors $\mathbf{x}_1$, $\mathbf{x}_2$, $\mathbf{x}_3$, $\mathbf{x}_4$ are single-digit integers ($-9 \le x_{ij} \le 9$ for $1 \le i, j \le 4$) then $B \le \sqrt{4 \cdot 9^2} = 18$. We get this upper bound on the number of passes through the loop in Step (4) of the LLL algorithm:

$$-\frac{24 \log 18}{\log \alpha} + 3.$$

For $\alpha = 0.75, 0.9, 0.99, 0.999$ we obtain the upper bounds 244, 661, 6905, 69337 (taking the floor of the real number). Comparing these numbers to the 23 iterations with $\alpha = 1$ required by the 4 pseudorandom vectors in Example 4.9, we see that the upper bound of Theorem 4.19 is very weak.

Definition 4.21. We recall the **big O notation**. Suppose that $f(x)$ and $g(x)$ are functions with the same domain, which is a subset of the real numbers. The statement $f(x) = O(g(x))$ means that for sufficiently large values of x, the quantity $f(x)$ is at most a constant multiple of the quantity $g(x)$, in absolute value. That is, $f(x) = O(g(x))$ if and only if there exists a real number $c > 0$ and a real number x_0 such that $|f(x)| \le c|g(x)|$ for all $x \ge x_0$. For further information, see Graham et al. [50].

Theorem 4.22. *For a fixed value of the parameter α, the number of times the LLL algorithm passes through the loop in Step (4) is $O(n^2 \log B)$, and the number of arithmetic operations performed by the algorithm is $O(n^4 \log B)$.*

Proof. The first claim follows immediately from Theorem 4.19. For the second claim, we observe that

- The initial computation of the Gram-Schmidt orthogonalization in Step (2) requires $O(n^3)$ arithmetic operations.

- The initial reduction in Step (4)(a) requires $O(n)$ operations, the $k-2$ reductions in Step (4)(b)(i) require $O(n^2)$ operations, and the exchange in Step (4)(b)(iii) requires $O(n)$ operations. Thus one pass through the loop in Step (4) requires $O(n^2)$ operations, and so the entire loop in Step (4) requires $O(n^4 \log B)$ operations.

Since $O(n^3)$ is dominated by $O(n^4 \log B)$, this completes the proof. $\qquad \square$

Theorem 4.23. *The binary lengths of the integers arising in the LLL algorithm are $O(n \log B)$.*

Proof. We will find upper bounds on the quantities $|\mathbf{y}_i|$, $|\mathbf{y}_i^*|$ and $|\mu_{ij}|$. At the start of the algorithm, we have $\mathbf{y}_i = \mathbf{x}_i$ for $1 \le i \le n$. After the initial computation of the Gram-Schmidt orthogonalization, Theorem 3.4 shows that

$$|\mathbf{y}_i^*| \le |\mathbf{y}_i| = |\mathbf{x}_i| \le B.$$

Equation (4.5) implies

$$d_i \le B^{2i} \le B^{2n} \quad \text{and so} \quad \log d_i = O(n \log B).$$

Using Proposition 3.11 we see that B^{2n} is an upper bound for the denominators of the components of the vectors $\mathbf{y}_i^*$ and the quantities μ_{ij}, and so these denominators are also $O(n \log B)$. (Recall that by assumption the components of the vectors $\mathbf{y}_i$ are integers.)

It remains to find upper bounds for the numerators. We first show that during the reduction, Steps (4)(a) and (4)(b)(i), we have

$$|\mathbf{y}_\ell| \le n^{1/2} B \ (1 \le \ell \le k-1), \qquad |\mathbf{y}_k| \le n(2B)^n. \tag{4.6}$$

These bounds hold at the start of the algorithm; and the exchange in Step (4)(b)(iii) does not affect the lengths $|\mathbf{y}_i|$ for $1 \le i \le n$. It will therefore follow that these bounds hold throughout the algorithm. So we may assume that these bounds hold immediately before the reduction. The vectors $\mathbf{y}_i$ for $i \ne k$ are not changed by the reduction, so it suffices to consider $\mathbf{y}_k$. We write

$$\mu_k = \max_{1 \le \ell \le k} |\mu_{k\ell}|.$$

Since

$$\mathbf{y}_k = \sum_{\ell=1}^{k} \mu_{k\ell} \mathbf{y}_\ell^*,$$

we get

$$|\mathbf{y}_k|^2 = \sum_{\ell=1}^{k} |\mu_{k\ell}|^2 |\mathbf{y}_\ell^*|^2 \le n\mu_k^2 B^2,$$

and hence

$$|\mathbf{y}_k| \le n^{1/2} \mu_k B. \tag{4.7}$$

At the end of the reduction, we have $\mu_{kk} = 1$ and $|\mu_{k\ell}| \le \frac{1}{2}$ for $1 \le \ell \le k-1$, so we have proved the first part of equation (4.6). Part (c) of Proposition 3.11 says that at the start of the reduction we have

$$|\mu_{k\ell}| \le d_{\ell-1}^{1/2} |\mathbf{y}_k| \quad (1 \le \ell < k).$$

Therefore

$$1 \le \mu_k \le \left(\max_{1 \le \ell < k} d_{\ell-1}^{1/2} \right) |\mathbf{y}_k| \le B^{k-2} n^{1/2} B \le n^{1/2} B^{n-1}. \tag{4.8}$$

We now get

$$\left| \mu_{kj} - \lceil \mu_{k\ell} \rceil \mu_{\ell j} \right| \le |\mu_{kj}| + \left| \lceil \mu_{k\ell} \rceil \right| |\mu_{\ell j}| \le \mu_k + \left(\mu_k + \tfrac{1}{2} \right) \tfrac{1}{2} \le 2\mu_k,$$

since $|\mu_{\ell j}| \le \frac{1}{2}$ for $1 \le j < \ell$. Thus for each ℓ in Step (4)(b)(i), the value of μ_k increases by at most a factor of 2; hence during the reduction μ_k increases by at most $2^{k-1} \le 2^{n-1}$. Combining this with equation (4.8) we see that throughout the algorithm we have

$$\mu_k \le n^{1/2} (2B)^{n-1}. \tag{4.9}$$

Now using equation (4.7) we get

$$|\mathbf{y}_k|^2 \le n(2B)^n,$$

so we have proved the second part of equation (4.6).

We can now complete the proof. Recall that $|\mathbf{z}|_\infty \le |\mathbf{z}|$ for any vector $\mathbf{z} \in \mathbb{R}^n$, where $|\mathbf{z}|_\infty$ is the max-norm (the maximum of the absolute values of the components). That is, the Euclidean norm is an upper bound for the absolute values of the components. Equation (4.6) already gives bounds on $|\mathbf{y}_i|$, and these imply that the (integral) components of $\mathbf{y}_i$ are $O(n \log B)$. Proposition 3.11 states that the vector $d_{i-1}\mathbf{y}_i^*$ has integral components for $1 \le i \le n$, and that the quantity $d_i\mu_{ij}$ is an integer for $1 \le j \le i \le n$. At the start of this proof we noted that $|\mathbf{y}_i^*| \le B$ and that $d_i \le B^{2n}$; hence the integral components of $d_{i-1}\mathbf{y}_i^*$ are less than or equal to B^{2n+1}, which has binary length $O(n \log B)$. Equation (4.9) implies that

$$|d_i\mu_{ij}| \le B^{2n} n^{1/2} (2B)^{n-1} = 2^{n-1} n^{1/2} B^{3n-1},$$

which is also $O(n \log B)$. This completes the proof. $\square$

Remark 4.24. In our analysis of the LLL algorithm, we have made the simplifying assumption that the original lattice basis vectors have integral components. This assumption can be removed, as follows. In view of the proof of Lemma 4.18, it suffices to show that the Gram determinants d_i have a positive lower bound which depends only on the lattice L. This is implied by a basic theorem in the geometry of numbers: by Cassels [22] (Chapter II, Theorem I, page 31) the lattice L contains a nonzero vector $\mathbf{z}$ with

$$|\mathbf{z}|^2 \leq \left(\frac{4}{3}\right)^{(i-1)/2} d_i^{1/i} \quad (1 \leq i \leq n),$$

and hence

$$d_i \geq \left(\frac{3}{4}\right)^{i(i-1)/2} |\mathbf{z}|^2 \quad (1 \leq i \leq n),$$

where $\mathbf{z}$ is a shortest nonzero vector in L.

The LLL algorithm can be applied with only trivial modifications (see the Maple code in Figures 4.3 and 4.4) to arbitrary linearly independent vectors $\mathbf{x}_1, \mathbf{x}_2, \ldots, \mathbf{x}_m$ in $\mathbb{R}^n$ with $m \leq n$. We conclude this section with an example.

Example 4.25. Let L be the 6-dimensional lattice in $\mathbb{R}^9$ whose original basis consists of the rows $\mathbf{x}_1, \mathbf{x}_2, \ldots, \mathbf{x}_6$ of the matrix X:

$$X = \begin{bmatrix} 4 & 9 & 3 & -5 & -5 & -1 & 7 & -1 & -5 \\ -2 & -8 & -7 & -1 & -3 & 6 & -3 & 9 & 8 \\ 1 & -3 & -2 & 3 & 9 & 7 & 2 & 7 & -2 \\ -5 & 6 & 4 & -2 & -2 & -7 & -2 & -9 & 1 \\ 1 & -2 & -2 & 7 & 7 & -3 & -9 & -5 & -4 \\ 7 & 1 & -4 & 3 & -2 & 9 & 9 & 7 & 6 \end{bmatrix}$$

The squared lengths of these original basis vectors are

$$232, \quad 317, \quad 210, \quad 220, \quad 238, \quad 326.$$

The LLL algorithm with $\alpha = 1$ performs the 25 reduce and exchange steps summarized in Figure 4.5. The final basis consists of the rows $\mathbf{y}_1, \mathbf{y}_2, \ldots, \mathbf{y}_6$ of the matrix Y:

$$Y = \begin{bmatrix} -4 & 3 & 2 & 1 & 7 & 0 & 0 & -2 & -1 \\ 3 & -1 & -6 & 1 & -1 & 2 & -5 & 3 & -1 \\ -2 & 4 & -2 & -5 & -1 & 5 & 4 & 6 & 2 \\ 1 & 9 & -4 & 3 & 2 & 4 & 2 & -1 & 5 \\ 2 & -5 & 2 & 1 & 3 & 5 & 7 & 6 & 0 \\ 3 & 11 & -1 & -3 & 1 & 1 & 2 & 0 & -7 \end{bmatrix}$$

The squared lengths of these reduced basis vectors are

$$84, \quad 87, \quad 131, \quad 157, \quad 153, \quad 195.$$

This is an especially impressive example of the power of the LLL algorithm: the longest reduced vector is shorter than the shortest original vector.

iteration 1	reduce	$k = 2$	$\ell = 1$	$\lceil \mu_{k\ell} \rceil = -1$
iteration 2	exchange	$k = 3$		
iteration 3	exchange	$k = 2$		
iteration 4	exchange	$k = 4$		
iteration 5	exchange	$k = 3$		
iteration 6	reduce	$k = 2$	$\ell = 1$	$\lceil \mu_{k\ell} \rceil = -1$
iteration 7	exchange	$k = 2$		
iteration 8	exchange	$k = 4$		
iteration 9	exchange	$k = 3$		
iteration 10	reduce	$k = 2$	$\ell = 1$	$\lceil \mu_{k\ell} \rceil = -1$
iteration 11	reduce	$k = 4$	$\ell = 3$	$\lceil \mu_{k\ell} \rceil = -1$
iteration 12	reduce	$k = 4$	$\ell = 2$	$\lceil \mu_{k\ell} \rceil = 1$
iteration 13	exchange	$k = 5$		
iteration 14	exchange	$k = 4$		
iteration 15	reduce	$k = 3$	$\ell = 2$	$\lceil \mu_{k\ell} \rceil = -1$
iteration 16	exchange	$k = 3$		
iteration 17	reduce	$k = 2$	$\ell = 1$	$\lceil \mu_{k\ell} \rceil = 1$
iteration 18	reduce	$k = 4$	$\ell = 2$	$\lceil \mu_{k\ell} \rceil = 1$
iteration 19	reduce	$k = 4$	$\ell = 1$	$\lceil \mu_{k\ell} \rceil = 1$
iteration 20	reduce	$k = 5$	$\ell = 4$	$\lceil \mu_{k\ell} \rceil = 1$
iteration 21	reduce	$k = 5$	$\ell = 3$	$\lceil \mu_{k\ell} \rceil = -1$
iteration 22	exchange	$k = 6$		
iteration 23	reduce	$k = 5$	$\ell = 4$	$\lceil \mu_{k\ell} \rceil = 1$
iteration 24	exchange	$k = 5$		
iteration 25	reduce	$k = 4$	$\ell = 1$	$\lceil \mu_{k\ell} \rceil = -1$

FIGURE 4.5
Reduce and exchange operations for Example 4.25

4.4   The closest vector problem

In this section we follow Babai [12] to show how the LLL algorithm can be used to find a good approximate solution to the closest vector problem (CVP): Given a vector $\mathbf{z} \in \mathbb{R}^n$ and a lattice $L \subset \mathbb{R}^n$ with basis vectors $\mathbf{x}_1, \mathbf{x}_2, \ldots, \mathbf{x}_n$, find a vector $\mathbf{y} \in L$ for which the distance $|\mathbf{z} - \mathbf{y}|$ is as small as possible.

We consider Babai's "nearest plane" algorithm. Let $\mathbf{x}_1^*, \mathbf{x}_2^*, \ldots, \mathbf{x}_n^*$ be the Gram-Schmidt orthogonalization of the lattice basis $\mathbf{x}_1, \mathbf{x}_2, \ldots, \mathbf{x}_n$. Let $U \subset \mathbb{R}^n$ be the hyperplane (subspace of dimension $n-1$) with basis $\mathbf{x}_1, \mathbf{x}_2, \ldots, \mathbf{x}_{n-1}$; note that we omit the last vector $\mathbf{x}_n$. Let $L^{(n-1)} = U \cap L$ be the corresponding sublattice of L; thus $L^{(n-1)} \subset U$ is the lattice with basis $\mathbf{x}_1, \mathbf{x}_2, \ldots, \mathbf{x}_{n-1}$. We consider translations of the hyperplane U by lattice vectors

$\mathbf{y} \in L$; that is, subsets of $\mathbb{R}^n$ of the form

$$U + \mathbf{y} = \{ \mathbf{u} + \mathbf{y} \mid \mathbf{u} \in U \} \qquad (\mathbf{y} \in L).$$

Given an arbitrary vector $\mathbf{z} \in \mathbb{R}^n$, the nearest plane algorithm says that we should find the vector $\mathbf{y} \in L$ for which the (orthogonal) distance from $\mathbf{z}$ to the translated hyperplane $U + \mathbf{y}$ is as small as possible.

For this we use the following recursive procedure. We write $\mathbf{z}$ as a linear combination (with real coefficients) of the GSO vectors:

$$\mathbf{z} = a_1 \mathbf{x}_1^* + a_2 \mathbf{x}_2^* + \cdots + a_{n-1} \mathbf{x}_{n-1}^* + a_n \mathbf{x}_n^* \qquad (a_1, a_2, \ldots, a_{n-1}, a_n \in \mathbb{R}).$$

We write $\lceil a_n \rfloor \in \mathbb{Z}$ for the nearest integer to a_n, and define

$$\mathbf{w} = \lceil a_n \rfloor \mathbf{x}_n, \qquad \mathbf{z}^* = a_1 \mathbf{x}_1^* + a_2 \mathbf{x}_2^* + \cdots + a_{n-1} \mathbf{x}_{n-1}^* + \mathbf{w}.$$

Then $\mathbf{z}^*$ is the orthogonal projection of $\mathbf{z}$ onto the translated hyperplane $U + \mathbf{w}$. We clearly have $\mathbf{z}^* - \mathbf{w} \in U$, and we recursively find the vector $\mathbf{y}^{(n-1)} \in L^{(n-1)}$ closest to $\mathbf{z}^* - \mathbf{w}$. (We have reduced the dimension of the problem from n to $n-1$.) We then set $\mathbf{y} = \mathbf{y}^{(n-1)} + \mathbf{w}$. For the case $n = 1$, which is the basis of the recursion, we are simply finding the closest integer multiple of one nonzero real number to another real number.

Theorem 4.26. Babai's Theorem. (Babai [12], page 4) *Suppose that the basis* $\mathbf{x}_1, \mathbf{x}_2, \ldots, \mathbf{x}_n$ *of the lattice* $L \subset \mathbb{R}^n$ *is LLL-reduced with standard reduction parameter* $\alpha = \frac{3}{4}$. *Let* $\mathbf{z} \in \mathbb{R}^n$ *be an arbitrary vector. Then the lattice vector* $\mathbf{y} \in L$ *produced by the nearest plane algorithm on input* $\mathbf{z}$ *satisfies*

$$|\mathbf{z} - \mathbf{y}| \le 2^{n/2} |\mathbf{z} - \mathbf{u}|,$$

where $\mathbf{u} \in L$ *is the closest lattice vector to* $\mathbf{z} \in \mathbb{R}^n$.

Proof. For $n = 1$, the nearest plane algorithm clearly finds the closest lattice vector; as mentioned above, this is simply the closest multiple of one nonzero real number to another real number.

Now suppose that $n \ge 2$; we use induction on n. We have

$$|\mathbf{z} - \mathbf{z}^*| \le \frac{1}{2} |\mathbf{x}_n^*|, \tag{4.10}$$

$$|\mathbf{z} - \mathbf{z}^*| \le |\mathbf{z} - \mathbf{u}|, \tag{4.11}$$

since $|\mathbf{x}_n^*|$ is the distance between two consecutive hyperplanes $U + \mathbf{y}$, and $|\mathbf{z} - \mathbf{z}^*|$ is the distance from $\mathbf{z}$ to the nearest such hyperplane. Inequality (4.10) trivially implies

$$|\mathbf{z} - \mathbf{z}^*|^2 \le \frac{1}{4} |\mathbf{x}_n^*|^2,$$

and then induction (corresponding to the recursion in the algorithm) implies

$$|\mathbf{z} - \mathbf{y}|^2 \le \frac{1}{4} \left(|\mathbf{x}_1^*|^2 + |\mathbf{x}_2^*|^2 + \cdots + |\mathbf{x}_n^*|^2 \right).$$

Proposition 4.6 (with $\beta = 2$ since $\alpha = \frac{3}{4}$) gives

$$\frac{1}{4}\left(|\mathbf{x}_1^*|^2 + |\mathbf{x}_2^*|^2 + \cdots + |\mathbf{x}_n^*|^2\right) \leq \frac{1}{4}\left(2^{n-1} + 2^{n-2} + \cdots + 1\right)|\mathbf{x}_n^*|^2$$
$$= \frac{1}{4}\left(2^n - 1\right)|\mathbf{x}_n^*|^2$$
$$< 2^{n-2}|\mathbf{x}_n^*|^2.$$

Combining these inequalities gives

$$|\mathbf{z} - \mathbf{y}| \leq 2^{n/2-1}|\mathbf{x}_n^*|. \tag{4.12}$$

We now consider two cases, corresponding to whether the closest lattice vector $\mathbf{u} \in L$ does or does not belong to the translated hyperplane $U + \mathbf{w}$.

Case 1 ($\mathbf{u} \in U + \mathbf{w}$): In this case $\mathbf{u} - \mathbf{w}$ is the closest vector in the sublattice $L^{(n-1)}$ to the vector $\mathbf{z}^* - \mathbf{w} \in U$. Therefore the inductive hypothesis gives

$$|\mathbf{z}^* - \mathbf{y}| = |\mathbf{z}^* - \mathbf{w} - \mathbf{y}^{(n-1)}|$$
$$\leq 2^{(n-1)/2}|\mathbf{z}^* - \mathbf{w} - (\mathbf{u} - \mathbf{w})|$$
$$= 2^{(n-1)/2}|\mathbf{z}^* - \mathbf{u}|$$
$$\leq 2^{(n-1)/2}|\mathbf{z} - \mathbf{u}|.$$

Combining this with inequality (4.11) gives

$$|\mathbf{z} - \mathbf{y}|^2 = |\mathbf{z} - \mathbf{z}^*|^2 + |\mathbf{z}^* - \mathbf{y}|^2$$
$$\leq |\mathbf{z} - \mathbf{u}|^2 + 2^{n-1}|\mathbf{z} - \mathbf{u}|^2$$
$$\leq 2^n|\mathbf{z} - \mathbf{u}|^2.$$

Hence $|\mathbf{z} - \mathbf{y}| \leq 2^{n/2}|\mathbf{z} - \mathbf{u}|$, as required.

Case 2 ($\mathbf{u} \notin U + \mathbf{w}$): In this case we must have

$$|\mathbf{z} - \mathbf{u}| \geq \frac{1}{2}|\mathbf{x}_n^*|.$$

Comparing this with inequality (4.12) we again obtain $|\mathbf{z} - \mathbf{y}| \leq 2^{n/2}|\mathbf{z} - \mathbf{u}|$, and this completes the proof. $\qquad\square$

4.5 Projects

Project 4.1. Write a report and give a seminar talk on Cassels [22], Chapter II, Theorem I, page 31: If L is any lattice in $\mathbb{R}^n$ with basis $\mathbf{x}_1$, $\mathbf{x}_2$, $\ldots$, $\mathbf{x}_n$

(we do not assume that the basis vectors have integral components), then L contains a nonzero vector $\mathbf{z}$ for which

$$|\mathbf{z}|^2 \le \left(\frac{4}{3}\right)^{(i-1)/2} d_i^{1/i} \quad (1 \le i \le n),$$

where d_i is the i-th Gram determinant for the basis $\mathbf{x}_1, \mathbf{x}_2, \ldots, \mathbf{x}_n$.

Project 4.2. Choose a set A of reduction parameters for the LLL algorithm; for example,

$$A = \{\, 0.35,\, 0.4,\, 0.45,\, 0.5,\, 0.55,\, 0.6,\, 0.65,\, 0.7,\, 0.75,\, 0.8,\, 0.85,\, 0.9,\, 0.95 \,\}.$$

Fix a dimension $n \ge 3$ and let L be the lattice in $\mathbb{R}^n$ with basis $\mathbf{x}_1, \mathbf{x}_2, \ldots, \mathbf{x}_n$. Apply the LLL algorithm to L for each value $\alpha \in A$ and denote the resulting α-reduced basis of L by $\mathbf{y}_1^{(\alpha)}, \mathbf{y}_2^{(\alpha)}, \ldots, \mathbf{y}_n^{(\alpha)}$. Let $s(\alpha)$ be the length of the shortest vector in this α-reduced basis:

$$s(\alpha) = \min\left(|\mathbf{y}_1^{(\alpha)}|,\, |\mathbf{y}_2^{(\alpha)}|,\, \ldots,\, |\mathbf{y}_n^{(\alpha)}| \right).$$

Let $s(L)$ be the length of the shortest vector obtained over all values $\alpha \in A$:

$$s(L) = \min_{\alpha \in A} s(\alpha).$$

Define $\alpha(L)$ to be the smallest $\alpha \in A$ which produces the shortest vector:

$$\alpha(L) = \min \{\, \alpha \in A \mid s(\alpha) = s(L) \,\}.$$

We can think of $\alpha(L)$ as the "optimal value of the reduction parameter" for the lattice L: that is, the smallest α which produces the shortest vector.

Choose a set of Λ of lattices in $\mathbb{R}^n$; for example, 1000 executions of the Maple command

```
X := LinearAlgebra[RandomMatrix]( n, n, generator = -9..9 );
```

will produce 1000 lattice bases in $\mathbb{R}^n$ with single-digit components. (The rows of the matrix are the basis vectors; there is a small chance that some of these matrices will be singular, and these matrices should be ignored.) For each $L \in \Lambda$ compute $\alpha(L)$, and the expected value of $\alpha(L)$ over all $L \in \Lambda$:

$$E(\Lambda) = \frac{1}{|\Lambda|} \sum_{L \in \Lambda} \alpha(L).$$

We can think of $E(\Lambda)$ as the "expected value of the reduction parameter" for lattices of dimension n (of course, this is relative to the chosen set A of parameter values and the chosen set Λ of n-dimensional lattices).

Study the behavior of the quantity $E(\Lambda)$ as a function of the dimension n.

Project 4.3. Lovász and Scarf [92] have generalized the LLL algorithm by replacing the Euclidean norm on $\mathbb{R}^n$ by an arbitrary norm. For example, one might want to reduce a lattice basis with respect to the max-norm (the maximum of the absolute values of the components of the vectors) or the one-norm (the sum of the absolute values of the components of the vectors). Write a report and present a seminar talk on this paper. Implement this algorithm on a computer and test it on numerous examples.

Project 4.4. Write a report and present a seminar talk on the average-case behavior of the LLL algorithm, based on the papers by Daudé and Vallée [35] and Nguyen and Stehlé [107]; see also the survey by Vallée and Vera [142].

Project 4.5. Study the behavior of the LLL algorithm with the limiting value $\alpha = 1$ of the reduction parameter, based on the paper by Akhavi [10]. Write a report and present a seminar talk on this topic.

Project 4.6. Two recent survey papers on floating-point versions of the LLL algorithm are Schnorr [126] and Stehlé [133]. Write a report and present a seminar talk on this topic.

Project 4.7. Write a computer program to implement Babai's nearest plane algorithm for the closest vector problem discussed in Section 4.4.

Project 4.8. The paper of Babai [12] discusses another algorithm which provides a good approximate solution to the closest vector problem. In the notation of Section 4.4, we first write the arbitrary vector $\mathbf{z} \in \mathbb{R}^n$ as a linear combination (with real coefficients) of the lattice basis vectors:

$$\mathbf{z} = a_1 \mathbf{x}_1 + a_2 \mathbf{x}_2 + \cdots + a_n \mathbf{x}_n \qquad (a_1, a_2, \ldots, a_n \in \mathbb{R}).$$

We then replace each real coefficient a_i by the nearest integer $\lceil a_i \rfloor$ to obtain a lattice vector $\mathbf{w}$:

$$\mathbf{w} = \lceil a_1 \rfloor \mathbf{x}_1 + \lceil a_2 \rfloor \mathbf{x}_2 + \cdots + \lceil a_n \rfloor \mathbf{x}_n \qquad (\lceil a_1 \rfloor, \lceil a_2 \rfloor, \ldots, \lceil a_n \rfloor \in \mathbb{Z}).$$

The proof that $\mathbf{w}$ is a good approximation to the closest lattice vector to $\mathbf{z}$ requires a geometric result on the shape of certain "Lovász-reduced" parallelipipeds in $\mathbb{R}^n$. Write a report and present a seminar talk on this second solution to the closest vector problem.

Project 4.9. Write a report and present a seminar talk on the applications of the LLL algorithm to integer programming, based on the survey paper by Aardal and Eisenbrand [2].

Project 4.10. Write a report and present a seminar talk on the applications of the closest vector problem in wireless communication. Basic references are the papers by Viterbo and Boutros [146], Agrell et al. [4], and Mow [104].

4.6 Exercises

Exercise 4.1. Let X be an $n \times n$ matrix with integer entries and determinant ± 1. Prove that the rows of X span the same lattice in $\mathbb{R}^n$ as the rows of the identity matrix I_n.

Exercise 4.2. Let L (respectively M) be the lattice in $\mathbb{R}^3$ with basis consisting of the rows of the matrix X (respectively Y) below. Prove that $L \neq M$:

$$X = \begin{bmatrix} -9 & -9 & -1 \\ 7 & 6 & -7 \\ 6 & -5 & -4 \end{bmatrix}, \qquad Y = \begin{bmatrix} -8 & 3 & 7 \\ 6 & 2 & -2 \\ -2 & 8 & -8 \end{bmatrix}.$$

Exercise 4.3. Let L (respectively M) be the lattice in $\mathbb{R}^3$ with basis consisting of the rows of the matrix X (respectively Y) below. Prove that $L = M$:

$$X = \begin{bmatrix} 4 & 2 & -9 \\ -1 & 8 & 6 \\ 6 & -6 & 4 \end{bmatrix}, \qquad Y = \begin{bmatrix} -168 & 602 & 58 \\ 157 & -564 & -57 \\ 594 & -2134 & -219 \end{bmatrix}.$$

Exercise 4.4. Find lattices L, M in $\mathbb{R}^2$ with $\det(L) = \det(M)$ but $L \neq M$. Do the same for $\mathbb{R}^3$ and $\mathbb{R}^4$.

Exercise 4.5. Find a 4×4 matrix $C = (c_{ij})$ simultaneously satisfying the following conditions:

$$c_{ij} \in \mathbb{Z} \text{ for all } i, j; \qquad |c_{ij}| \geq 100 \text{ for all } i, j; \qquad \det(C) = 1.$$

Exercise 4.6. (a) For the matrices X and Y below, find a matrix C such that $\det(C) = \pm 1$ and $Y = CX$:

$$X = \begin{bmatrix} 3 & -5 & 9 & 4 \\ 1 & -3 & 5 & 4 \\ 2 & -1 & -1 & 6 \\ 3 & 2 & 2 & 5 \end{bmatrix}, \qquad Y = \begin{bmatrix} 3 & 0 & -2 & 2 \\ -1 & -1 & 1 & 4 \\ 1 & 3 & 3 & -1 \\ -2 & 2 & -4 & 0 \end{bmatrix}.$$

(b) Show that the rows of X and Y are bases for the same lattice in $\mathbb{R}^4$.
(c) Show that the basis X is not α-reduced for any α ($\frac{1}{4} < \alpha < 1$).
(d) Show that the basis Y is α-reduced for every α ($\frac{1}{4} < \alpha < 1$).

Exercise 4.7. (a) Trace the LLL algorithm with the standard value $\alpha = \frac{3}{4}$ on the lattice L consisting of all integral linear combinations of the rows $\mathbf{x}_1$, $\mathbf{x}_2$, $\mathbf{x}_3$ of the matrix X:

$$X = \begin{bmatrix} 4 & 5 & 1 \\ 4 & 8 & 2 \\ 6 & 2 & 6 \end{bmatrix}$$

(b) Verify that the longest vector in the reduced basis is shorter than the shortest vector in the original basis.
(c) Are the results any different for $\alpha = \frac{1}{2}$?
(d) Are the results any different for $\alpha = 1$?

Exercise 4.8. Trace the LLL algorithm with $\alpha = 1$ on the lattice M consisting of all integral linear combinations of the rows $\mathbf{y}_1$, $\mathbf{y}_2$, $\mathbf{y}_3$, $\mathbf{y}_4$ of the matrix F of Example 4.1. Compare the results with the matrix X of Example 4.1.

Exercise 4.9. (a) Write a computer program (in a language other than Maple) to implement the LLL algorithm of Figure 4.1.
(b) Test your program with $\alpha = \frac{3}{4}$ on the lattice consisting of all integral linear combinations of the rows of the matrix X:

$$
X = \begin{bmatrix}
8 & -3 & -3 & -9 & 1 & 9 & -3 & -9 \\
-7 & 5 & 1 & 1 & 9 & -3 & -4 & -2 \\
-5 & 2 & 1 & -3 & -4 & 5 & 5 & 4 \\
-4 & 9 & -6 & -5 & -7 & 2 & -1 & 5 \\
-5 & 0 & 2 & 2 & 0 & 5 & 6 & -5 \\
-8 & -2 & 3 & 5 & -1 & 7 & 7 & 4 \\
3 & -9 & 3 & -7 & 3 & 2 & -3 & 2 \\
-4 & -2 & -8 & 6 & 0 & 4 & -9 & 7
\end{bmatrix}
$$

(c) Are the results any different for $\alpha = \frac{1}{2}$? What can you say about the shortest and longest vectors in the reduced basis?
(d) Are the results any different for $\alpha = 1$? What can you say about the shortest and longest vectors in the reduced basis?

Exercise 4.10. Redo Exercise 4.7, keeping track of the loop invariant D at each step of the algorithm.

Exercise 4.11. Redo Exercise 4.8, keeping track of the loop invariant D at each step of the algorithm.

Exercise 4.12. Redo Exercise 4.9, keeping track of the loop invariant D at each step of the algorithm.

Exercise 4.13. Prove Lemma 4.13.

Exercise 4.14. Make the necessary changes in the LLL algorithm of Figure 4.1 so that it applies to m linearly independent vectors in $\mathbb{R}^n$ for any $m \leq n$.

Exercise 4.15. Trace the algorithm of Exercise 4.14 with $\alpha = 1$ on the lattice L consisting of all integral linear combinations of the rows of the matrix X:

$$
X = \begin{bmatrix}
4 & 2 & 7 & -4 & -3 & -6 \\
-9 & 2 & -7 & 9 & -3 & -4 \\
3 & -2 & 8 & -6 & 5 & -2
\end{bmatrix}
$$

Exercise 4.16. Write a computer program (in a language other than Maple) to implement the algorithm of Exercise 4.14. Test your program for various values of α on the lattice consisting of all integral linear combinations of the rows of the matrix X:

$$
X = \begin{bmatrix}
8 & 8 & 7 & 5 & 1 & 2 & 6 & 0 & 6 & 3 & 1 & 4 & 5 & 9 & 3 & 4 \\
5 & 7 & 8 & 3 & 2 & 2 & 1 & 5 & 5 & 3 & 5 & 3 & 9 & 2 & 7 & 0 \\
4 & 9 & 0 & 3 & 0 & 9 & 6 & 5 & 9 & 4 & 5 & 9 & 2 & 6 & 8 & 2 \\
1 & 4 & 6 & 5 & 4 & 1 & 2 & 7 & 5 & 2 & 7 & 9 & 8 & 9 & 9 & 9 \\
3 & 8 & 1 & 3 & 6 & 5 & 2 & 9 & 3 & 4 & 4 & 6 & 1 & 4 & 0 & 5 \\
6 & 1 & 1 & 5 & 1 & 4 & 1 & 7 & 5 & 2 & 0 & 7 & 6 & 2 & 8 & 2 \\
2 & 5 & 5 & 1 & 1 & 5 & 7 & 0 & 5 & 0 & 6 & 3 & 9 & 7 & 8 & 1 \\
1 & 7 & 3 & 3 & 8 & 6 & 4 & 3 & 8 & 4 & 4 & 6 & 4 & 1 & 8 & 4
\end{bmatrix}
$$

Exercise 4.17. Modify the discussion of the closest vector problem in Section 4.4 so that the general reduction paramater α is used in place of the standard value $\alpha = \frac{3}{4}$. In particular, how does Theorem 4.26 change?

5

Deep Insertions

CONTENTS

In this chapter we consider a modification of the exchange step of the original LLL algorithm: instead of transposing the vectors in positions k and $k-1$, we cyclically permute the vectors in positions i, $i+1$, ..., k where i is the smallest index for which vectors i and k satisfy the exchange condition. In other words, we put vector k into position i, and put vectors i to $k-1$ into positions $i+1$ to k. (If $i = k-1$ then we have the exchange step of the original LLL algorithm.) This idea, called a **deep insertion** of vector k into position i, was introduced in 1994 by Schnorr and Euchner [127]. A more detailed exposition is given in Cohen [26], Section §2.6.2.

5.1 Modifying the exchange condition

One problem with the method of deep insertions compared to the original LLL algorithm is that it is more difficult to keep track of the Gram-Schmidt orthogonalization of the lattice basis vectors. Recall that in the original LLL algorithm, we compute the GSO once at the start of the computation, and thereafter we merely update the GSO data after each reduce or exchange step by modifying the square lengths γ_i^* of orthogonal basis vectors and the projection coefficients μ_{ij}. This is possible since, as we saw in Chapter 3, there are simple formulas for calculating the effect on γ_i^* and μ_{ij} of size-reducing a basis vector or transposing the basis vectors in positions k and $k-1$, so there is no need to recompute the GSO. However, with a deep insertion of vector k into position i for $i < k-1$, the formulas become much more complicated; see Section 5.3. Therefore, in the LLL algorithm with deep insertions, we recompute the Gram-Schmidt orthogonalization of the lattice basis vectors up to vector k at the start of each iteration of the main loop.

To start, let's recall the exchange condition in the original LLL algorithm. The basic idea is to check whether transposing the vectors in positions k and $k-1$ and then recomputing the GSO will give us a shorter vector $\mathbf{x}_{k-1}^*$. The exchange is performed if and only if

$$|\mathbf{x}_k^*|^2 + \mu_{k,k-1}^2 |\mathbf{x}_{k-1}^*|^2 < \alpha |\mathbf{x}_{k-1}^*|^2; \tag{5.1}$$

that is, the exchange produces a new GSO vector $\widehat{\mathbf{x}}_{k-1}^*$ which is substantially shorter (depending on the parameter α) than the old GSO vector $\mathbf{x}_{k-1}^*$.

Suppose that at some point in the computation, the current ordered basis of the lattice consists of the following m vectors in $\mathbb{R}^n$:

$$\mathbf{x}_1, \ldots, \mathbf{x}_{i-1}, \mathbf{x}_i, \mathbf{x}_{i+1}, \ldots, \mathbf{x}_{k-1}, \mathbf{x}_k, \mathbf{x}_{k+1}, \ldots, \mathbf{x}_m.$$

Computing the Gram-Schmidt orthogonalization produces

$$\mathbf{x}_i^* = \mathbf{x}_i - \sum_{j=1}^{i-1} \mu_{ij} \mathbf{x}_j^* \quad (i = 1, \ldots, m),$$

or equivalently,

$$\mathbf{x}_i = \mathbf{x}_i^* + \sum_{j=1}^{i-1} \mu_{ij} \mathbf{x}_j^* \quad (i = 1, \ldots, m),$$

which gives

$$|\mathbf{x}_i|^2 = |\mathbf{x}_i^*|^2 + \sum_{j=1}^{i-1} \mu_{ij}^2 |\mathbf{x}_j^*|^2 \quad (i = 1, \ldots, m), \tag{5.2}$$

since $\mathbf{x}_1^*, \ldots, \mathbf{x}_m^*$ are orthogonal. If we now perform a deep insertion of $\mathbf{x}_k$ into position i, then the new ordered basis of the lattice consists of the vectors

$$\mathbf{x}_1, \ldots, \mathbf{x}_{i-1}, \mathbf{x}_k, \mathbf{x}_i, \ldots, \mathbf{x}_{k-2}, \mathbf{x}_{k-1}, \mathbf{x}_{k+1}, \ldots, \mathbf{x}_m.$$

Computing the GSO of this ordered basis gives the same vectors $\mathbf{x}_1^*, \ldots, \mathbf{x}_{i-1}^*$ but a new vector in position i:

$$\widehat{\mathbf{x}}_i^* = \mathbf{x}_k - \sum_{j=1}^{i-1} \mu_{kj} \mathbf{x}_j^*,$$

or equivalently,

$$\mathbf{x}_k = \widehat{\mathbf{x}}_i^* + \sum_{j=1}^{i-1} \mu_{kj} \mathbf{x}_j^*,$$

which gives

$$|\mathbf{x}_k|^2 = |\widehat{\mathbf{x}}_i^*|^2 + \sum_{j=1}^{i-1} \mu_{kj}^2 |\mathbf{x}_j^*|^2.$$

In particular, for the GSO vector in position i we have

$$|\widehat{\mathbf{x}}_i^*|^2 = |\mathbf{x}_k|^2 - \sum_{j=1}^{i-1} \mu_{kj}^2 |\mathbf{x}_j^*|^2. \tag{5.3}$$

In the special case $i = k-1$ we obtain

$$|\widehat{\mathbf{x}}_{k-1}^*|^2 = |\mathbf{x}_k|^2 - \sum_{j=1}^{k-2} \mu_{kj}^2 |\mathbf{x}_j^*|^2 = |\mathbf{x}_k^*|^2 + \sum_{j=1}^{k-1} \mu_{kj}^2 |\mathbf{x}_j^*|^2 - \sum_{j=1}^{k-2} \mu_{kj}^2 |\mathbf{x}_j^*|^2$$

where we have used equation (5.2) with $i = k$. This simplifies to

$$|\widehat{\mathbf{x}}_{k-1}^*|^2 = |\mathbf{x}_k^*|^2 + \mu_{k,k-1}^2 |\mathbf{x}_{k-1}^*|^2,$$

which is the quantity we have seen before in the exchange condition of the original LLL algorithm: the left side of the inequality (5.1). In the special cases $i = 1, 2, \ldots, k-1$ equation (5.3) gives the following results:

$$|\widehat{\mathbf{x}}_1^*|^2 = |\mathbf{x}_k|^2,$$
$$|\widehat{\mathbf{x}}_2^*|^2 = |\mathbf{x}_k|^2 - \mu_{k1}^2 |\mathbf{x}_1^*|^2,$$
$$|\widehat{\mathbf{x}}_3^*|^2 = |\mathbf{x}_k|^2 - \mu_{k1}^2 |\mathbf{x}_1^*|^2 - \mu_{k2}^2 |\mathbf{x}_2^*|^2,$$

$$\vdots$$

$$|\widehat{\mathbf{x}}_{k-1}^*|^2 = |\mathbf{x}_k|^2 - \mu_{k1}^2 |\mathbf{x}_1^*|^2 - \cdots - \mu_{k,k-2}^2 |\mathbf{x}_{k-2}^*|^2.$$

These equations tells us that in order to determine the position i into which we should do a deep insertion of $\mathbf{x}_k$, we should initialize a new variable $C = |\mathbf{x}_k|^2$, repeatedly subtract $\mu_{ki}^2 |\mathbf{x}_i^*|^2$ for $i = 1, 2, \ldots$, and stop as soon as soon as the **deep exchange condition** is satisfied:

$$C < \alpha |\mathbf{x}_i^*|^2, \qquad \text{equivalently} \qquad |\widehat{\mathbf{x}}_i^*|^2 < \alpha |\mathbf{x}_i^*|^2.$$

This means that the new GSO vector $\widehat{\mathbf{x}}_i^*$ in position i is sufficiently shorter (depending on the parameter α) than the old GSO vector $\mathbf{x}_i^*$ in position i. It is important to note that the initial value of C is the squared length of the lattice basis vector in position k, not the GSO vector in position k.

If the deep exchange condition is not satisfied for any value $1 \le i \le k-1$ then we do not perform a deep insertion; just as in the original LLL algorithm, we increment the index variable k and return to the top of the main loop.

We can now present the complete LLL algorithm with deep insertions; see Figure 5.1. There is a separate `reduce` procedure but not a separate `exchange` procedure: the original exchange procedure, which transposed two consecutive vectors and updated the GSO, has been replaced by a deep insertion and a recomputation of the GSO inside the main loop. Furthermore, the algorithm performs a complete size-reduction of $\mathbf{x}_k$ before starting to test the deep exchange condition; since this condition depends on all $\mathbf{x}_i$ for $i < k$, it is not

- *Input:* An $m \times n$ $(m \leq n)$ integer matrix X of rank m, and a reduction parameter $\alpha \in \mathbb{R}$ in the range $\frac{1}{4} < \alpha < 1$. The rows of X form a basis for an m-dimensional lattice L in $\mathbb{R}^n$.

- *Output*: An $m \times n$ integer matrix Y containing an α-reduced basis of L.

- Procedure reduce(k, ℓ):

  If $|\mu_{k\ell}| > 1/2$ then
  - (1) Set $r \leftarrow \lceil \mu_{k\ell} \rfloor$.
  - (2) Set $\mathbf{y}_k \leftarrow \mathbf{y}_k - r\mathbf{y}_\ell$.
  - (3) For $j = 1, 2, \ldots, \ell-1$ do: Set $\mu_{kj} \leftarrow \mu_{kj} - r\mu_{\ell j}$.
  - (4) Set $\mu_{k\ell} \leftarrow \mu_{k\ell} - r$.

- *Main loop*:
  - (1) Set $Y \leftarrow X$.
  - (2) Set $k \leftarrow 2$.
  - (3) While $k \leq m$ do:

    (recompute the GSO up to index k)
    - (a) For $i = 1, 2, \ldots, k$ do:

      Set $\mathbf{y}_i^* \leftarrow \mathbf{y}_i$.
      For $j = 1, 2, \ldots, i-1$ do:
        Set $\mu_{ij} \leftarrow (\mathbf{y}_i \cdot \mathbf{y}_j^*)/\gamma_j^*$. Set $\mathbf{y}_i^* \leftarrow \mathbf{y}_i^* - \mu_{ij}\mathbf{y}_j^*$.
      Set $\gamma_i^* \leftarrow \mathbf{y}_i^* \cdot \mathbf{y}_i^*$.

      (perform complete size-reduction of $\mathbf{y}_k$)
    - (b) For $\ell = k-1, \ldots, 2, 1$ do: Call reduce(k, ℓ).

      (find position i for deep insertion of $\mathbf{y}_k$)
    - (c) Set $C \leftarrow \mathbf{y}_k \cdot \mathbf{y}_k$. Set $i \leftarrow 1$. Set startagain $\leftarrow$ false.
    - (d) While $i < k$ and not startagain do:

      (check deep exchange condition)
      If $C \geq \alpha\gamma_i^*$ then
        Set $C \leftarrow C - \mu_{ki}^2 \gamma_i^*$. Set $i \leftarrow i + 1$.
      (deep insertion of $\mathbf{y}_k$ into position i)
      else
        Set $\mathbf{z} \leftarrow \mathbf{y}_k$. For $j = k, \ldots, i+1$ do: Set $\mathbf{y}_j \leftarrow \mathbf{y}_{j-1}$.
        Set $\mathbf{y}_i \leftarrow \mathbf{z}$.
        Set $k \leftarrow \max(i-1, 2)$.
        Set startagain $\leftarrow$ true.
    - (e) Set $k \leftarrow k+1$.

FIGURE 5.1
LLL algorithm with deep insertions

sufficient to merely reduce $\mathbf{x}_k$ using $\mathbf{x}_{k-1}$ as in the original LLL algoritmhm. The "goto" statements of the original algorithms presented by Schnorr and Euchner [127] and Cohen [26] (§2.6.2) have been removed, and a new Boolean variable `startagain` has been introduced.

5.2 Examples of deep insertion

We now consider some examples. When we say `exchange`(k) we simply mean a "deep insertion" of vector k into position $k-1$. We use the notation $k \rightarrow i$ to indicate a deep insertion of vector k into position i.

Example 5.1. Consider the lattice in $\mathbb{R}^3$ spanned by the rows of the matrix

$$\begin{bmatrix} 9 & 2 & 7 \\ 8 & 6 & 1 \\ 3 & 2 & 6 \end{bmatrix}$$

The original LLL algorithm with $\alpha = \frac{3}{4}$ does 5 passes through the main loop:

`reduce`$(2,1)$, `exchange`(2) $\qquad \begin{bmatrix} -1 & 4 & -6 \\ 9 & 2 & 7 \\ 3 & 2 & 6 \end{bmatrix}$

`reduce`$(2,1)$ $\qquad \begin{bmatrix} -1 & 4 & -6 \\ 8 & 6 & 1 \\ 3 & 2 & 6 \end{bmatrix}$

`exchange`(3) $\qquad \begin{bmatrix} -1 & 4 & -6 \\ 3 & 2 & 6 \\ 8 & 6 & 1 \end{bmatrix}$

`reduce`$(2,1)$ $\qquad \begin{bmatrix} -1 & 4 & -6 \\ 2 & 6 & 0 \\ 8 & 6 & 1 \end{bmatrix}$

`reduce`$(3,2)$, `reduce`$(3,1)$ $\qquad \begin{bmatrix} -1 & 4 & -6 \\ 2 & 6 & 0 \\ 3 & -2 & -5 \end{bmatrix}$

Row 3 is the shortest reduced lattice vector with square length 38.

The LLL algorithm ($\alpha = \frac{3}{4}$) with deep insertions also does 5 passes through the main loop, but performs one deep insertion:

`reduce`$(2,1)$, `exchange`(2) $\qquad \begin{bmatrix} -1 & 4 & -6 \\ 9 & 2 & 7 \\ 3 & 2 & 6 \end{bmatrix}$

$$\text{reduce}(3,1), \text{exchange}(3) \qquad \begin{bmatrix} -1 & 4 & -6 \\ 2 & 6 & 0 \\ 9 & 2 & 7 \end{bmatrix}$$

$$\text{reduce}(3,2), \text{reduce}(3,1), \text{deep insertion } 3 \to 1 \qquad \begin{bmatrix} 3 & -2 & -5 \\ -1 & 4 & -6 \\ 2 & 6 & 0 \end{bmatrix}$$

$$\text{reduce}(3,2), \text{reduce}(3,1), \text{exchange}(3) \qquad \begin{bmatrix} 3 & -2 & -5 \\ 6 & 0 & 1 \\ -1 & 4 & -6 \end{bmatrix}$$

$$\text{reduce}(3,2), \text{reduce}(3,1) \qquad \begin{bmatrix} 3 & -2 & -5 \\ 6 & 0 & 1 \\ 2 & 6 & 0 \end{bmatrix}$$

Row 3 is the shortest reduced lattice vector with square length 37.

Example 5.2. Consider the lattice in $\mathbb{R}^3$ spanned by the rows of the matrix

$$\begin{bmatrix} 83 & 29 & 21 \\ 99 & 45 & 96 \\ 2 & 65 & 31 \end{bmatrix}$$

The original LLL algorithm with $\alpha = \frac{3}{4}$ does 2 passes through the main loop; the first pass does reduce(2,1) and the second pass does nothing. We obtain the following reduced basis, in which row 3 (squared length 5190) is the shortest lattice vector; this has not changed from the original basis:

$$\begin{bmatrix} 83 & 29 & 21 \\ 16 & 16 & 75 \\ 2 & 65 & 31 \end{bmatrix}$$

The LLL algorithm ($\alpha = \frac{3}{4}$) with deep insertions does 4 passes through the main loop, including one deep insertion:

$$\text{reduce}(2,1) \qquad \begin{bmatrix} 83 & 29 & 21 \\ 16 & 16 & 75 \\ 2 & 65 & 31 \end{bmatrix}$$

$$\text{deep insertion } 3 \to 1 \qquad \begin{bmatrix} 2 & 65 & 31 \\ 83 & 29 & 21 \\ 16 & 16 & 75 \end{bmatrix}$$

$$\text{reduce}(3,1), \text{exchange}(3) \qquad \begin{bmatrix} 2 & 65 & 31 \\ 14 & -49 & 44 \\ 83 & 29 & 21 \end{bmatrix}$$

$$\text{reduce}(3,1) \qquad \begin{bmatrix} 2 & 65 & 31 \\ 14 & -49 & 44 \\ 81 & -36 & -10 \end{bmatrix}$$

Row 2 (squared length 4533) is the shortest lattice vector; its squared length is $\approx 87.34\%$ of that of the shortest vector found by the original LLL algorithm.

Example 5.3. Consider the lattice in $\mathbb{R}^3$ spanned by the rows of the matrix

$$\begin{bmatrix} -270 & 983 & -834 \\ -725 & -979 & 143 \\ 929 & -612 & -27 \end{bmatrix}$$

The original LLL algorithm with $\alpha = \frac{3}{4}$ does 2 passes through the main loop; the first does reduce(2,1) and the second does nothing. We obtain the following reduced basis, in which row 3 (squared length 1238314) is the shortest vector; this has not changed from the original basis:

$$\begin{bmatrix} -270 & 983 & -834 \\ -995 & 4 & -691 \\ 929 & -612 & -27 \end{bmatrix}$$

The LLL algorithm ($\alpha = \frac{3}{4}$) with deep insertions does 4 passes through the main loop, including two deep insertions:

reduce$(2,1)$	$\begin{bmatrix} -270 & 983 & -834 \\ -995 & 4 & -691 \\ 929 & -612 & -27 \end{bmatrix}$
deep insertion $3 \to 1$	$\begin{bmatrix} 929 & -612 & -27 \\ -270 & 983 & -834 \\ -995 & 4 & -691 \end{bmatrix}$
reduce$(2,1)$, deep insertion $3 \to 1$	$\begin{bmatrix} -66 & -608 & -718 \\ 929 & -612 & -27 \\ -270 & 983 & -834 \end{bmatrix}$
reduce$(3,2)$	$\begin{bmatrix} -66 & -608 & -718 \\ 929 & -612 & -27 \\ 659 & 371 & -861 \end{bmatrix}$

Row 1 (squared length 889544) is the shortest lattice vector; its squared length is $\approx 71.84\%$ of that of the shortest vector found by the original LLL algorithm.

Example 5.4. Consider the lattice in $\mathbb{R}^4$ spanned by the rows of the matrix

$$\begin{bmatrix} 84 & 3 & 34 & 17 \\ 20 & 48 & 66 & 19 \\ 69 & 14 & 63 & 78 \\ 28 & 72 & 36 & 57 \end{bmatrix}$$

The original LLL algorithm ($\alpha = \frac{3}{4}$) does 3 passes through the main loop, involving 6 reductions and no exchanges. We obtain the following reduced basis, in which row 4 (squared length 5635) is the shortest vector:

$$\begin{bmatrix} 84 & 3 & 34 & 17 \\ -64 & 45 & 32 & 2 \\ -35 & -37 & -37 & 42 \\ 43 & 61 & 7 & -4 \end{bmatrix}$$

The LLL algorithm ($\alpha = \frac{3}{4}$) with deep insertions does 7 passes through the main loop, including 9 reductions, 2 exchanges, and 2 deep insertions; it produces the following reduced basis, in which row 1 (squared length 2984) is the shortest vector:

$$\begin{bmatrix} 8 & 24 & -30 & 38 \\ -15 & 11 & 29 & 61 \\ 43 & 61 & 7 & -4 \\ -41 & 58 & -27 & -21 \end{bmatrix}$$

Deep insertion gives a shortest vector whose length is $\approx 52.95\%$ of that of the shortest vector found by the original LLL algorithm.

5.3   Updating the GSO

For the original LLL algorithm, we worked out formulas for updating the Gram-Schmidt orthogonalization after the exchange of lattice basis vectors $\mathbf{y}_k$ and $\mathbf{y}_{k-1}$: the GSO data consists of the squared lengths γ_i^* of the orthogonal basis vectors $\mathbf{y}_i^*$, and the orthogonal projection coefficients μ_{ij}. We expressed the GSO data for the new lattice basis (after the exchange) in terms of the old lattice basis (before the exchange). Recall that the exchange of $\mathbf{y}_k$ and $\mathbf{y}_{k-1}$ can be regarded as the simplest case of deep insertion: the insertion of vector $\mathbf{y}_k$ into position $k-1$. In this section, we extend the formulas for updating the GSO data to the next case: the insertion of vector $\mathbf{y}_k$ into position $k-2$.

Before the insertion, the lattice basis consists of the vectors denoted by

$$\mathbf{y}_1, \ \cdots, \ \mathbf{y}_{k-2}, \ \mathbf{y}_{k-1}, \ \mathbf{y}_k, \ \cdots, \ \mathbf{y}_m.$$

After the insertion, the lattice basis consists of the vectors

$$\mathbf{y}_1, \ \cdots, \ \mathbf{y}_k, \ \mathbf{y}_{k-2}, \ \mathbf{y}_{k-1}, \ \cdots, \ \mathbf{y}_m.$$

To make it easier to distinguish this new lattice basis from the old lattice basis, we will also denote the new basis by

$$\mathbf{z}_1, \ \cdots, \ \mathbf{z}_{k-2}, \ \mathbf{z}_{k-1}, \ \mathbf{z}_k, \ \cdots, \ \mathbf{z}_m.$$

That is,

$$\mathbf{z}_{k-2} = \mathbf{y}_k, \quad \mathbf{z}_{k-1} = \mathbf{y}_{k-2}, \quad \mathbf{z}_k = \mathbf{y}_{k-1}, \quad \mathbf{z}_i = \mathbf{y}_i \ (i \neq k-2, k-1, k).$$

We use asterisks for the Gram-Schmidt orthogonal basis: before the insertion, the Gram-Schmidt basis is $\mathbf{y}_1^*, \ldots, \mathbf{y}_m^*$ with coefficients μ_{ij}; after the insertion, the Gram-Schmidt basis is $\mathbf{z}_1^*, \ldots, \mathbf{z}_m^*$ with coefficients ν_{ij}. Since $\mathbf{z}_i = \mathbf{y}_i$ for $i < k-2$, it is clear that $\mathbf{z}_i^* = \mathbf{y}_i^*$ for $i < k-2$. Since $\mathbf{z}_i = \mathbf{y}_i$ for $i > k$, and the subspaces spanned by $\mathbf{y}_1, \ldots, \mathbf{y}_i$ and $\mathbf{z}_1, \ldots, \mathbf{z}_i$ are the same for $i > k$, it is clear that $\mathbf{z}_i^* = \mathbf{y}_i^*$ for $i > k$. So we only have to find formulas for the three vectors $\mathbf{z}_{k-2}^*, \mathbf{z}_{k-1}^*, \mathbf{z}_k^*$ in terms of the vectors $\mathbf{y}_1^*, \ldots, \mathbf{y}_m^*$.

Although the following calculations are not necessary for understanding and implementing the LLL algorithm with deep insertions, working out these formulas is an excellent way to develop a better understanding of how much the Gram-Schmidt process depends on the order of the original basis vectors.

Vector $\mathbf{z}_{k-2}^*$. By definition of the Gram-Schmidt basis, we have

$$\mathbf{z}_{k-2}^* = \mathbf{z}_{k-2} - \sum_{i=1}^{k-3} \frac{\mathbf{z}_{k-2} \cdot \mathbf{z}_i^*}{\mathbf{z}_i^* \cdot \mathbf{z}_i^*} \mathbf{z}_i^*. \tag{5.4}$$

Rewriting this in terms of the $\mathbf{y}_i$ and $\mathbf{y}_i^*$ we get

$$\mathbf{z}_{k-2}^* = \mathbf{y}_k - \sum_{i=1}^{k-3} \frac{\mathbf{y}_k \cdot \mathbf{y}_i^*}{\mathbf{y}_i^* \cdot \mathbf{y}_i^*} \mathbf{y}_i^* = \mathbf{y}_k - \sum_{i=1}^{k-3} \mu_{ki} \mathbf{y}_i^*.$$

Subtracting and adding two more terms gives

$$\mathbf{z}_{k-2}^* = \mathbf{y}_k - \sum_{i=1}^{k-1} \mu_{ki} \mathbf{y}_i^* + \mu_{k,k-1} \mathbf{y}_{k-1}^* + \mu_{k,k-2} \mathbf{y}_{k-2}^*,$$

and this gives the formula for $\mathbf{z}_{k-2}^*$ in terms of the $\mathbf{y}_i^*$:

$$\mathbf{z}_{k-2}^* = \mathbf{y}_k^* + \mu_{k,k-1} \mathbf{y}_{k-1}^* + \mu_{k,k-2} \mathbf{y}_{k-2}^*. \tag{5.5}$$

Since the Gram-Schmidt basis vectors are orthogonal, we get

$$|\mathbf{z}_{k-2}^*|^2 = |\mathbf{y}_k^*|^2 + \mu_{k,k-1}^2 |\mathbf{y}_{k-1}^*|^2 + \mu_{k,k-2}^2 |\mathbf{y}_{k-2}^*|^2. \tag{5.6}$$

Equation (5.5) gives

$$\mathbf{y}_{k-2} \cdot \mathbf{z}_{k-2}^* = \mathbf{y}_{k-2} \cdot \mathbf{y}_k^* + \mu_{k,k-1} \mathbf{y}_{k-2} \cdot \mathbf{y}_{k-1}^* + \mu_{k,k-2} \mathbf{y}_{k-2} \cdot \mathbf{y}_{k-2}^*.$$

Since $\mathbf{y}_j \cdot \mathbf{y}_i^* = 0$ for $j < i$ we have

$$\mathbf{y}_{k-2} \cdot \mathbf{z}_{k-2}^* = \mu_{k,k-2} \mathbf{y}_{k-2} \cdot \mathbf{y}_{k-2}^*.$$

But $\mathbf{y}_{k-2} = \mathbf{y}^*_{k-2} + \cdots$ where the dots represent a linear combination of the $\mathbf{y}^*_j$ for $j < k-2$, and so

$$\mathbf{y}_{k-2} \cdot \mathbf{z}^*_{k-2} = \mu_{k,k-2}|\mathbf{y}^*_{k-2}|^2. \tag{5.7}$$

From equation (5.4) we also find that

$$\nu_{k-2,i} = \frac{\mathbf{z}_{k-2} \cdot \mathbf{z}^*_i}{\mathbf{z}^*_i \cdot \mathbf{z}^*_i} = \frac{\mathbf{y}_k \cdot \mathbf{y}^*_i}{\mathbf{y}^*_i \cdot \mathbf{y}^*_i} = \mu_{ki} \quad (i < k-2).$$

Vector $\mathbf{z}^*_{k-1}$. By definition of the Gram-Schmidt basis, we have

$$\mathbf{z}^*_{k-1} = \mathbf{z}_{k-1} - \sum_{i=1}^{k-2} \frac{\mathbf{z}_{k-1} \cdot \mathbf{z}^*_i}{\mathbf{z}^*_i \cdot \mathbf{z}^*_i} \mathbf{z}^*_i. \tag{5.8}$$

Since $\mathbf{z}_{k-1} = \mathbf{y}_{k-2}$ we get

$$\mathbf{z}^*_{k-1} = \mathbf{y}_{k-2} - \sum_{i=1}^{k-3} \frac{\mathbf{y}_{k-2} \cdot \mathbf{z}^*_i}{\mathbf{z}^*_i \cdot \mathbf{z}^*_i} \mathbf{z}^*_i - \frac{\mathbf{y}_{k-2} \cdot \mathbf{z}^*_{k-2}}{\mathbf{z}^*_{k-2} \cdot \mathbf{z}^*_{k-2}} \mathbf{z}^*_{k-2}.$$

Rewriting the sum in terms of the $\mathbf{y}^*_i$ gives

$$\mathbf{z}^*_{k-1} = \mathbf{y}_{k-2} - \sum_{i=1}^{k-3} \frac{\mathbf{y}_{k-2} \cdot \mathbf{y}^*_i}{\mathbf{y}^*_i \cdot \mathbf{y}^*_i} \mathbf{y}^*_i - \frac{\mathbf{y}_{k-2} \cdot \mathbf{z}^*_{k-2}}{\mathbf{z}^*_{k-2} \cdot \mathbf{z}^*_{k-2}} \mathbf{z}^*_{k-2},$$

and this simplifies to

$$\mathbf{z}^*_{k-1} = \mathbf{y}^*_{k-2} - \frac{\mathbf{y}_{k-2} \cdot \mathbf{z}^*_{k-2}}{\mathbf{z}^*_{k-2} \cdot \mathbf{z}^*_{k-2}} \mathbf{z}^*_{k-2}.$$

Using equations (5.5) and (5.7) we obtain

$$\mathbf{z}^*_{k-1} = \mathbf{y}^*_{k-2} - \frac{\mu_{k,k-2}|\mathbf{y}^*_{k-2}|^2}{\mathbf{z}^*_{k-2} \cdot \mathbf{z}^*_{k-2}} \left(\mathbf{y}^*_k + \mu_{k,k-1}\mathbf{y}^*_{k-1} + \mu_{k,k-2}\mathbf{y}^*_{k-2} \right),$$

and this simplifies to

$$\mathbf{z}^*_{k-1} = \left(1 - \frac{\mu^2_{k,k-2}|\mathbf{y}^*_{k-2}|^2}{\mathbf{z}^*_{k-2} \cdot \mathbf{z}^*_{k-2}} \right) \mathbf{y}^*_{k-2} - \frac{\mu_{k,k-2}|\mathbf{y}^*_{k-2}|^2}{\mathbf{z}^*_{k-2} \cdot \mathbf{z}^*_{k-2}} \left(\mathbf{y}^*_k + \mu_{k,k-1}\mathbf{y}^*_{k-1} \right).$$

Rewriting this gives

$$\mathbf{z}^*_{k-1} = \frac{1}{\mathbf{z}^*_{k-2} \cdot \mathbf{z}^*_{k-2}} \times$$

$$\left(\left(\mathbf{z}^*_{k-2} \cdot \mathbf{z}^*_{k-2} - \mu^2_{k,k-2}|\mathbf{y}^*_{k-2}|^2 \right) \mathbf{y}^*_{k-2} - \mu_{k,k-2}|\mathbf{y}^*_{k-2}|^2 \left(\mathbf{y}^*_k + \mu_{k,k-1}\mathbf{y}^*_{k-1} \right) \right).$$

Using equation (5.6) we obtain

$$\mathbf{z}^*_{k-1} = \frac{1}{\mathbf{z}^*_{k-2} \cdot \mathbf{z}^*_{k-2}} \times$$

$$\left(\left(|\mathbf{y}^*_k|^2 + \mu^2_{k,k-1}|\mathbf{y}^*_{k-1}|^2 \right)\mathbf{y}^*_{k-2} - \mu_{k,k-2}|\mathbf{y}^*_{k-2}|^2 \left(\mathbf{y}^*_k + \mu_{k,k-1}\mathbf{y}^*_{k-1} \right) \right).$$

Another application of equation (5.6) gives the formula we want expressing $\mathbf{z}^*_{k-1}$ in terms of the $\mathbf{y}^*_i$:

$$\mathbf{z}^*_{k-1} = \begin{cases} \dfrac{1}{|\mathbf{y}^*_k|^2 + \mu^2_{k,k-1}|\mathbf{y}^*_{k-1}|^2 + \mu^2_{k,k-2}|\mathbf{y}^*_{k-2}|^2} \times \\[2mm] \left(\left(|\mathbf{y}^*_k|^2 + \mu^2_{k,k-1}|\mathbf{y}^*_{k-1}|^2 \right)\mathbf{y}^*_{k-2} \right. \\[2mm] \left. \qquad -\mu_{k,k-2}|\mathbf{y}^*_{k-2}|^2 \left(\mathbf{y}^*_k + \mu_{k,k-1}\mathbf{y}^*_{k-1} \right) \right). \end{cases} \tag{5.9}$$

From this we can easily write down the formula for $|\mathbf{z}^*_{k-1}|^2$. Equation (5.8) also shows that for $i \leq k-2$ we have

$$\nu_{k-1,i} = \frac{\mathbf{z}_{k-1} \cdot \mathbf{z}^*_i}{\mathbf{z}^*_i \cdot \mathbf{z}^*_i} = \frac{\mathbf{y}_{k-2} \cdot \mathbf{z}^*_i}{\mathbf{z}^*_i \cdot \mathbf{z}^*_i} = \begin{cases} \dfrac{\mathbf{y}_{k-2} \cdot \mathbf{y}^*_i}{\mathbf{y}^*_i \cdot \mathbf{y}^*_i} & (i < k-2) \\[3mm] \dfrac{\mathbf{y}_{k-2} \cdot \mathbf{z}^*_{k-2}}{\mathbf{z}^*_{k-2} \cdot \mathbf{z}^*_{k-2}} & (i = k-2) \end{cases}$$

Using equations (5.6) and (5.7) we obtain

$$\nu_{k-1,i} = \begin{cases} \mu_{k-2,i} & (i < k-2) \\[3mm] \dfrac{\mu_{k,k-2}|\mathbf{y}^*_{k-2}|^2}{|\mathbf{y}^*_k|^2 + \mu^2_{k,k-1}|\mathbf{y}^*_{k-1}|^2 + \mu^2_{k,k-2}|\mathbf{y}^*_{k-2}|^2} & (i = k-2) \end{cases}$$

Vector $\mathbf{z}^*_k$. By definition of the Gram-Schmidt basis, we have

$$\mathbf{z}^*_k = \mathbf{z}_k - \sum_{i=1}^{k-1} \frac{\mathbf{z}_k \cdot \mathbf{z}^*_i}{\mathbf{z}^*_i \cdot \mathbf{z}^*_i}\mathbf{z}^*_i. \tag{5.10}$$

Since $\mathbf{z}_k = \mathbf{y}_{k-1}$ and $\mathbf{z}_i = \mathbf{y}_i$ for $i < k-2$, we get

$$\mathbf{z}^*_k = \mathbf{y}_{k-1} - \sum_{i=1}^{k-3} \frac{\mathbf{y}_{k-1} \cdot \mathbf{y}^*_i}{\mathbf{y}^*_i \cdot \mathbf{y}^*_i}\mathbf{y}^*_i - \frac{\mathbf{y}_{k-1} \cdot \mathbf{z}^*_{k-2}}{\mathbf{z}^*_{k-2} \cdot \mathbf{z}^*_{k-2}}\mathbf{z}^*_{k-2} - \frac{\mathbf{y}_{k-1} \cdot \mathbf{z}^*_{k-1}}{\mathbf{z}^*_{k-1} \cdot \mathbf{z}^*_{k-1}}\mathbf{z}^*_{k-1}.$$

Subtracting and adding the same term gives

$$\mathbf{z}^*_k = \mathbf{y}_{k-1} - \sum_{i=1}^{k-2} \frac{\mathbf{y}_{k-1} \cdot \mathbf{y}^*_i}{\mathbf{y}^*_i \cdot \mathbf{y}^*_i}\mathbf{y}^*_i + \frac{\mathbf{y}_{k-1} \cdot \mathbf{y}^*_{k-2}}{\mathbf{y}^*_{k-2} \cdot \mathbf{y}^*_{k-2}}\mathbf{y}^*_{k-2}$$

$$-\frac{\mathbf{y}_{k-1} \cdot \mathbf{z}_{k-2}^*}{\mathbf{z}_{k-2}^* \cdot \mathbf{z}_{k-2}^*}\mathbf{z}_{k-2}^* - \frac{\mathbf{y}_{k-1} \cdot \mathbf{z}_{k-1}^*}{\mathbf{z}_{k-1}^* \cdot \mathbf{z}_{k-1}^*}\mathbf{z}_{k-1}^*.$$

By definition of $\mathbf{y}_{k-1}^*$ we have

$$\mathbf{z}_k^* = \mathbf{y}_{k-1}^* + \frac{\mathbf{y}_{k-1} \cdot \mathbf{y}_{k-2}^*}{\mathbf{y}_{k-2}^* \cdot \mathbf{y}_{k-2}^*}\mathbf{y}_{k-2}^* - \frac{\mathbf{y}_{k-1} \cdot \mathbf{z}_{k-2}^*}{\mathbf{z}_{k-2}^* \cdot \mathbf{z}_{k-2}^*}\mathbf{z}_{k-2}^* - \frac{\mathbf{y}_{k-1} \cdot \mathbf{z}_{k-1}^*}{\mathbf{z}_{k-1}^* \cdot \mathbf{z}_{k-1}^*}\mathbf{z}_{k-1}^*.$$

Since $\mathbf{y}_{k-1} = \mathbf{y}_{k-1}^* + \mu_{k-1,k-2}\mathbf{y}_{k-2}^* + \cdots$ where the dots represent a linear combination of the $\mathbf{y}_i^*$ for $i < k-2$, we get

$$\mathbf{y}_{k-1} \cdot \mathbf{y}_{k-2}^* = \mu_{k-1,k-2}|\mathbf{y}_{k-2}^*|^2,$$

and hence

$$\mathbf{z}_k^* = \mathbf{y}_{k-1}^* + \mu_{k-1,k-2}\mathbf{y}_{k-2}^* - \frac{\mathbf{y}_{k-1} \cdot \mathbf{z}_{k-2}^*}{\mathbf{z}_{k-2}^* \cdot \mathbf{z}_{k-2}^*}\mathbf{z}_{k-2}^* - \frac{\mathbf{y}_{k-1} \cdot \mathbf{z}_{k-1}^*}{\mathbf{z}_{k-1}^* \cdot \mathbf{z}_{k-1}^*}\mathbf{z}_{k-1}^*. \qquad (5.11)$$

Using equation (5.5) we find that

$$\mathbf{y}_{k-1} \cdot \mathbf{z}_{k-2}^* = \left[\mathbf{y}_{k-1}^* + \mu_{k-1,k-2}\mathbf{y}_{k-2}^* + \cdots\right] \cdot \left[\mathbf{y}_k^* + \mu_{k,k-1}\mathbf{y}_{k-1}^* + \mu_{k,k-2}\mathbf{y}_{k-2}^*\right],$$

and hence

$$\mathbf{y}_{k-1} \cdot \mathbf{z}_{k-2}^* = \mu_{k,k-1}|\mathbf{y}_{k-1}^*|^2 + \mu_{k-1,k-2}\mu_{k,k-2}|\mathbf{y}_{k-2}^*|^2. \qquad (5.12)$$

Using equation (5.9) we can obtain a similar expression for $\mathbf{y}_{k-1} \cdot \mathbf{z}_{k-1}^*$. If we use this expression and equation (5.12) in equation (5.11), and then use equations (5.5) and (5.9) for $\mathbf{z}_{k-2}^*$ and $\mathbf{z}_{k-1}^*$, together with the corresponding equations for $|\mathbf{z}_{k-2}^*|^2$ and $|\mathbf{z}_{k-1}^*|^2$, then we obtain a (very complicated) formula for $\mathbf{z}_k^*$, from which we can derive a formula for $|\mathbf{z}_k^*|^2$. The remaining details of this calculation, together with the formulas for ν_{ki} ($i < k$), are left to the reader (Exercise 5.16).

5.4   Projects

Project 5.1. Implement the LLL algorithm with deep insertions on a computer, and test your program on a number of examples. For example, generate 1000 pseudorandom 10-dimensional lattices in $\mathbb{R}^{10}$. For each lattice, compare the results produced by three basis reduction algorithms:

(*i*) the original LLL algorithm with $\alpha = 1$

(*ii*) the original LLL algorithm with $\alpha = \frac{3}{4}$ and deep insertions

(*iii*) the original LLL algorithm with $\alpha = 1$ and deep insertions

Discuss the following questions:

(a) Which algorithm gives the shortest lattice vectors in general?

(b) Which algorithm uses the least running time in general?

(c) Which algorithm is the best in general assuming that shorter vectors are better but longer running times are worse?

Extend your computations to dimensions 20, 30, 40, 50, ...

Project 5.2. A number of papers provide experimental data on the performance of the LLL algorithm, including the version with deep insertions. Write a report and present a seminar talk on this topic, based on the following references: Backes and Wetzel [13], Gama and Nguyen [45], Schnorr [126].

5.5 Exercises

For Exercises 5.1 to 5.14, use the LLL algorithm ($\alpha = \frac{3}{4}$) with deep insertions to find a reduced basis of the lattice spanned by the rows of the matrix, and compare the results with those obtained from the original LLL algorithm.

Exercise 5.1.
$$\begin{bmatrix} 8 & 0 & 7 \\ 4 & 8 & 3 \\ 0 & 4 & 8 \end{bmatrix}$$

Exercise 5.2.
$$\begin{bmatrix} 1 & 5 & 2 \\ 2 & 1 & 5 \\ 4 & 8 & 0 \end{bmatrix}$$

Exercise 5.3.
$$\begin{bmatrix} 7 & -5 & -1 \\ 7 & 3 & 1 \\ -1 & 6 & -4 \end{bmatrix}$$

Exercise 5.4.
$$\begin{bmatrix} 5 & -9 & -4 \\ 2 & 4 & 9 \\ 2 & 4 & -7 \end{bmatrix}$$

Exercise 5.5.
$$\begin{bmatrix} 85 & 56 & 42 \\ 55 & 15 & 5 \\ 63 & 60 & 57 \end{bmatrix}$$

Exercise 5.6.

$$\begin{bmatrix} -68 & 36 & 69 \\ 58 & 54 & -50 \\ 38 & -47 & 51 \end{bmatrix}$$

Exercise 5.7.

$$\begin{bmatrix} 4 & 8 & 5 & 6 \\ 3 & 7 & 8 & 1 \\ 4 & 8 & 0 & 3 \\ 4 & 3 & 9 & 5 \end{bmatrix}$$

Exercise 5.8.

$$\begin{bmatrix} 5 & -1 & 6 & 4 \\ -4 & 8 & 6 & 2 \\ 4 & 0 & -6 & -3 \\ -1 & -3 & 6 & -2 \end{bmatrix}$$

Exercise 5.9.

$$\begin{bmatrix} 58 & 75 & 53 & 76 \\ 68 & 93 & 4 & 47 \\ 84 & 85 & 62 & 32 \\ 62 & 49 & 23 & 21 \end{bmatrix}$$

Exercise 5.10.

$$\begin{bmatrix} 87 & 58 & 5 & 55 \\ 82 & -26 & 57 & -72 \\ 44 & -76 & -62 & -37 \\ 49 & -60 & -9 & 66 \end{bmatrix}$$

Exercise 5.11.

$$\begin{bmatrix} 933 & 416 & 972 & 709 \\ 505 & 857 & 375 & 213 \\ 622 & 473 & 540 & 558 \\ 181 & 765 & 407 & 503 \end{bmatrix}$$

Exercise 5.12.

$$\begin{bmatrix} -592 & 221 & -280 & -596 \\ 446 & 403 & 580 & 164 \\ 869 & -734 & 456 & -707 \\ 454 & 214 & 244 & -755 \end{bmatrix}$$

Exercise 5.13.

$$\begin{bmatrix} 6 & 7 & 3 & 1 & 9 \\ 9 & 2 & 9 & 1 & 1 \\ 1 & 4 & 9 & 4 & 3 \\ 4 & 6 & 1 & 8 & 1 \\ 1 & 5 & 3 & 6 & 2 \end{bmatrix}$$

Exercise 5.14.

$$\begin{bmatrix} -2 & -1 & 1 & 0 & 3 \\ 1 & 9 & -1 & 9 & -2 \\ -5 & -7 & 3 & 0 & 3 \\ 1 & 5 & 0 & -2 & -1 \\ -7 & 1 & 5 & 2 & -2 \end{bmatrix}$$

Exercise 5.15. For 2-dimensional lattices, show that the LLL algorithm with deep insertions never finds a shorter vector than the original LLL algorithm.

Exercise 5.16. Complete the calculations in Section 5.3: work out an explicit formula for $\mathbf{z}_k^*$ in terms of the $\mathbf{y}_i^*$, the corresponding formula for $|\mathbf{z}_k^*|^2$, and the equations for ν_{ki} $(i < k)$.

6

Linearly Dependent Vectors

CONTENTS

In this chapter we consider a modification of the LLL algorithm in which the input vectors are allowed to be linearly dependent. This algorithm, called the **modified LLL algorithm**, or **MLLL algorithm** for short, was introduced by Pohst [118] in 1987. In the simplest case, we are given m vectors $\mathbf{x}_1$, $\mathbf{x}_2$, ..., $\mathbf{x}_m$ in an $(m-1)$-dimensional lattice $L \subset \mathbb{R}^n$ where we assume that the first $m-1$ vectors $\mathbf{x}_1$, $\mathbf{x}_2$, ..., $\mathbf{x}_{m-1}$ are linearly independent. This special case, which is essentially the inductive step in the general case, is discussed in Pohst and Zassenhaus [119], §3.3. A similar algorithm is discussed in Cohen [26], §2.6.4. One drawback of these original presentations is the large number of "goto" statements, which make the logical structure of the algorithm difficult to follow. The discussion in this chapter follows the more structured version of Pohst's original algorithm given by Sims [131], §8.7; his single "goto" statement is easily replaced by a Boolean variable. Other closely related algorithms are presented in Grötschel et al. [51], §5.4; Buchmann and Pohst [20]; Schnorr and Euchner [127]; and Hobby [63].

6.1   Embedding dependent vectors

Following Pohst [118], pages 123–124, we first describe how the original LLL algorithm can be adapted to process linearly dependent vectors.

Let L be a lattice in $\mathbb{R}^n$; hence $\dim(L) \leq n$. The elements of L will be regarded as row vectors. We define

$$M = \min \{ \, |\mathbf{x}|^2 \in L \mid \mathbf{x} \neq 0 \, \}.$$

We consider m vectors $\mathbf{x}_1$, $\mathbf{x}_2$, ..., $\mathbf{x}_m$ in L, which are assumed to be linearly dependent. We use the following trick of converting these vectors into linearly independent vectors by embedding them in a vector space of higher dimension. Let $\mathbf{e}_1$, $\mathbf{e}_1$, ..., $\mathbf{e}_m$ be the standard basis vectors in $\mathbb{R}^m$; then the vectors $[\mathbf{e}_1, \mathbf{x}_1]$, $[\mathbf{e}_2, \mathbf{x}_2]$, ..., $[\mathbf{e}_m, \mathbf{x}_m]$ in $\mathbb{R}^{m+n}$ are linearly independent (Exercise 6.1). We also introduce a positive integral exponent ℓ; this is a scaling factor whose significance will be made clear shortly.

As usual, α is the reduction parameter and $\beta = 4/(4\alpha-1)$. We define vectors $\mathbf{y}_1$, $\mathbf{y}_2$, ..., $\mathbf{y}_m$ in $\mathbb{R}^{m+n}$ by the formula

$$\mathbf{y}_i = [\mathbf{e}_i, \beta^\ell \mathbf{x}_i] \qquad (1 \le i \le m).$$

Since $\mathbf{y}_1$, $\mathbf{y}_2$, ..., $\mathbf{y}_m$ are linearly independent, we consider the lattice $\widehat{L}$ in $\mathbb{R}^{m+n}$ with $\mathbf{y}_1$, $\mathbf{y}_2$, ..., $\mathbf{y}_m$ as basis. We define

$$N = \min\{\, |\mathbf{y}|^2 \in \widehat{L} \mid \mathbf{y} \ne \mathbf{0} \,\}.$$

We apply the LLL algorithm to the basis $\mathbf{y}_1$, $\mathbf{y}_2$, ..., $\mathbf{y}_m$ of the lattice $\widehat{L}$, and obtain a reduced basis $\mathbf{z}_1$, $\mathbf{z}_2$, ..., $\mathbf{z}_m$ of $\widehat{L}$. By Theorem 4.7 we have

$$|\mathbf{z}_1|^2 \le \beta^{m-1} N. \tag{6.1}$$

(For the standard value $\alpha = \frac{3}{4}$ we obtain $|\mathbf{z}_1|^2 \le 2^{m-1} N$.)

Since $\mathbf{z}_1 \in \widehat{L}$ we have

$$\mathbf{z}_1 = [\mathbf{c}, \beta^\ell \mathbf{x}], \qquad \mathbf{c} \in \mathbb{Z}^m, \qquad \mathbf{x} \in L.$$

Therefore

$$|\mathbf{z}_1|^2 = |\mathbf{c}|^2 + \beta^{2\ell} |\mathbf{x}|^2,$$

and this gives

$$|\mathbf{c}|^2 + \beta^{2\ell} |\mathbf{x}|^2 \le \beta^{m-1} N.$$

We want to choose the exponent ℓ large enough so that this last inequality will imply $\mathbf{x} = 0$; that is, so that $\mathbf{x} \ne \mathbf{0}$ gives a contradiction. Now $\mathbf{x} \ne \mathbf{0}$ implies $|\mathbf{x}|^2 \ge M$, and so it suffices to choose ℓ so that

$$\beta^{2\ell} M \ge \beta^{m-1} N.$$

Since $\mathbf{x}_1$, $\mathbf{x}_2$, ..., $\mathbf{x}_m$ in L are linearly dependent, there is a nontrivial linear dependence relation, say

$$c_1 \mathbf{x}_1 + c_2 \mathbf{x}_2 + \cdots + c_m \mathbf{x}_m = 0, \qquad [c_1, c_2, \ldots, c_m] \ne \mathbf{0}.$$

We write

$$C = \max\{|c_1|, \ldots, |c_m|\}, \qquad C \ge 1.$$

Then clearly $\widehat{L}$ contains the nonzero vector

$$[c_1, c_2, \ldots, c_m, 0, \ldots, 0] \in \widehat{L},$$

and considering the length of this vector shows that

$$N \leq mC^2.$$

The condition on ℓ now becomes

$$\beta^{2\ell}M \geq \beta^{m-1}mC^2. \tag{6.2}$$

Solving this inequality for ℓ gives (Exercise 6.2)

$$\ell \geq \frac{1}{2}\left(m - 1 + \log_\beta \frac{mC^2}{M}\right). \tag{6.3}$$

If ℓ satisfies this inequality, then $\mathbf{x} = \mathbf{0}$ and so $\mathbf{z}_1 = [c, 0, \ldots, 0]$.

This approach has one major disadvantage: it may be very difficult to determine the numbers M and C, the length of a shortest nonzero vector in the original lattice L, and the length of the coefficient vector of a shortest nontrivial linear dependence relation among $\mathbf{x}_1$, $\mathbf{x}_2$, $\ldots$, $\mathbf{x}_m$. If we cannot determine M and C, then we cannot calculate the lower bound on ℓ from inequality (4.3).

We present one example to illustrate this approach; further examples may be found in Exercises 6.3 to 6.5.

Example 6.1. One simple way to choose m linearly dependent vectors in $\mathbb{R}^n$ is to fix $m > n$ and then use a computer algebra system to generate random vectors. For example, if we set $m = 6$ and $n = 4$, and use the Maple command

```
LinearAlgebra[RandomMatrix]( 6, 4, generator = -9..9 );
```

then we obtain a matrix such as

$$X = \begin{bmatrix} \mathbf{x}_1 \\ \mathbf{x}_2 \\ \mathbf{x}_3 \\ \mathbf{x}_4 \\ \mathbf{x}_5 \\ \mathbf{x}_6 \end{bmatrix} = \begin{bmatrix} 5 & 1 & 6 & 0 \\ -8 & -3 & 2 & -1 \\ -5 & 9 & -4 & 4 \\ -1 & -8 & 9 & -7 \\ -5 & -6 & 2 & -7 \\ -4 & 6 & 0 & -8 \end{bmatrix}$$

We use the standard parameters $\alpha = \frac{3}{4}$ and $\beta = 2$. We scale the dependent vectors $\mathbf{x}_1, \ldots, \mathbf{x}_6$ in $\mathbb{R}^4$ and embed them as independent vectors in $\mathbb{R}^{10}$:

$$\begin{bmatrix} 1 & 0 & 0 & 0 & 0 & 0 & 5 \cdot 2^\ell & 2^\ell & 6 \cdot 2^\ell & 0 \\ 0 & 1 & 0 & 0 & 0 & 0 & -8 \cdot 2^\ell & -3 \cdot 2^\ell & 2 \cdot 2^\ell & -2^\ell \\ 0 & 0 & 1 & 0 & 0 & 0 & -5 \cdot 2^\ell & 9 \cdot 2^\ell & -4 \cdot 2^\ell & 4 \cdot 2^\ell \\ 0 & 0 & 0 & 1 & 0 & 0 & -2^\ell & -8 \cdot 2^\ell & 9 \cdot 2^\ell & -7 \cdot 2^\ell \\ 0 & 0 & 0 & 0 & 1 & 0 & -5 \cdot 2^\ell & -6 \cdot 2^\ell & 2 \cdot 2^\ell & -7 \cdot 2^\ell \\ 0 & 0 & 0 & 0 & 0 & 1 & -4 \cdot 2^\ell & 6 \cdot 2^\ell & 0 & -8 \cdot 2^\ell \end{bmatrix}$$

These vectors form a basis of the lattice $\widehat{L}$ in $\mathbb{R}^{10}$. We use a trial-and-error approach to find the smallest value of ℓ that produces a zero row in the right block after the LLL algorithm is used to reduce the basis.

For $\ell = 5$ the LLL algorithm takes 73 steps (reduce and exchange) and gives this α-reduced basis of $\widehat{L}$:

$$\begin{bmatrix}
-12 & -4 & 3 & 13 & -12 & 1 & 0 & 32 & 32 & 32 \\
23 & 15 & -13 & -28 & 16 & 2 & 0 & 32 & 0 & 32 \\
-19 & -6 & 4 & 20 & -19 & 2 & 0 & 32 & 0 & -32 \\
-4 & 9 & -10 & -1 & -13 & 6 & 0 & 32 & -32 & 32 \\
1 & -18 & 19 & 8 & 17 & -10 & 32 & 0 & 0 & -32 \\
9 & -12 & 14 & -2 & 22 & -9 & -32 & 32 & 0 & 0
\end{bmatrix}$$

None of the rows in the right block is zero.

For $\ell = 6$ the LLL algorithm takes 76 steps and gives this α-reduced basis:

$$\begin{bmatrix}
35 & 19 & -16 & -41 & 28 & 1 & 0 & 0 & -64 & 0 \\
-27 & -6 & 3 & 27 & -29 & 4 & 0 & 0 & -64 & 0 \\
23 & 15 & -13 & -28 & 16 & 2 & 0 & 64 & 0 & 64 \\
19 & 6 & -4 & -20 & 19 & -2 & 0 & -64 & 0 & 64 \\
24 & -3 & 6 & -20 & 33 & -8 & 64 & 64 & 0 & 0 \\
29 & -24 & 29 & -14 & 58 & -21 & 0 & 0 & 0 & 0
\end{bmatrix}$$

One row in the right block is zero, but it is not the first row.

For $\ell = 7$ the LLL algorithm takes 87 steps and gives this α-reduced basis:

$$\begin{bmatrix}
29 & -24 & 29 & -14 & 58 & -21 & 0 & 0 & 0 & 0 \\
-33 & -49 & 48 & 54 & 1 & -18 & 0 & 0 & 0 & 0 \\
-27 & -6 & 3 & 27 & -29 & 4 & 0 & 0 & -128 & 0 \\
23 & 15 & -13 & -28 & 16 & 2 & 0 & 128 & 0 & 128 \\
-19 & -22 & 21 & 28 & -5 & -7 & 128 & 0 & 0 & 128 \\
5 & -21 & 23 & 6 & 25 & -13 & -128 & -128 & 0 & 0
\end{bmatrix}$$

Now the first row, in fact the first two rows, in the right block are zero: we have found the minimal value of the exponent ℓ.

6.2   The modified LLL algorithm

We now present Pohst's modified LLL algorithm following the more structured approach of Sims [131], §8.7. As in the original LLL algorithm, we have a reduction parameter α satisfying $\frac{1}{4} < \alpha < 1$. As before, the nearest integer to the real number x will be denoted $\lceil x \rfloor$; recall that if x has the form $x = n + \frac{1}{2}$ for some $n \in \mathbb{Z}$ then we round down, not up: $\lceil x \rfloor = n$.

The original spanning vectors $\mathbf{x}_1, \mathbf{x}_2, \ldots, \mathbf{x}_m$ for the lattice $L \subset \mathbb{R}^n$ are the rows of the $m \times n$ matrix X. The reduced set of spanning vectors $\mathbf{y}_1$, $\mathbf{y}_2, \ldots, \mathbf{y}_m$ are the rows of the $m \times n$ matrix Y; when the MLLL algorithm terminates, the nonzero rows of Y are a basis for the lattice L. The rows of the $m \times n$ matrix Y^* contain the Gram-Schmidt orthogonalization of the rows of Y. As before, the $m \times m$ lower-triangular matrix μ contains the Gram-Schmidt coefficients μ_{ij}, and the $m \times 1$ vector γ^* contains the squared lengths of the rows of Y^*.

The fundamental difference between the LLL and MLLL algorithms is that the latter algorithm has a more complicated control structure resulting from the necessity of dealing with linearly dependent vectors. This requires that the single index variable from LLL must be replaced by more than one index variable in MLLL. We cannot compute the Gram-Schmidt orthogonalization of the original lattice basis all at once at the start of the algorithm, for the simple reason that we do not start with a linearly independent set. This means that we must perform one step of the GSO at a time, checking at each step whether one of the GSO vectors has become 0, which indicates a linear dependence relation among the lattice vectors. If the current GSO vector is not zero, then we perform lattice basis reduction on the current subset of linearly independent vectors. We distinguish the following indices:

$g =$ row index in Y for the current vector in the GSO computation,

$k =$ row index in Y for the current vector in the LLL computation,

$\ell =$ row index in Y for the last vector in the current LLL computation,

$z =$ row index in Y for last vector not known to be zero.

Apart from this change of notation, the reduce and exchange procedures are essentially the same as in the original LLL algorithm; see Figure 6.1. The most important new features are that the exchange procedure has two arguments, and the quantities ν and δ are calculated in the main loop of the algorithm, which is displayed in Figure 6.2. To make it easier to trace the execution, we adopt an idea of Sims [131] and indicate the positions in the algorithm where the vectors $\mathbf{y}_i$ are modified by the symbols (P1), (P2), (P3), (P4), (P5). (Note also that an alternative method of computing the new Gram-Schmidt coefficients using 2×2 matrices has been used in the exchange procedure.) To save space but to keep the logical structure clear, we have occasionally used "fi" and "od" to indicate the end of an "if" or a "do" statement.

Example 6.2. If we let $n = 1$ then the MLLL algorithm computes the GCD of a set of integers using the version of the Euclidean algorithm with least absolute remainders. For example, let $m = 3$ and consider the integers 6, 10, 15. At the start of the MLLL algorithm, we have

$$Y = \begin{bmatrix} 6 \\ 10 \\ 15 \end{bmatrix}$$

- Procedure reduce(k, p):

    If $|\mu_{kp}| > \frac{1}{2}$ then

    - Set $r \leftarrow \lceil \mu_{kp} \rfloor$.
    - Set $\mathbf{y}_k \leftarrow \mathbf{y}_k - r\mathbf{y}_p$.
    - For $j = 1, 2, \ldots, p-1$ do: Set $\mu_{kj} \leftarrow \mu_{kj} - r\mu_{pj}$.
    - Set $\mu_{kp} \leftarrow \mu_{kp} - r$.

- Procedure exchange(k, ℓ):

    - Set $\mu_{k,k-1} \leftarrow \nu\gamma_{k-1}^*/\delta$.
    - Set $\gamma_k^* \leftarrow \gamma_{k-1}^*\gamma_k^*/\delta$.
    - For $j = k+1, \ldots, \ell$ do:

$$\text{Set } \begin{bmatrix} \mu_{j,k-1} \\ \mu_{jk} \end{bmatrix} \leftarrow \begin{bmatrix} 1 & \mu_{k,k-1} \\ 0 & 1 \end{bmatrix} \begin{bmatrix} 0 & 1 \\ 1 & -\nu \end{bmatrix} \begin{bmatrix} \mu_{j,k-1} \\ \mu_{jk} \end{bmatrix}.$$

FIGURE 6.1
Reduce and exchange procedures for the MLLL algorithm

The first modification occurs at (P2) where row 2 is reduced using row 1:

$$Y = \begin{bmatrix} 6 \\ -2 \\ 15 \end{bmatrix}$$

The next modification occurs at (P5) where rows 1 and 2 are exchanged:

$$Y = \begin{bmatrix} -2 \\ 6 \\ 15 \end{bmatrix}$$

At (P2) row 2 is reduced using row 1:

$$Y = \begin{bmatrix} -2 \\ 0 \\ 15 \end{bmatrix}$$

A zero row has appeared and so (P4) swaps rows 2 and 3:

$$Y = \begin{bmatrix} -2 \\ 15 \\ 0 \end{bmatrix}$$

- *Input*: An $m \times n$ integer matrix X whose rows $\mathbf{x}_1, \mathbf{x}_2, \ldots \mathbf{x}_m$, which are not necessarily linearly dependent, span a lattice $L \subset \mathbb{R}^n$.

- *Output*: An $m \times n$ integer matrix Y in which the nonzero rows $\mathbf{y}_i$ form a basis for the lattice L.

- Set $Y \leftarrow X$, $z \leftarrow m$, $g \leftarrow 1$.

- While $g \leq z$ do:

    - If $\mathbf{y}_g = \mathbf{0}$ then

        If $g < z$ then swap $\mathbf{y}_g$ and $\mathbf{y}_z$ fi. Set $z \leftarrow z-1$.   (P1)

        else

        Set $\mathbf{y}_g^* \leftarrow \mathbf{y}_g$.

        For $j = 1, \ldots, g-1$ do: Set $\mu_{gj} \leftarrow (\mathbf{y}_g \cdot \mathbf{y}_j^*)/\gamma_j^*$, $\mathbf{y}_g^* \leftarrow \mathbf{y}_g^* - \mu_{gj}\mathbf{y}_j^*$.

        Set $\gamma_g^* \leftarrow \mathbf{y}_g^* \cdot \mathbf{y}_g^*$.

    - If $g = 1$ then set $g \leftarrow 2$ else

        * Set $\ell \leftarrow g$, $k \leftarrow g$, startagain $\leftarrow$ false.
        * While $k \leq \ell$ and not startagain do:
            · reduce$(k, k-1)$. Set $\nu \leftarrow \mu_{k,k-1}$, $\delta \leftarrow \gamma_k^* + \nu^2\gamma_{k-1}^*$.   (P2)
            · If $\delta \geq \alpha\gamma_{k-1}^*$ then

                For $p = k-2, \ldots, 1$ do: reduce(k, p) od. Set $k \leftarrow k+1$.   (P3)

                else

                If $\mathbf{y}_k = 0$ then

                    If $k < z$ then swap $\mathbf{y}_k$ and $\mathbf{y}_z$.   (P4)

                    Set $z \leftarrow z-1$, $g \leftarrow k$, startagain $\leftarrow$ true.

                else

                    If $\delta \neq 0$ then exchange(k, ℓ) fi. Set $\gamma_{k-1}^* \leftarrow \delta$.

                    Swap $\mathbf{y}_k$ and $\mathbf{y}_{k-1}$. If $\gamma_{k-1}^* = 0$ then set $\ell \leftarrow k-1$ fi.   (P5)

                    For $j = 1, \ldots, k-2$ do swap $\mu_{k-1,j}$ and μ_{kj}.

                    Set $\mathbf{y}_{k-1}^* \leftarrow \mathbf{y}_{k-1}$.

                    For $j = 1, \ldots, k-2$ do set $\mathbf{y}_{k-1}^* \leftarrow \mathbf{y}_{k-1}^* - \mu_{k-1,j}\mathbf{y}_j^*$.

                    If $k \leq \ell$ then

                        Set $\mathbf{y}_k^* \leftarrow \mathbf{y}_k$. For $j = 1, \ldots, k-1$ do: Set $\mathbf{y}_k^* \leftarrow \mathbf{y}_k^* - \mu_{kj}\mathbf{y}_j^*$.

                    If $k > 2$ then set $k \leftarrow k-1$.

            · If not startagain then set $g \leftarrow g+1$.

FIGURE 6.2

Main loop of the MLLL algorithm

At (P2) row 2 is reduced using row 1:

$$Y = \begin{bmatrix} -2 \\ -1 \\ 0 \end{bmatrix}$$

At (P5) rows 1 and 2 are exchanged:

$$Y = \begin{bmatrix} -1 \\ -2 \\ 0 \end{bmatrix}$$

At (P2) row 2 is reduced using row 1:

$$Y = \begin{bmatrix} -1 \\ 0 \\ 0 \end{bmatrix}$$

Another zero row has appeared, and the algorithm terminates. The nonzero entry in row 1 is the GCD of 6, 10, 15 (up to a sign).

Example 6.3. We consider these three row vectors in $\mathbb{R}^2$:

$$Y = \begin{bmatrix} 5 & 2 \\ 4 & 1 \\ -9 & 6 \end{bmatrix}$$

(P2) reduces row 2 using row 1:

$$Y = \begin{bmatrix} 5 & 2 \\ -1 & -1 \\ -9 & 6 \end{bmatrix}$$

(P5) exchanges rows 1 and 2:

$$Y = \begin{bmatrix} -1 & -1 \\ 5 & 2 \\ -9 & 6 \end{bmatrix}$$

(P2) reduces row 2 using row 1:

$$Y = \begin{bmatrix} -1 & -1 \\ 1 & -2 \\ -9 & 6 \end{bmatrix}$$

(P2) reduces row 3 using row 2:

$$Y = \begin{bmatrix} -1 & -1 \\ 1 & -2 \\ -4 & -4 \end{bmatrix}$$

(P5) exchanges rows 2 and 3:

$$Y = \begin{bmatrix} -1 & -1 \\ -4 & -4 \\ 1 & -2 \end{bmatrix}$$

(P2) reduces row 2 using row 1:

$$Y = \begin{bmatrix} -1 & -1 \\ 0 & 0 \\ 1 & -2 \end{bmatrix}$$

A zero row has appeared, and (P4) swaps it to the last row:

$$Y = \begin{bmatrix} -1 & -1 \\ 1 & -2 \\ 0 & 0 \end{bmatrix}$$

Finally, (P2) reduces row 2 using row 1; this changes Y^* and μ but not Y. The two nonzero rows of the last matrix form a reduced basis of the lattice spanned by the rows of the original matrix.

Example 6.4. In the previous two examples, the operations in positions (P1) and (P4) were never executed. This example uses all five positions (P1), (P2), (P3), (P4), (P5). We start with five vectors in $\mathbb{R}^3$:

$$Y = \begin{bmatrix} 0 & 0 & 3 \\ 2 & 0 & 1 \\ 3 & 1 & 0 \\ 0 & 0 & 0 \\ 0 & 2 & 3 \end{bmatrix}$$

The trace of the execution of the MLLL algorithm is displayed in Figure 6.3.

6.3 Projects

Project 6.1. Implement the MLLL algorithm in your favorite computer language, and test your program on a large number of pseudorandom integer matrices of many different sizes.

Project 6.2. An algorithm similar to the MLLL algorithm has been proposed by Hobby [63], with an application to electronic typesetting. Write a report and give a seminar talk on this paper.

			row 1	row 2	row 3	row 4	row 5
P2	reduce	2, 1	$[0, 0, 3]$	$[2, 0, 1]$	$[3, 1, 0]$	$[0, 0, 0]$	$[0, 2, 3]$
P5	exchange	2, 1	$[2, 0, 1]$	$[0, 0, 3]$	$[3, 1, 0]$	$[0, 0, 0]$	$[0, 2, 3]$
P2	reduce	2, 1	$[2, 0, 1]$	$[-2, 0, 2]$	$[3, 1, 0]$	$[0, 0, 0]$	$[0, 2, 3]$
P2	reduce	3, 2	$[2, 0, 1]$	$[-2, 0, 2]$	$[3, 1, 0]$	$[0, 0, 0]$	$[0, 2, 3]$
P5	exchange	3, 2	$[2, 0, 1]$	$[3, 1, 0]$	$[-2, 0, 2]$	$[0, 0, 0]$	$[0, 2, 3]$
P2	reduce	2, 1	$[2, 0, 1]$	$[1, 1, -1]$	$[-2, 0, 2]$	$[0, 0, 0]$	$[0, 2, 3]$
P5	exchange	2, 1	$[1, 1, -1]$	$[2, 0, 1]$	$[-2, 0, 2]$	$[0, 0, 0]$	$[0, 2, 3]$
P2	reduce	2, 1	$[1, 1, -1]$	$[2, 0, 1]$	$[-2, 0, 2]$	$[0, 0, 0]$	$[0, 2, 3]$
P2	reduce	3, 2	$[1, 1, -1]$	$[2, 0, 1]$	$[-2, 0, 2]$	$[0, 0, 0]$	$[0, 2, 3]$
P5	exchange	3, 2	$[1, 1, -1]$	$[-2, 0, 2]$	$[2, 0, 1]$	$[0, 0, 0]$	$[0, 2, 3]$
P2	reduce	2, 1	$[1, 1, -1]$	$[-1, 1, 1]$	$[2, 0, 1]$	$[0, 0, 0]$	$[0, 2, 3]$
P2	reduce	3, 2	$[1, 1, -1]$	$[-1, 1, 1]$	$[2, 0, 1]$	$[0, 0, 0]$	$[0, 2, 3]$
P3	reduce	3, 1	$[1, 1, -1]$	$[-1, 1, 1]$	$[2, 0, 1]$	$[0, 0, 0]$	$[0, 2, 3]$
P1	swap	4, 5	$[1, 1, -1]$	$[-1, 1, 1]$	$[2, 0, 1]$	$[0, 2, 3]$	$[0, 0, 0]$
P2	reduce	4, 3	$[1, 1, -1]$	$[-1, 1, 1]$	$[2, 0, 1]$	$[-2, 2, 2]$	$[0, 0, 0]$
P5	exchange	4, 3	$[1, 1, -1]$	$[-1, 1, 1]$	$[-2, 2, 2]$	$[2, 0, 1]$	$[0, 0, 0]$
P2	reduce	3, 2	$[1, 1, -1]$	$[-1, 1, 1]$	$[0, 0, 0]$	$[2, 0, 1]$	$[0, 0, 0]$
P4	swap	3, 4	$[1, 1, -1]$	$[-1, 1, 1]$	$[2, 0, 1]$	$[0, 0, 0]$	$[0, 0, 0]$
P2	reduce	3, 2	$[1, 1, -1]$	$[-1, 1, 1]$	$[2, 0, 1]$	$[0, 0, 0]$	$[0, 0, 0]$
P3	reduce	3, 1	$[1, 1, -1]$	$[-1, 1, 1]$	$[2, 0, 1]$	$[0, 0, 0]$	$[0, 0, 0]$

FIGURE 6.3
Trace of the execution of the MLLL algorithm for Example 6.4

6.4 Exercises

Exercise 6.1. Let e_1, e_2, ..., e_m be the standard basis vectors in $\mathbb{R}^m$, and let x_1, x_2, ..., x_m be arbitrary vectors in $\mathbb{R}^n$. Prove that the vectors $[e_1, x_1]$, $[e_2, x_2]$, ..., $[e_m, x_m]$ are linearly independent in $\mathbb{R}^{m+n}$.

Exercise 6.2. Prove that inequality (6.2) implies inequality (6.3), and conversely.

Exercise 6.3. Consider these three row vectors in $\mathbb{R}^2$:

$$\begin{bmatrix} x_1 \\ x_2 \\ x_3 \end{bmatrix} = \begin{bmatrix} 1 & -8 \\ -3 & -6 \\ 9 & 6 \end{bmatrix}$$

Apply the method of Section 6.1 to embed these vectors as linearly independent vectors in $\mathbb{R}^5$. Using $\alpha = \frac{3}{4}$ ($\beta = 2$), and following Example 6.1, show that $\ell = 2$ does not make any row of the right block equal to zero, but that $\ell = 3$ makes the first row of the right block equal to zero.

Exercise 6.4. Use a computer algebra system to redo Example 6.1 with the limiting value $\alpha = 1$ $(\beta = \frac{4}{3})$. Show that $\ell = 13$ does not make any row of the right block equal to zero, that $\ell = 14$ makes the second row equal to zero but not the first, and that $\ell = 15$ makes the first and third rows equal to zero.

Exercise 6.5. Consider these four row vectors in $\mathbb{R}^3$:

$$
\begin{bmatrix} x_1 \\ x_2 \\ x_3 \\ x_4 \end{bmatrix} = \begin{bmatrix} -32 & 8 & 44 \\ -74 & 69 & 92 \\ -4 & 99 & -31 \\ 27 & 29 & 67 \end{bmatrix}
$$

Following the method of Section 6.1 with $\alpha = \frac{3}{4}$ $(\beta = 2)$, find

(i) the smallest ℓ which makes one row of the right block zero,

(ii) the smallest ℓ which makes the first row equal to zero.

Does there exist a sufficiently large value of ℓ which makes two rows of the right block equal to zero? Why or why not?

Exercise 6.6. Following Example 6.2, use the MLLL algorithm to calculate the GCD of 6, 10 and 14.

Exercise 6.7. Repeat Exercise 6.6 for the other five permutations, namely

$$
\begin{bmatrix} 6 \\ 14 \\ 10 \end{bmatrix}, \quad \begin{bmatrix} 10 \\ 6 \\ 14 \end{bmatrix}, \quad \begin{bmatrix} 10 \\ 14 \\ 6 \end{bmatrix}, \quad \begin{bmatrix} 14 \\ 6 \\ 10 \end{bmatrix}, \quad \begin{bmatrix} 14 \\ 10 \\ 6 \end{bmatrix}.
$$

How does the trace of the algorithm change for the different permutations?

Exercise 6.8. Use MLLL to calculate the GCD of 105, 70, 42 and 30.

Exercise 6.9. Apply the MLLL algorithm to these three vectors in $\mathbb{R}^2$:

$$
\begin{bmatrix} 1 & -8 \\ -3 & -6 \\ 9 & 6 \end{bmatrix}
$$

Exercise 6.10. Use the MLLL algorithm to show that every integer vector in $\mathbb{R}^2$ is an integral linear combination of the four rows of this matrix:

$$
\begin{bmatrix} -3 & -4 \\ -6 & -8 \\ -7 & 5 \\ 5 & 9 \end{bmatrix}
$$

Exercise 6.11. Use the MLLL algorithm to find a reduced basis for the lattice in $\mathbb{R}^3$ spanned by the rows of this matrix:

$$\begin{bmatrix} 7 & 5 & 7 \\ 5 & 1 & 6 \\ 3 & 3 & 2 \\ 0 & 4 & 4 \end{bmatrix}$$

Exercise 6.12. Use the MLLL algorithm to find a reduced basis for the lattice in $\mathbb{R}^3$ spanned by the rows of this matrix:

$$\begin{bmatrix} 1 & 2 & 3 \\ 2 & 3 & 4 \\ 3 & 4 & 5 \\ 4 & 5 & 6 \end{bmatrix}$$

Exercise 6.13. Prove or disprove: Position (P1) of the MLLL algorithm is executed if and only if the original set of vectors contains the zero vector.

Exercise 6.14. Prove that the MLLL algorithm terminates in polynomial time. (See Cohen [26], page 96, and Sims [131], page 374.)

Exercise 6.15. The original version of the MLLL algorithm in Pohst [118] (see also Pohst and Zassenhaus [119], §3.3, and Cohen [26], §2.6.4) kept track of linear dependence relations among the original vectors. That is, if the original $m \times n$ matrix A has rank r, then the algorithm also outputs an $n \times n$ integer matrix H such that

(i) the first $n-r$ rows of H are linearly independent,

(ii) the last r rows of H are zero,

(iii) $HA = O$.

Modify the MLLL algorithm so that it computes such a matrix H. Explain how this can be used to produce a basis of integer vectors for the nullspace of an integer matrix.

Exercise 6.16. Show how the MLLL algorithm can be used to solve the following problem: Let U be a subspace of $\mathbb{Q}^n$, and let $L = U \cap \mathbb{Z}^n$ be the lattice of integer vectors in U. Find a basis for L. (See Hobby [63].)

7

The Knapsack Problem

CONTENTS

In this chapter we present our first application of lattice basis reduction to cryptography: we use the LLL algorithm to break a knapsack cryptosystem.

7.1 The subset-sum problem

Definition 7.1. Suppose that we are given a set of n distinct positive integers

$$A = \{\, a_1, a_2, \ldots, a_n \,\},$$

together with another positive integer s. The **subset-sum problem** (also called the **knapsack problem**) asks whether there exists a subset

$$I \subseteq \{\, 1, 2, \ldots, n \,\},$$

for which the sum of the corresponding elements of A is exactly s:

$$\sum_{i \in I} a_i = s. \tag{7.1}$$

We think of s as the size of a knapsack, and the numbers a_i as the sizes of items to be placed in the knapsack; the question is then whether the knapsack can be filled exactly by some subset of the items.

We reformulate this problem as follows. We define

$$x_i = \begin{cases} 1 & \text{if } i \in I \\ 0 & \text{if } i \notin I \end{cases} \qquad (i = 1, 2, \ldots, n).$$

This is the characteristic function of the subset I. Equation (7.1) becomes

$$\sum_{i=1}^{n} x_i a_i = s. \tag{7.2}$$

The subset-sum problem then asks whether there exists a vector

$$[x_1, x_2, \ldots, x_n] \in \{0,1\}^n,$$

which satisfies equation (7.2). This general problem is NP-complete (see Garey and Johnson [46]) but some special cases are easy to solve.

Example 7.2. If $a_i = 2^{i-1}$ for $i = 1, 2, \ldots, n$ then it is easy to see that a solution exists if and only if $0 \leq s \leq 2^n - 1$, and the vector $[x_1, x_2, \ldots, x_n]$ is just the binary representation of s in reverse order:

$$s = x_1 + 2x_2 + 4x_3 + \cdots + 2^{n-1} x_n = \sum_{i=1}^{n} x_i 2^{i-1}.$$

Definition 7.3. The sequence $a_1, a_2, \ldots, a_n$ is called **superincreasing** if

$$a_i > \sum_{j=1}^{i-1} a_j \qquad (i = 2, 3, \ldots, n).$$

That is, each term is greater than the sum of the previous terms. The sequence $a_i = 2^{i-1}$ from Example 7.2 is superincreasing.

The subset-sum problem is easy to solve recursively for any superincreasing sequence, starting with x_n and working backwards. Since

$$a_n > a_1 + a_2 + \cdots + a_{n-1},$$

we must set

$$x_n = \begin{cases} 1 & \text{if } s > a_1 + a_2 + \cdots + a_{n-1} \\ 0 & \text{if } s \leq a_1 + a_2 + \cdots + a_{n-1} \end{cases}$$

Put more simply,

$$x_n = 1 \quad \Longleftrightarrow \quad s > \sum_{j=1}^{n-1} a_j.$$

We can now remove a_n and reduce the problem to a smaller subset-sum problem in which the sequence is $a_1, a_2, \ldots, a_{n-1}$ and the sum is $s - x_n a_n$:

$$\sum_{i=1}^{n-1} x_i a_i = s - x_n a_n.$$

Reasoning as before we see that

$$x_{n-1} = 1 \quad \Longleftrightarrow \quad s - x_n a_n > \sum_{j=1}^{n-2} a_j.$$

Repeating this process quickly produces the entire solution $x_n, x_{n-1}, \ldots, x_1$. More precisely, we have

$$x_i = 1 \quad \Longleftrightarrow \quad s - \sum_{j=i+1}^{n} x_j a_j > \sum_{j=1}^{i-1} a_j \quad (i = n, n-1, \ldots, 1).$$

Example 7.4. Let $n = 15$, and define the superincreasing sequence $b_1, b_2, \ldots, b_n$ as follows:

- let b_1 be the smallest prime number greater than 2^n;

- for $i = 2, 3, \ldots, n$, let b_i be the smallest prime number greater than $2b_{i-1}$.

We obtain the following superincreasing sequence:

$$
\begin{array}{lll}
b_1 = 32771, & b_2 = 65543, & b_3 = 131101, \\
b_4 = 262187, & b_5 = 524387, & b_6 = 1048759, \\
b_7 = 2097523, & b_8 = 4195057, & b_9 = 8390143, \\
b_{10} = 16780259, & b_{11} = 33560539, & b_{12} = 67121039, \\
b_{13} = 134242091, & b_{14} = 268484171, & b_{15} = 536968403.
\end{array}
$$

Consider the subset-sum problem with sequence $b_1, b_2, \ldots, b_n$ and sum

$$s = 891221976.$$

Using the algorithm given above, we easily compute

$$x_1 x_2 \cdots x_{15} = 100111000101011,$$

which corresponds to the solution

$$b_1 + b_4 + b_5 + b_6 + b_{10} + b_{12} + b_{14} + b_{15} = s.$$

7.2 Knapsack cryptosystems

Definition 7.5. Merkle and Hellman [95] used the subset-sum problem as the basis for a cryptographic scheme known as a **knapsack cryptosystem**.

The receiver of the encrypted message secretly chooses a **private key**, which is a superincreasing sequence $b_1, b_2, \ldots, b_n$ satisfying

$$b_1 \approx 2^n, \qquad b_i > \sum_{j=1}^{i-1} b_j \ (i = 2, 3, \ldots, n), \qquad b_n \approx 2^{2n}.$$

The receiver also secretly chooses positive integers m (the **modulus**) and w (the **multiplier**) satisfying

$$m > \sum_{j=1}^{n} b_j, \qquad 0 < w < m, \qquad \gcd(m, w) = 1.$$

The last condition guarantees that w is invertible modulo m. Finally, the receiver chooses a permutation π of the integers $1, 2, \ldots, n$ and then computes the **public key**, which is the sequence $a_1, a_2, \ldots, a_n$ defined by

$$a_i \equiv w b_{\pi(i)} \pmod{m}, \qquad 0 < a_i < m. \tag{7.3}$$

Note that $a_i \neq 0$ for all i, since otherwise

$$b_{\pi(i)} \equiv w^{-1} a_i \equiv 0 \pmod{m},$$

contradicting the choice of m. The quantities $m, w, b_1, b_2, \ldots, b_n$ and π are kept private, whereas the quantities $a_1, a_2, \ldots, a_n$ are made public.

Example 7.6. We continue from Example 7.4. For the modulus m, we take the smallest prime greater than $b_1 + b_2 + \cdots + b_n$:

$$m = 1073903977.$$

For the multiplier, we take $\lceil m/n \rfloor$, the nearest integer to m/n; this is (almost) the average of $b_1, b_2, \ldots, b_n$:

$$w = 71593598.$$

We randomly choose a permutation π of the integers $1, 2, \ldots, 15$:

$$\pi = [7, 12, 1, 15, 9, 14, 2, 8, 11, 5, 10, 4, 13, 6, 3].$$

We then apply equation (7.3) to obtain the following public key:

$$
\begin{array}{lll}
a_1 = 929737936, & a_2 = 970987227, & a_3 = 787514290, \\
a_4 = 322163533, & a_5 = 926801380, & a_6 = 662236970, \\
a_7 = 572718201, & a_8 = 499197496, & a_9 = 270712809, \\
a_{10} = 142942483, & a_{11} = 994479591, & a_{12} = 143064843, \\
a_{13} = 724883274, & a_{14} = 285884973, & a_{15} = 71532418.
\end{array}
$$

We continue with the general discussion. Suppose that the sender has divided the message to be encrypted into blocks, each of which is represented as a bit string (0-1 vector) of length n: $x_1 x_2 \cdots x_n$. The sender uses the public key to encrypt each block of the message as a positive integer:

$$s = \sum_{i=1}^{n} x_i a_i.$$

The receiver uses the private quantities m and w to compute

$$t = w^{-1} s \;(\mathrm{mod}\; m), \qquad 0 \le t < m.$$

We now observe that

$$t = w^{-1} s = w^{-1} \sum_{i=1}^{n} x_i a_i = \sum_{i=1}^{n} x_i w^{-1} a_i = \sum_{i=1}^{n} x_i b_{\pi(i)}.$$

This is a subset-sum problem with a superincreasing sequence (permuted, but easily unpermuted). The receiver of the encrypted message can easily solve this subset-sum problem and recover the block of the original message.

Example 7.7. We continue from Example 7.6. There are 26 letters in the English alphabet, and so there are $26^3 = 17576$ triples of letters, which we order lexicographically:

$$aaa,\; aab,\; aac,\; \ldots,\; aba,\; abb,\; abc,\; \ldots,\; zza,\; zzb,\; zzc,\; \ldots,\; zzz.$$

More precisely, the triple $\alpha\beta\gamma$ precedes the triple $\alpha'\beta'\gamma'$ if and only if

- α precedes α' (in alphabetical order), or

- $\alpha = \alpha'$ and β precedes β', or

- $\alpha = \alpha'$ and $\beta = \beta'$ and γ precedes γ'.

If $\alpha\beta\gamma$ is a triple of letters then we define the corresponding triple of integers $(\delta_1, \delta_2, \delta_3)$ with $1 \le \delta_1, \delta_2, \delta_3 \le 26$ to be the indices of α, β, γ in the alphabet. The index Δ of the triple $\alpha\beta\gamma$ in the list of all triples is given by this formula:

$$\Delta = (\delta_1 - 1)26^2 + (\delta_2 - 1)26 + (\delta_3 - 1) + 1.$$

Since $2^{14} = 16384 < 26^3 < 32768 = 2^{15}$, every triple $\alpha\beta\gamma$ can be represented by a bit string of length 15, namely the binary numeral for its index:

$$\alpha\beta\gamma \longmapsto \Delta = \sum_{i=1}^{15} x_i \, 2^{15-i} \longmapsto x_1 x_2 \cdots x_{15}$$

We now convert the binary strings into subset sums using the public key:

$$s = \sum_{i=1}^{15} x_i a_i.$$

α	β	γ	δ_1	δ_2	δ_3	Δ	$x_1 x_2 \cdots x_{14} x_{15}$
i	t	h	9	20	8	5909	001011100010101
i	n	k	9	14	11	5756	001011001111100
c	o	m	3	15	13	1728	000011011000000
p	u	t	16	21	20	10679	010100110110111
e	r	a	5	18	1	3146	000110001001010
l	g	e	12	7	5	7596	001110110101100
b	r	a	2	18	1	1118	000010001011110
i	s	r	9	19	18	5893	001011100000101
e	a	l	5	1	12	2715	000101010011011
l	y	n	12	25	14	8073	001111110001001
i	c	e	9	3	5	5464	001010101011000

FIGURE 7.1

Encrypted message for Example 7.7

For example, consider the message

i think computer algebra is really nice

We ignore the spaces and break this message into 3-letter blocks:

ith, ink, com, put, era, lge, bra, isr, eal, lyn, ice

Applying the encryption algorithm, we obtain the results in Figure 7.1. The original message has been transformed into the following ciphertext $s_1, s_2, \ldots, s_{11}$:

4740166124,	4652635640,	2358948655,	4584789196,
1948627538,	4119285500,	3345826870,	3745686533,
2978559824,	3985229131,	3695291114.	

The receiver can now decrypt the message using the modulus m, the multiplier w, and the private key $b_1, b_2, \ldots, b_{15}$.

The problem with knapsack cryptosystems is that they can be easily broken; this was demonstrated by Shamir [129]. A method for breaking this cryptosystem using the LLL algorithm was given by Lagarias and Odlyzko [85].

Given the public key $a_1, a_2, \ldots, a_n$ and the ciphertext s, we construct the following $(n+1) \times (n+1)$ matrix M:

$$M = \begin{bmatrix} 1 & 0 & 0 & \cdots & 0 & -a_1 \\ 0 & 1 & 0 & \cdots & 0 & -a_2 \\ 0 & 0 & 1 & \cdots & 0 & -a_3 \\ \vdots & \vdots & \vdots & \vdots & \vdots & \vdots \\ 0 & 0 & 0 & \cdots & 1 & -a_n \\ 0 & 0 & 0 & \cdots & 0 & s \end{bmatrix}$$

It is clear that the rows $M_1, \ldots, M_{n+1}$ of the matrix M are linearly independent, and so they form a basis of a lattice $L \subset \mathbb{R}^n$. By assumption, we know that there must exist $x_1, x_2, \ldots, x_n \in \{0, 1\}$ for which

$$x_1 a_1 + x_2 a_2 + \cdots + x_n a_n = s.$$

Therefore the lattice L contains the vector

$$V = x_1 M_1 + x_2 M_2 + \cdots + x_n M_n + M_{n+1} = [\, x_1, \, x_2, \, \ldots, \, x_n, \, 0 \,].$$

This is a very short vector, since its length is at most $\sqrt{n}$, whereas the original lattice basis vectors $M_1, \ldots, M_{n+1}$ are very long. (In Example 7.6 the numbers $a_1, a_2, \ldots, a_n$ each have 8 or 9 digits.) So we can apply the LLL algorithm to the rows of the matrix M, and look for a reduced basis vector with components in the set $\{0, 1\}$. If we find such a vector, we need to check that the corresponding subset sum does indeed equal the ciphertext. We also need to consider the possibility that the LLL algorithm might produce a reduced basis vector with the opposite sign; that is,

$$V = -x_1 M_1 - x_2 M_2 - \cdots - x_n M_n - M_{n+1} = [\, -x_1, \, -x_2, \, \ldots, \, -x_n, \, 0 \,].$$

In this case we replace each component by its negative, and check that the subset sum equals the ciphertext. The LLL algorithm does not necessarily produce a reduced basis vector of the required form, even though we know that such a vector must exist. In this case, the decryption attempt fails.

Example 7.8. We continue from Example 7.7. The first block of the message is represented by the ciphertext 4740166124. The corresponding matrix M is displayed in Figure 7.2. We apply the LLL algorithm with the standard parameter $\alpha = \frac{3}{4}$ to the rows of M, and obtain the reduced basis consisting of the rows of the matrix displayed in Figure 7.3. Row 10 of the reduced matrix gives the 0-1 vector 0010111000101010, which corresponds to the triple *ith*, the first block of the message.

The next two blocks are decrypted correctly: for the second block, row 14 of the reduced matrix is 0010110011111000, corresponding to *ink*; for the third block, row 1 of the reduced matrix is 0000110110000000, corresponding to *com*.

The fourth block of the message is represented by the ciphertext 4584789196. We apply the LLL algorithm with $\alpha = \frac{3}{4}$ to the rows of the corresponding matrix M, and obtain the reduced basis consisting of the rows of the matrix in Figure 7.4. None of the rows of the reduced matrix has all of its entries either in $\{0, 1\}$ or in $\{0, -1\}$. The decryption attempt fails.

For the fifth block, row 14 of the reduced matrix in Figure 7.5 is the negative of the 0-1 vector 0001100010010100, corresponding to *era*.

Continuing, we obtain the following results for the 11 blocks of the message: success for blocks 1, 2, 3, 5, 6, 7, 9, 11 and failure for blocks 4, 8, 10.

$$
\begin{bmatrix}
1 & . & . & . & . & . & . & . & . & . & . & . & . & . & . & -929737936 \\
. & 1 & . & . & . & . & . & . & . & . & . & . & . & . & . & -970987227 \\
. & . & 1 & . & . & . & . & . & . & . & . & . & . & . & . & -787514290 \\
. & . & . & 1 & . & . & . & . & . & . & . & . & . & . & . & -322163533 \\
. & . & . & . & 1 & . & . & . & . & . & . & . & . & . & . & -926801380 \\
. & . & . & . & . & 1 & . & . & . & . & . & . & . & . & . & -662236970 \\
. & . & . & . & . & . & 1 & . & . & . & . & . & . & . & . & -572718201 \\
. & . & . & . & . & . & . & 1 & . & . & . & . & . & . & . & -499197496 \\
. & . & . & . & . & . & . & . & 1 & . & . & . & . & . & . & -270712809 \\
. & . & . & . & . & . & . & . & . & 1 & . & . & . & . & . & -142942483 \\
. & . & . & . & . & . & . & . & . & . & 1 & . & . & . & . & -994479591 \\
. & . & . & . & . & . & . & . & . & . & . & 1 & . & . & . & -143064843 \\
. & . & . & . & . & . & . & . & . & . & . & . & 1 & . & . & -724883274 \\
. & . & . & . & . & . & . & . & . & . & . & . & . & 1 & . & -285884973 \\
. & . & . & . & . & . & . & . & . & . & . & . & . & . & 1 & -71532418 \\
. & . & . & . & . & . & . & . & . & . & . & . & . & . & . & 4740166124
\end{bmatrix}
$$

FIGURE 7.2
The original lattice basis for block 1

Example 7.9. We repeat the same computations as in Example 7.8, except that we use the reduction parameter $\alpha = \frac{99}{100}$ in the LLL algorithm. We now obtain success in all 11 blocks: lattice basis reduction has correctly decrypted the entire message. The new reduced matrix for block 4 (which produced a decryption failure for $\alpha = \frac{3}{4}$) is displayed in Figure 7.6; now row 3 is the 0-1 vector 0101001101101110 corresponding to *put*.

7.3 Projects

Project 7.1. Write a report and present a seminar talk on the "rise and fall" of knapsack cryptosystems, based on the original papers by Merkle and Hellman [95], Shamir [129], Lagarias and Odlyzko [85], and Odlyzko [113]. There is also a very readable survey paper by Odlyzko [114].

Project 7.2. Write your own a software package to encrypt and decrypt messages using the methods described in this chapter.

7.4   Exercises

Exercise 7.1. Consider the following knapsack cryptosystem. Start with the extended alphabet consisting of the 26 letters together with the space, for a total of 27 symbols. There are $27^2 = 729$ ordered pairs of symbols; these pairs are ordered lexicographically and identified with the binary representations (using ten bits) of the numbers from 1 to 729. Messages are broken into a sequence of pairs and each pair is treated separately. Encryption and decryption are performed as described in this chapter. Suppose that you are an eavesdropper who knows the public key:

$$747852, \quad 315290, \quad 551065, \quad 108243, \quad 157491,$$
$$526441, \quad 105165, \quad 735540, \quad 524902, \quad 321446.$$

You have intercepted the following ciphertext:

1919750,	1264751,	1841466,	2989048,	1638010,
1707060,	1480519,	1377919,	971520,	3669902,
1427372,	2255047,	2872802,	581024,	2534735,
1501039,	1069298,	2077241,	2458503,	1815303,
1819407,	1292966,	2392531,	1682436,	526441,
2713977,	429689,	1069298.		

Use lattice basis reduction to recover the original message.

Exercise 7.2. Same as Exercise 7.1 for this ciphertext:

1477954,	1759386,	948435,	1946631,	1585684,
840705,	2458503,	1873785,	2132953,	840705,
1556443,	2236579,	1174463,	2637027,	2337948,
2392531,	1319642,	1427885,	2423619,	1841466,
2458503,	1184210,	1407160,	1372789,	2137262,
1833771,	581024,	1129011,	846348,	899392,
213408,	1930523,	1274293,	3345583,	2288597,
1218786.				

Exercise 7.3. Same as Exercise 7.1 for this ciphertext:

2715516,	2686275,	2077241,	2397148,	1104387,
2175737,	1895126,	1632367,	1657504,	2302756,
2610043,	1523098,	551065,	2255047,	1331954,
2772767,	2131106,	1739379,	1218786,	1069298.

Exercise 7.4. Same as Exercise 7.1 for this ciphertext:

213408,	1930523,	1272754,	581024,	948435,
1377919,	956130,	1261981,	1272754,	429689,
1069298,	971520,	2458503,	1184210,	2358673,
1010508,	1377919,	1345292,	2137262,	1261981,
2182406,	2132953,	1682436,	526441,	2713977,
429689,	1728606,	1867629,	1450457,	2028506,
2776871,	2008294,	2637027.		

Exercise 7.5. Consider the following knapsack cryptosystem. Start with the extended alphabet consisting of the 26 letters together with the space, for a total of 27 symbols. There are $27^3 = 19683$ ordered triples of symbols; these triples are ordered lexicographically and identified with the binary representations (using 15 bits) of the numbers from 1 to 19683. Messages are broken into a sequence of triples and each triple is treated separately. Encryption and decryption are performed as described in this chapter. Suppose that you are an eavesdropper who knows the public key:

644824022,	644404566,	653213142,	680058326,	286344257,
286619525,	646501846,	286409797,	644273486,	644266932,
286357365,	304236677,	290814085,	287458437,	644299702.

You have intercepted the following ciphertext:

4326972602,	4357553566,	4109958741,	2866800403,	2471727472,
2439743952,	1610872689,	4074853333,	3066083510,	3025999246,
2370505312,	3764062656,	4686854923,	2734162737,	3392933671.

Use lattice basis reduction to recover the original message. You may need to increase the value of the reduction parameter α.

Exercise 7.6. Same as Exercise 7.5 for this ciphertext:

3475549025,	3545757657,	3738488948,	5358661763,	2512239930,
4123768019,	4423967432,	1610872689,	2435418312,	4370316388,
2154356579,	864892047,	4427985034,	3115470084,	2799724583,
3205707740,	878314639,	3205832266.		

Exercise 7.7. Same as Exercise 7.5 for this ciphertext:

3092943986,	2439743952,	2818881925,	5008112338,	2901595589,
2489279084,	4071189647,	3814393006,	3851226486,	1581589417,
3205707740,	878314639,	2901595589,	2202095915,	3478924335,

3491350719, 3832154346, 2902696661, 3152436828, 2514868084,
3120129978, 3084010884, 3822152942, 4325681464, 4036125747,
3392933671.

Exercise 7.8. Same as Exercise 7.5 for this ciphertext:

3475549025, 3545757657, 4383856952, 2165885065, 2507049162,
3854555918, 3397390391, 1586072353, 4370316388, 2154356579,
864892047, 4427985034, 3115470084, 2799724583, 3205707740,
878314639, 2901595589, 4066739481, 3785291062, 3146463950,
3192534200.

$$
\begin{bmatrix}
2 & -1 & -2 & 0 & -1 & 0 & 1 & 1 & 2 & 0 & 0 & 0 & 0 & 0 & 0 & 0 \\
1 & 0 & 0 & -1 & 0 & -2 & 1 & 0 & -1 & 0 & -1 & -1 & 1 & 0 & 0 & -1 \\
-1 & 0 & -1 & 0 & 1 & -1 & 0 & 2 & -1 & -1 & 0 & 1 & 1 & 0 & 0 & -1 \\
0 & 1 & 2 & 1 & -2 & -1 & 0 & 1 & 0 & 1 & -1 & 0 & 0 & 0 & 0 & 2 \\
-2 & 0 & 3 & -1 & -1 & -1 & 0 & 0 & -1 & -1 & -1 & 0 & 0 & 0 & 0 & 2 \\
0 & 1 & -1 & -1 & -1 & -2 & 0 & 2 & 2 & 0 & 1 & 0 & -1 & 0 & 0 & 0 \\
0 & 0 & 1 & -1 & 1 & -1 & 0 & -2 & -1 & 0 & 0 & 0 & -2 & -1 & 0 & 0 \\
1 & 0 & -2 & 0 & 0 & 0 & -1 & -1 & 0 & 2 & 2 & 0 & 0 & 0 & -1 & 0 \\
1 & 0 & 1 & 0 & 0 & 0 & 1 & -1 & 0 & -1 & 0 & 0 & -1 & 0 & 1 & -1 \\
0 & 0 & 1 & 0 & 1 & 1 & 1 & 0 & 1 & 0 & -1 & -1 & 1 & 0 & -1 & 0 \\
1 & 0 & 0 & -1 & 1 & -1 & 1 & 2 & -1 & 1 & 1 & 1 & -1 & -1 & 0 & -1 \\
0 & -2 & 1 & 0 & 0 & 0 & -1 & 1 & -1 & -1 & -2 & 1 & 0 & 0 & 0 & -1 \\
1 & 0 & 1 & 0 & 0 & 1 & -2 & 0 & -1 & 0 & 0 & 1 & 1 & 1 & 0 & -1 \\
1 & 1 & 1 & 0 & 0 & 0 & -1 & -1 & -1 & 1 & -1 & -1 & -1 & -1 & 1 & -1 \\
-1 & 1 & 1 & -1 & -1 & 0 & 0 & -1 & -1 & 2 & -1 & 0 & -1 & -1 & 2 & -1 \\
-133 & -99 & -15 & 528 & 272 & -278 & -63 & -129 & 302 & 78 & 0 & -78 & -17 & -122 & 102 & -40
\end{bmatrix}
$$

FIGURE 7.3
The reduced basis ($\alpha = 3/4$) for block 1

$$
\begin{bmatrix}
2 & -1 & -2 & 0 & -1 & 0 & 1 & 1 & 2 & 0 & 0 & 0 & 0 & 0 & 0 & 0 \\
1 & 0 & 0 & -1 & 0 & -2 & 1 & 0 & 1 & 0 & -1 & 1 & 1 & 0 & 0 & 1 \\
-1 & 0 & -1 & 0 & 1 & -1 & 0 & 2 & -1 & -1 & 0 & 1 & 1 & 0 & 0 & -1 \\
0 & 1 & 2 & 1 & -2 & -1 & 0 & 1 & 0 & 1 & -1 & 0 & 0 & 0 & 0 & 2 \\
-2 & 0 & 3 & -1 & -1 & -1 & 0 & 0 & 1 & 1 & 1 & 0 & 0 & 0 & 0 & 2 \\
0 & 1 & 1 & 1 & 1 & -2 & 0 & 2 & 2 & 0 & 0 & 0 & -1 & 0 & 0 & 0 \\
0 & 0 & 1 & -1 & 1 & -1 & 1 & -2 & -1 & 0 & 2 & 0 & -2 & -1 & 0 & 0 \\
1 & 0 & -1 & 0 & 0 & 0 & 1 & -1 & 0 & 2 & 0 & 0 & 0 & 0 & 1 & 0 \\
1 & 0 & -2 & 1 & 0 & 0 & 1 & 1 & 0 & -1 & -1 & 0 & -1 & 0 & -1 & 1 \\
1 & 0 & 1 & -1 & 1 & 0 & 1 & 1 & 2 & -1 & 1 & -1 & 1 & 0 & 0 & 0 \\
1 & 0 & 1 & 0 & 1 & -1 & -1 & 2 & 1 & 1 & -2 & -1 & -1 & 1 & 0 & 1 \\
0 & -2 & 0 & 0 & 0 & 0 & -2 & 1 & -1 & 1 & 0 & 1 & 0 & 0 & 1 & -1 \\
1 & 0 & 1 & 0 & 0 & 1 & 1 & 0 & 1 & 0 & -1 & 1 & -1 & -1 & -1 & 1 \\
1 & 0 & 1 & 0 & 0 & 0 & 0 & -1 & 0 & 1 & -1 & 1 & 1 & 0 & 2 & -1 \\
-1 & 1 & 1 & 0 & -1 & 0 & 0 & -1 & -1 & 2 & 0 & 0 & 1 & -1 & -1 & 1 \\
-181 & -148 & -60 & 486 & 208 & -303 & -91 & -155 & 273 & 67 & -58 & -83 & -57 & -133 & 92 & -37
\end{bmatrix}
$$

FIGURE 7.4
The reduced basis ($\alpha = 3/4$) for block 4

$$
\begin{bmatrix}
2 & -1 & -2 & 0 & -1 & 0 & 1 & 1 & 2 & 0 & 0 & 0 & 0 & 0 & 0 & 0 \\
1 & 0 & 0 & -1 & 0 & -2 & 1 & 0 & -1 & -1 & -1 & -1 & -1 & 0 & 0 & 1 \\
-1 & 0 & -1 & 0 & 1 & -1 & 0 & 2 & -1 & 1 & 0 & -1 & 1 & 0 & 0 & -1 \\
0 & 1 & 2 & 1 & -2 & -1 & 0 & 1 & 0 & 1 & -1 & 0 & 0 & 0 & 0 & 2 \\
-2 & 0 & 3 & -1 & -1 & -1 & 0 & 0 & 1 & 0 & 1 & 0 & 0 & 0 & 0 & 2 \\
0 & 1 & -1 & -1 & -1 & -2 & 0 & 2 & 0 & 0 & 0 & 0 & 0 & 0 & 0 & 0 \\
0 & 0 & 1 & -1 & 1 & -1 & 0 & -2 & -2 & 0 & 2 & 0 & 0 & 0 & 0 & 0 \\
1 & 0 & -2 & 0 & 0 & 0 & -1 & -1 & -1 & -1 & 0 & 0 & 0 & -1 & 1 & 0 \\
1 & 0 & 1 & -1 & 0 & 0 & 1 & -1 & 0 & 1 & -1 & 0 & 0 & 0 & 0 & -1 \\
1 & 1 & 0 & 1 & 1 & 0 & 1 & 2 & 0 & 1 & -2 & -1 & -1 & 0 & 0 & -2 \\
-1 & 0 & -2 & 0 & 1 & -1 & 0 & 0 & -1 & 0 & 2 & 0 & 1 & 0 & 0 & -1 \\
2 & 0 & 1 & -1 & 0 & 0 & -1 & 1 & -1 & 0 & -1 & -1 & 1 & 0 & 0 & 1 \\
0 & 0 & 1 & 0 & -1 & -1 & 0 & 0 & 0 & 1 & 0 & -1 & 1 & -1 & 0 & 0 \\
0 & 0 & 0 & -1 & -1 & 0 & 0 & 2 & 1 & 0 & 0 & -1 & 1 & -1 & 0 & -1 \\
0 & 0 & 1 & 0 & 0 & 0 & 0 & -1 & 0 & 0 & -1 & 1 & 0 & -1 & 2 & 0 \\
1 & 0 & 1 & -1 & -1 & 0 & -1 & 0 & -1 & 1 & -1 & -1 & -1 & -1 & 0 & -1 \\
-176 & -169 & -118 & 117 & -48 & -180 & -101 & -113 & 54 & 3 & -141 & -44 & -109 & -79 & 21 & -14
\end{bmatrix}
$$

FIGURE 7.5
The reduced basis ($\alpha = 3/4$) for block 5

$$
\begin{bmatrix}
1 & 0 & 0 & 1 & 0 & 0 & 1 & -1 & 0 & -1 & -1 & 0 & -1 & 0 & 1 & -1 \\
0 & 0 & 0 & 0 & 0 & 0 & 0 & 0 & 0 & -2 & 0 & -1 & 0 & 1 & 2 & 0 \\
0 & 1 & 1 & 1 & 0 & 0 & 1 & 1 & 0 & 2 & 1 & 0 & 1 & 1 & 1 & 0 \\
1 & 0 & 0 & 0 & 1 & 0 & 1 & -1 & 0 & 0 & 0 & 0 & 0 & -1 & 0 & 0 \\
0 & 2 & 1 & 1 & 1 & 0 & -1 & 0 & -1 & -1 & 0 & 0 & 2 & 0 & 0 & 0 \\
0 & -2 & 0 & 0 & 0 & 0 & 2 & 0 & 0 & 1 & 0 & -1 & 1 & 0 & -1 & -1 \\
-1 & 1 & -1 & 1 & -1 & -1 & -1 & 0 & -2 & 0 & 1 & 0 & -1 & 0 & -1 & 0 \\
2 & 1 & 0 & 0 & -1 & -1 & 0 & 1 & 0 & 1 & 2 & 1 & 0 & 0 & -1 & -1 \\
1 & 0 & 1 & 1 & 0 & 1 & 1 & 0 & 1 & 0 & 0 & 0 & -1 & 0 & 0 & 1 \\
1 & 0 & 0 & 0 & 0 & -1 & -1 & 1 & 1 & 0 & -1 & 0 & 0 & 0 & 0 & 1 \\
2 & -1 & -1 & 0 & 0 & 0 & 0 & 0 & -1 & -1 & 0 & 0 & -1 & 0 & -1 & 1 \\
0 & 1 & 1 & 0 & 0 & 0 & 0 & -1 & 0 & 1 & 1 & 0 & 0 & 0 & 0 & -1 \\
2 & 0 & 0 & 0 & 1 & 1 & 0 & 0 & 1 & 0 & 0 & -2 & 0 & 2 & -2 & -1 \\
1 & 1 & 0 & -2 & -1 & 0 & -2 & -1 & 1 & 1 & -1 & 1 & -1 & 1 & 0 & -1 \\
0 & -1 & 0 & 0 & 0 & 1 & 0 & 0 & 1 & 0 & 2 & 1 & 0 & 0 & 1 & 0 \\
1 & 0 & -1 & -2 & -1 & -1 & -1 & -1 & 1 & 1 & -1 & -1 & -1 & 1 & 0 & -1 \\
-182 & -150 & -59 & 485 & 207 & -303 & -91 & -154 & 274 & 65 & -58 & -83 & -59 & -134 & 92 & -38
\end{bmatrix}
$$

FIGURE 7.6
The reduced basis ($\alpha = 99/100$) for block 4

8

Coppersmith's Algorithm

CONTENTS

At a cryptography conference in 1996, Coppersmith [30] (see also [31]) showed how lattice basis reduction could be used to find small roots of modular polynomial equations in one variable. Coppersmith's algorithm implies that RSA cryptosystems with low public exponents can be broken in polynomial time using the LLL algorithm. In this chapter, we present the basic ideas underlying this algorithm, following Coppersmith [32, 33] and Howgrave-Graham [66]. The great impact of these ideas on cryptography is discussed in the survey papers by Coupé et al. [34], Nguyen and Stern [108, 109], and May [94].

8.1 Introduction to the problem

We consider a monic integer polynomial $p(x)$ of degree d in the variable x:

$$p(x) = x^d + p_{d-1}x^{d-1} + \cdots + p_1 x + p_0 \qquad (p_0, p_1, \ldots, p_{d-1} \in \mathbb{Z}). \qquad (8.1)$$

(The term "monic" simply means that $p_d = 1$.) We consider a modulus M; this is a large composite integer, and we make no assumptions about its prime factors. Our goal is to find integer solutions x_0 of this modular equation:

$$p(x_0) \equiv 0 \pmod{M}. \qquad (8.2)$$

If there is a solution $x_0 \in \mathbb{Z}$ for which $|x_0| < M^{1/d}$ then Coppersmith's algorithm finds this solution, and does so in polynomial time in $\log M$ and 2^d. The basic idea is to construct a matrix C using the coefficients of $p(x)$, and then to apply lattice basis reduction to the rows of C. From the reduced basis we construct another (non-modular) equation which has x_0 as a root.

We fix the reduction parameter in the LLL algorithm at its standard value $\alpha = \frac{3}{4}$. We write $\mathbf{x}_1, \mathbf{x}_2, \ldots, \mathbf{x}_n$ for an LLL-reduced basis of the lattice L in $\mathbb{R}^n$, and $\mathbf{x}_1^*, \mathbf{x}_2^*, \ldots, \mathbf{x}_n^*$ for its Gram-Schmidt orthogonalization. We know that the determinant $\det(L)$ of the lattice L can be expressed as

$$\det(L)^2 = \prod_{i=1}^{n} |\mathbf{x}_i^*|^2 = |\mathbf{x}_1^*|^2 |\mathbf{x}_2^*|^2 \cdots |\mathbf{x}_n^*|^2,$$

and that the output of the LLL algorithm satisfies

$$|\mathbf{x}_i^*|^2 \leq 2|\mathbf{x}_{i+1}^*|^2 \quad (i = 1, 2, \ldots, n-1).$$

From these two results (and induction on i) it follows that

$$\det(L)^2 \leq \left(2^{n-1}|\mathbf{x}_n^*|^2\right) \cdots \left(2^2|\mathbf{x}_n^*|^2\right)\left(2|\mathbf{x}_n^*|^2\right) = 2^{n(n-1)/2}|\mathbf{x}_n^*|^{2n}.$$

Rearranging this inequality and taking the $2n$-th root gives

$$|\mathbf{x}_n^*| \geq 2^{-(n-1)/4} \det(L)^{1/n}. \tag{8.3}$$

This is a lower bound on the length of the last vector in an LLL-reduced basis.

Now consider an arbitrary vector $\mathbf{v}$ in the lattice L:

$$\mathbf{v} = \sum_{i=1}^{n} v_i \mathbf{x}_i = v_1 \mathbf{x}_1 + v_2 \mathbf{x}_2 + \cdots + v_n \mathbf{x}_n \quad (v_1, v_2, \ldots, v_n \in \mathbb{Z}).$$

Each basis vector of the lattice can be written in terms the Gram-Schmidt orthogonalization of the basis:

$$\mathbf{x}_i = \mathbf{x}_i^* + \sum_{j=1}^{i-1} \mu_{ij} \mathbf{x}_j^*.$$

From this we see that if we express $\mathbf{v}$ as a linear combination of the orthogonal basis vectors $\mathbf{x}_i^*$ then the coefficient of $\mathbf{x}_n^*$ is simply v_n, and hence

$$|\mathbf{v}| \geq |v_n| \, |\mathbf{x}_n^*|. \tag{8.4}$$

Lemma 8.1. Coppersmith's Lemma. (Coppersmith [32], page 236) *Let $L \subset \mathbb{R}^n$ be a lattice of dimension n with determinant $\det(L)$, and let $\mathbf{x}_1, \mathbf{x}_2, \ldots, \mathbf{x}_n$ be an LLL-reduced basis of L (with $\alpha = \frac{3}{4}$). If an element $\mathbf{v} \in L$ satisfies*

$$|\mathbf{v}| < 2^{-(n-1)/4} \det(L)^{1/n},$$

then $\mathbf{v}$ belongs to the hyperplane spanned by $\mathbf{x}_1, \mathbf{x}_2, \ldots, \mathbf{x}_{n-1}$.

Proof. Using inequality (8.3), the hypothesis of the Lemma implies $|\mathbf{v}| < |\mathbf{x}_n^*|$. Using equation (8.4), this implies $v_n = 0$. □

8.2 Construction of the matrix

We start with the polynomial $p(x)$ of equation (8.1). In addition to the degree d of the polynomial, we use an auxiliary integer $h \geq 1$ to be determined later.

Definition 8.2. We introduce a family of dh polynomials $q_{ij}(x)$ as follows:

$$q_{ij}(x) = x^i\, p(x)^j \quad (\, i = 0, 1, \ldots, d-1; \; j = 1, 2, \ldots, h-1 \,).$$

Congruence (8.2) is equivalent to

$$p(x_0) = y_0 M \quad (y_0 \in \mathbb{Z}), \tag{8.5}$$

and hence

$$q_{ij}(x_0) \equiv 0 \pmod{M^j}.$$

We use the coefficients of the polynomials $q_{ij}(x)$ to construct a matrix C. The entries of this matrix depend on an approximation factor $\epsilon > 0$, and the quantity

$$X = \frac{1}{2} M^{(1/d)-\epsilon}, \tag{8.6}$$

which will be an upper bound for the absolute values $|x_0|$ of the solutions $x_0 \in \mathbb{Z}$ of congruence (8.2). For notational convenience, we write

$$t = \frac{1}{\sqrt{dh}}.$$

We will also use the symbols X and t to denote rational approximations of the exact values of these quantities; this slight abuse of notation needs to be kept in mind but should not cause any confusion.

Definition 8.3. The **Coppersmith matrix** C is upper triangular with $d(2h-1)$ rows and columns, and has the block structure given in Figure 8.1:

- The upper left block of size $dh \times dh$ is a diagonal matrix containing (rational approximations to) the quantities t/X^{k-1} for $k = 1, 2, \ldots, dh$.

- The lower right block of size $d(h-1) \times d(h-1)$ is a diagonal matrix containing the entries M^j for $j = 1, 2, \ldots, h-1$ where each entry occurs d times (in Figure 8.1, I_d denotes the $d \times d$ identity matrix).

- The upper right block of size $dh \times d(h-1)$ contains the coefficients of the polynomials $q_{ij}(x)$ as follows: The coefficient of x^{k-1} in $q_{ij}(x)$ is the entry in row k and column $d(h+j-1)+i+1$ of the matrix C. That is, column $d(j-1)+i+1$ of the upper right block contains the coefficients of $q_{ij}(x)$.

- The lower left block of size $d(h-1) \times dh$ is zero.

$$
\left[
\begin{array}{ccccc|c}
t & 0 & \cdots & 0 & 0 & \\
0 & t/X & \cdots & 0 & 0 & \text{coefficient of } x^{k-1} \text{ in } q_{ij}(x) \text{ is in} \\
\vdots & \vdots & \ddots & \vdots & \vdots & \\
0 & 0 & \cdots & t/X^{dh-2} & 0 & \text{row } k, \text{ column } d(h+j-1)+i+1 \\
0 & 0 & \cdots & 0 & t/X^{dh-1} & \\
\hline
& & & & & \begin{array}{ccccc}
MI_d & O & \cdots & O & O \\
O & M^2 I_d & \cdots & O & O \\
\vdots & \vdots & \ddots & \vdots & \vdots \\
O & O & \cdots & M^{h-2}I_d & O \\
O & O & \cdots & O & M^{h-1}I_d
\end{array} \\
& & O & & &
\end{array}
\right]
$$

FIGURE 8.1
Block structure of the Coppersmith matrix

We write L for the lattice in $\mathbb{Q}^{d(2h-1)}$ spanned by the rows of C.

The structure of the Coppersmith matrix will be made clearer by considering a few small examples; see Figure 8.2.

Definition 8.4. We consider the row vector $\mathbf{r}$ of dimension $d(2h-1)$ in which the first dh components are increasing powers of x_0 and the last $d(h-1)$ components are the negatives of products of powers of the integers x_0 and y_0 from equation (8.5):

$\mathbf{r} =$
$$
[1, x_0, x_0^2, \ldots, x_0^{dh-1}, \ -y_0, -y_0 x_0, \ldots, -y_0 x_0^{d-1}, \ldots, -y_0^{h-1}, \ldots, -y_0^{h-1} x_0^{d-1}].
$$

We also consider the vector-matrix product $\mathbf{s} = \mathbf{r}C$. Since every component $\mathbf{r}$ is an integer, the vector $\mathbf{s}$ belongs to the Coppersmith lattice L.

Lemma 8.5. *We have* $|\mathbf{s}| < 1$.

Proof. The first dh components of $\mathbf{s}$ are

$$
t, \quad t\left(\frac{x_0}{X}\right), \quad t\left(\frac{x_0}{X}\right)^2, \quad \ldots, \quad t\left(\frac{x_0}{X}\right)^{dh-1}.
$$

The last $d(h-1)$ components of $\mathbf{s}$ are all zero, by equation (8.5):

$$
q_{ij}(x_0) - x_0^i y_0^j M^j = x_0^i p(x_0)^j - x_0^i y_0^j M^j = x_0^i (y_0 M)^j - x_0^i y_0^j M^j = 0.
$$

Hence the Euclidean length of $\mathbf{s}$ satisfies

$$
|\mathbf{s}|^2 = \sum_{k=1}^{dh} \left(t\left(\frac{x_0}{X}\right)^{k-1} \right)^2.
$$

For $d = 3$ and $h = 2$ we write $p(x) = x^3 + ax^2 + bx + c$ and obtain

$$
\begin{bmatrix}
t & 0 & 0 & 0 & 0 & 0 & c & 0 & 0 \\
0 & t/X & 0 & 0 & 0 & 0 & b & c & 0 \\
0 & 0 & t/X^2 & 0 & 0 & 0 & a & b & c \\
0 & 0 & 0 & t/X^3 & 0 & 0 & 1 & a & b \\
0 & 0 & 0 & 0 & t/X^4 & 0 & 0 & 1 & a \\
0 & 0 & 0 & 0 & 0 & t/X^5 & 0 & 0 & 1 \\
0 & 0 & 0 & 0 & 0 & 0 & M & 0 & 0 \\
0 & 0 & 0 & 0 & 0 & 0 & 0 & M & 0 \\
0 & 0 & 0 & 0 & 0 & 0 & 0 & 0 & M
\end{bmatrix}
$$

For $d = 2$ and $h = 3$ we write $p(x) = x^2 + ax + b$ and obtain

$$
\begin{bmatrix}
t & 0 & 0 & 0 & 0 & 0 & b & 0 & b^2 & 0 \\
0 & t/X & 0 & 0 & 0 & 0 & a & b & 2ab & b^2 \\
0 & 0 & t/X^2 & 0 & 0 & 0 & 1 & a & a^2 + 2b & 2ab \\
0 & 0 & 0 & t/X^3 & 0 & 0 & 0 & 1 & 2a & a^2 + 2b \\
0 & 0 & 0 & 0 & t/X^4 & 0 & 0 & 0 & 1 & 2a \\
0 & 0 & 0 & 0 & 0 & t/X^5 & 0 & 0 & 0 & 1 \\
0 & 0 & 0 & 0 & 0 & 0 & M & 0 & 0 & 0 \\
0 & 0 & 0 & 0 & 0 & 0 & 0 & M & 0 & 0 \\
0 & 0 & 0 & 0 & 0 & 0 & 0 & 0 & M^2 & 0 \\
0 & 0 & 0 & 0 & 0 & 0 & 0 & 0 & 0 & M^2
\end{bmatrix}
$$

For $d = 3$ and $h = 3$ the upper right block of the matrix C is

$$
\begin{bmatrix}
c & 0 & 0 & c^2 & 0 & 0 \\
b & c & 0 & 2bc & c^2 & 0 \\
a & b & c & 2ac + b^2 & 2bc & c^2 \\
1 & a & b & 2ab + 2c & 2ac + b^2 & 2bc \\
0 & 1 & a & a^2 + 2b & 2ab + 2c & 2ac + b^2 \\
0 & 0 & 1 & 2a & a^2 + 2b & 2ab + 2c \\
0 & 0 & 0 & 1 & 2a & a^2 + 2b \\
0 & 0 & 0 & 0 & 1 & 2a \\
0 & 0 & 0 & 0 & 0 & 1
\end{bmatrix}
$$

FIGURE 8.2
Small examples of the Coppersmith matrix

$$\begin{bmatrix} t & & & & * & & \\ & \ddots & & & 1 & \ddots & \\ & & \ddots & & & \ddots & * \\ & & & t/X^{dh-1} & & & 1 \\ \hline & & & & M & & \\ & O & & & & \ddots & \\ & & & & & & M^{h-1} \end{bmatrix} \longmapsto \begin{bmatrix} & C_0 & & & O & \\ \hline & & & 1 & & \\ & * & & & \ddots & \\ & & & & & 1 \end{bmatrix}$$

$$C \longmapsto C'$$

FIGURE 8.3
Row operations transforming C to C'

By the definition of X we have $|x_0| < X$ and hence

$$|\mathbf{s}|^2 < \sum_{k=1}^{dh} t^2 = dh t^2 = 1,$$

since $t = 1/\sqrt{dh}$.   □

Since $p(x)$ is a monic polynomial, so is $q_{ij}(x)$ for all i and j. It follows that the bottom $d(h-1)$ rows of the upper right block of C form an upper triangular matrix with every diagonal entry equal to 1. Therefore we can perform elementary row operations on C and obtain a new matrix C' in which the upper right $dh \times d(h-1)$ block is the zero matrix and the lower right $d(h-1) \times d(h-1)$ block is the identity matrix; see Figure 8.3. We write C_0 for the upper left $dh \times dh$ block of C'. (We usually multiply C_0 by the least common multiple of the denominators of its entries, in order to obtain a matrix of integers.)

Definition 8.6. The **Coppersmith lattice** L_0 is the lattice in $\mathbb{Q}^{dh}$ spanned by the rows of C_0. Since the last $d(h-1)$ components of the vector $\mathbf{s}$ are zero, it follows that $\mathbf{s}$ belongs to L_0; in fact, $\mathbf{s}$ is a short nonzero element of L_0.

Example 8.7. For $d = 3$ and $h = 2$, Figure 8.4 shows the Coppersmith matric C, the reduced matrix C', and the matrix C_0 whose rows (after clearing the denominators) span the Coppersmith lattice.

$$C = \left[\begin{array}{cccc|cc} t & 0 & 0 & 0 & b & 0 \\ 0 & t/X & 0 & 0 & a & b \\ 0 & 0 & t/X^2 & 0 & 1 & a \\ 0 & 0 & 0 & t/X^3 & 0 & 1 \\ \hline 0 & 0 & 0 & 0 & M & 0 \\ 0 & 0 & 0 & 0 & 0 & M \end{array}\right]$$

$$C' = \left[\begin{array}{cccc|cc} t & 0 & -bt/X^2 & abt/X^3 & 0 & 0 \\ 0 & t/X & -at/X^2 & (a^2-b)t/X^3 & 0 & 0 \\ 0 & 0 & -tM/X^2 & atM/X^3 & 0 & 0 \\ 0 & 0 & 0 & -tM/X^3 & 0 & 0 \\ \hline 0 & 0 & t/X^2 & -at/X^3 & 1 & 0 \\ 0 & 0 & 0 & t/X^3 & 0 & 1 \end{array}\right]$$

$$C_0 = \left[\begin{array}{cccc} t & 0 & -bt/X^2 & abt/X^3 \\ 0 & t/X & -at/X^2 & (a^2-b)t/X^3 \\ 0 & 0 & -tM/X^2 & atM/X^3 \\ 0 & 0 & 0 & -tM/X^3 \end{array}\right]$$

FIGURE 8.4

Coppersmith matrices for $d = 3$ and $h = 2$

8.3 Determinant of the lattice

The elimination process which starts with C and ends with C' requires only two types of elementary row operation: interchanging two rows, and adding a multiple of one row to another row; it does not require multiplying a row by a nonzero scalar. It follows that the determinant of the matrix can change by at most a sign; that is, $\det(C') = \pm \det(C)$.

Lemma 8.8. *The lattice L_0 has the same determinant as the lattice L, namely*

$$\det(L_0) = \det(L) = \left(t\, X^{-(dh-1)/2} M^{(h-1)/2} \right)^{dh}.$$

Proof. The block structure of C' in Figure 8.3 shows that

$$\det(C_0) = \det(C_0) \cdot \det(I_{d(h-1)}) = \det C' = \pm \det C.$$

It follows that

$$\det(L_0) = |\det(C_0)| = |\det(C)| = \det(L).$$

Since C is upper triangular, its determinant is the product of its diagonal entries:

$$\det(C) = \prod_{k=1}^{dh} \frac{t}{X^{k-1}} \prod_{j=0}^{h-1} (M^j)^d = t^{dh} X^{-(dh-1)dh/2} M^{d(h-1)h/2}.$$

This quantity is clearly positive.   □

The lattice L_0 has dimension dh, and so the inequality of Lemma 8.1 says

$$|\mathbf{v}| < 2^{-(dh-1)/4} \det(L_0)^{1/dh}.$$

Using Lemma 8.8 we obtain

$$|\mathbf{v}| < 2^{-(dh-1)/4} t X^{-(dh-1)/2} M^{(h-1)/2}.$$

Lemma 8.5 gives $|\mathbf{s}| < 1$, and so the hypothesis of Lemma 8.1 applies to $\mathbf{s}$ if

$$1 \le 2^{-(dh-1)/4} t X^{-(dh-1)/2} M^{(h-1)/2}.$$

This inequality is equivalent to

$$X^{(dh-1)/2} \le 2^{-(dh-1)/4} t M^{(h-1)/2},$$

which gives the condition

$$X \le 2^{-1/2} t^{2/(dh-1)} M^{(h-1)/(dh-1)}.$$

Recalling that $t = 1/\sqrt{dh}$, we obtain

$$X \le 2^{-1/2} (dh)^{-1/(dh-1)} \cdot M^{(h-1)/(dh-1)}. \tag{8.7}$$

We use this inequality to determine the auxiliary integer $h \ge 1$; recall that h is the upper bound in Definition 8.2 for the powers of $p(x_0)$ in $q_{ij}(x)$.

We consider separately the two factors on the right side of inequality (8.7). We first want to ensure that

$$2^{-1/2} (dh)^{-1/(dh-1)} \ge \frac{1}{2}, \tag{8.8}$$

which is equivalent to

$$(dh)^{1/(dh-1)} \le \sqrt{2} \approx 1.414.$$

Consider the following table of approximate values of $n^{1/(n-1)}$:

n	2	3	4	5	6	7	8	9	10
$n^{1/(n-1)}$	2	1.732	1.587	1.495	1.431	1.383	1.346	1.316	1.292

Lemma 8.9. *If $n \geq 7$ then $n^{1/(n-1)} < \sqrt{2}$, and conversely for integers $n \geq 2$.*

Proof. Exercise 8.7. □

It follows that condition (8.8) holds if and only if $dh \geq 7$, or equivalently

$$h \geq \frac{7}{d}.$$

Referring to equation (8.6), we also want to ensure that

$$M^{(h-1)/(dh-1)} \geq M^{(1/d)-\epsilon}, \tag{8.9}$$

which (taking logarithms to the base M) is equivalent to

$$\frac{h-1}{dh-1} \geq \frac{1}{d} - \epsilon.$$

Lemma 8.10. *Inequality (8.9) holds if and only if*

$$h \geq \frac{d(\epsilon+1)-1}{d^2\epsilon}.$$

Proof. Exercise 8.8. □

Inequalities (8.8) and (8.9) both hold if and only if we define h as follows.

Definition 8.11. We choose an integer h such that

$$h \geq \max\left(\frac{7}{d}, \frac{d(\epsilon+1)-1}{d^2\epsilon}\right).$$

To give a rough idea of the behavior of the quantity on the right side, we include a short table of its ceiling (the smallest integer larger than the quantity):

$\epsilon\backslash d$	1	2	3	4	5	6	7	8	9	10
0.1	7	4	3	3	2	2	2	2	2	1
0.01	7	26	23	19	17	15	13	12	10	10
0.001	7	251	223	188	161	140	123	110	99	91
0.0001	7	2501	2223	1876	1601	1390	1225	1094	988	901

To summarize, we have the following situation. We start with a polynomial $p(x)$ of degree d and a modulus M. We choose an approximation factor $\epsilon > 0$, and we then choose the auxiliary integer h according to Definition 8.11. We construct the $d(2h-1) \times d(2h-1)$ matrix C according to Definition 8.3. We perform unimodular elementary row operations on C to obtain the matrix C', and extract its upper left $dh \times dh$ block C_0 (and clear the denominators). The rows of C_0 span a dh-dimensional lattice L_0 in $\mathbb{Q}^{dh}$, and this lattice contains a certain nonzero vector $\mathbf{s}$. If we then choose the upper bound X for the (absolute values of the) solutions x_0 of the congruence $p(x_0) \equiv 0 \pmod{M}$ so that

$$X \leq \frac{1}{2} M^{1/d - \epsilon},$$

then we will have

$$|\mathbf{s}| < 1 \leq 2^{-(dh-1)/4} \det(L_0)^{1/dh},$$

according to Lemma 8.1.

8.4   Application of the LLL algorithm

We now apply the LLL algorithm (still assuming $\alpha = \frac{3}{4}$) to the dh-dimensional Coppersmith lattice $L_0 \subset \mathbb{Q}^{dh}$ spanned by the rows of the $dh \times dh$ matrix C_0. We obtain an LLL-reduced basis of L_0 which we denote by $\mathbf{x}_1, \mathbf{x}_2, \ldots, \mathbf{x}_{dh}$. Let $\mathbf{s}_0$ be the vector obtained from the first dh components of the vector $\mathbf{s} = \mathbf{r}C$ of Definition 8.4; recall that the remaining $d(h-1)$ components of $\mathbf{s}$ are zero. By the previous sections, we know that $\mathbf{s}_0$ belongs to L_0. In fact, $\mathbf{s}_0$ is a short nonzero element of L_0 which belongs to the hyperplane spanned by the first $dh-1$ basis vectors $\mathbf{x}_1, \mathbf{x}_2, \ldots, \mathbf{x}_{dh-1}$. This is the hyperplane orthogonal to the Gram-Schmidt vector $\mathbf{x}_{dh}^*$; we can easily compute the components of $\mathbf{x}_{dh}^*$, say

$$\mathbf{x}_{dh}^* = [\, \nu_0, \; \nu_1, \; \ldots, \; \nu_{dh-1} \,].$$

The same components give an equation for the hyperplane containing $\mathbf{s}_0$. From the proof of Lemma 8.5 we know that the components of $\mathbf{s}_0$ are

$$t, \quad t\left(\frac{x_0}{X}\right), \quad t\left(\frac{x_0}{X}\right)^2, \quad \ldots, \quad t\left(\frac{x_0}{X}\right)^{dh-1},$$

and so we obtain the equation

$$\sum_{k=0}^{dh-1} \nu_k \, t\left(\frac{x_0}{X}\right)^k = 0 \qquad (\nu_0, \nu_1, \ldots, \nu_{dh-1} \in \mathbb{Q}).$$

This is a polynomial with rational coefficients and the root x_0, the desired integral solution of the original congruence (8.2). We clear denominators and cancel common factors to obtain a primitive polynomial with integral coefficients which has x_0 as a root:

$$\sum_{k=0}^{dh-1} n_k x_0^k = 0; \qquad n_0, n_1, \ldots, n_{dh-1} \in \mathbb{Z}; \qquad \gcd(n_0, n_1, \ldots, n_{dh-1}) = 1.$$

We have thus reduced our original problem to the well-understood problem of solving a polynomial equation in one variable with integer coefficients. (See for example the discussion of polynomial division over principal ideal domains in Knuth [78], §4.6.1.) We have proved the main result of this chapter.

Theorem 8.12. Coppersmith's Theorem. (Coppersmith [32], page 243)
*Let $p(x)$ be a monic polynomial of degree d with integer coefficients, and let M
be a given modulus of unknown factorization. Let X be a known upper bound
on the absolute value of the desired solution x_0 to the modular equation*

$$p(x_0) \equiv 0 \pmod{M}.$$

If $\epsilon > 0$ is an approximation factor, and if

$$X < \frac{1}{2} M^{1/d - \epsilon},$$

then we can determine x_0 in time polynomial in $(\log M, d, 1/\epsilon)$.

Corollary 8.13. *We make the same assumptions as in Theorem 8.12, but
without the approximation factor $\epsilon > 0$. If*

$$X < M^{1/d},$$

then we can determine x_0 in time polynomial in $(\log M, 2^d)$.

Proof. Coppersmith [32], Corollary 1, page 243. □

Example 8.14. We present the details of an example from Howgrave-Graham
[66], §4. We take $d = 2$ and consider the quadratic polynomial

$$p(x) = x^2 + 14x + 19.$$

We take $M = 35$ as the modulus, and to simplify the computations we set
$t = 1$ and $X = 2$. With these values we obtain the following Coppersmith
matrix C:

$$C = \left[\begin{array}{cccccc|cccc}
1 & 0 & 0 & 0 & 0 & 0 & 19 & 0 & 361 & 0 \\
0 & 1/2 & 0 & 0 & 0 & 0 & 14 & 19 & 532 & 361 \\
0 & 0 & 1/4 & 0 & 0 & 0 & 1 & 14 & 234 & 532 \\
0 & 0 & 0 & 1/8 & 0 & 0 & 0 & 1 & 28 & 234 \\
0 & 0 & 0 & 0 & 1/16 & 0 & 0 & 0 & 1 & 28 \\
0 & 0 & 0 & 0 & 0 & 1/32 & 0 & 0 & 0 & 1 \\
\hline
0 & 0 & 0 & 0 & 0 & 0 & 35 & 0 & 0 & 0 \\
0 & 0 & 0 & 0 & 0 & 0 & 0 & 35 & 0 & 0 \\
0 & 0 & 0 & 0 & 0 & 0 & 0 & 0 & 1225 & 0 \\
0 & 0 & 0 & 0 & 0 & 0 & 0 & 0 & 0 & 1225
\end{array}\right]$$

A sequence of unimodular elementary row operations produces the matrix C':

$$C' = \left[\begin{array}{cccccc|cccc}
1 & 0 & -19/4 & 133/4 & -3363/16 & 10507/8 & 0 & 0 & 0 & 0 \\
0 & 1/2 & -7/2 & 177/8 & -553/4 & 27605/32 & 0 & 0 & 0 & 0 \\
0 & 0 & -35/4 & 245/4 & -2765/8 & 3675/2 & 0 & 0 & 0 & 0 \\
0 & 0 & 0 & -35/8 & 245/4 & -9625/16 & 0 & 0 & 0 & 0 \\
0 & 0 & 0 & 0 & -1225/16 & 8575/8 & 0 & 0 & 0 & 0 \\
0 & 0 & 0 & 0 & 0 & -1225/32 & 0 & 0 & 0 & 0 \\
\hline
0 & 0 & 1/4 & -7/4 & 79/8 & -105/2 & 1 & 0 & 0 & 0 \\
0 & 0 & 0 & 1/8 & -7/4 & 275/16 & 0 & 1 & 0 & 0 \\
0 & 0 & 0 & 0 & 1/16 & -7/8 & 0 & 0 & 1 & 0 \\
0 & 0 & 0 & 0 & 0 & 1/32 & 0 & 0 & 0 & 1
\end{array}\right]$$

We clear denominators in the upper left block and obtain the matrix C_0:

$$C_0 = \left[\begin{array}{rrrrrr}
32 & 0 & -152 & 1064 & -6726 & 42028 \\
0 & 16 & -112 & 708 & -4424 & 27605 \\
0 & 0 & -280 & 1960 & -11060 & 58800 \\
0 & 0 & 0 & -140 & 1960 & -19250 \\
0 & 0 & 0 & 0 & -2450 & 34300 \\
0 & 0 & 0 & 0 & 0 & -1225
\end{array}\right]$$

We apply the LLL algorithm to the Coppersmith lattice L_0 spanned by the rows of C_0; the Maple command

```
convert( IntegerRelations[LLL](convert(C0,listlist)), Matrix ):
```

produces the reduced basis of L_0 given by the rows of this matrix:

$$\left[\begin{array}{rrrrrr}
0 & 160 & 0 & -60 & 0 & -100 \\
-64 & -64 & -88 & 80 & -72 & -51 \\
64 & -48 & 32 & 4 & -180 & 16 \\
128 & -80 & -48 & 16 & 116 & -13 \\
-32 & -96 & -16 & -132 & 90 & -108 \\
-64 & -32 & 248 & 96 & -30 & -141
\end{array}\right]$$

We apply Gram-Schmidt orthogonalization to the rows $\mathbf{x}_1, \mathbf{x}_2, \ldots, \mathbf{x}_6$ of this matrix, and obtain $\mathbf{x}_1^*, \mathbf{x}_2^*, \ldots, \mathbf{x}_6^*$. We retain only the last vector:

$$\mathbf{x}_6^* = \left[-\frac{117600}{17929}, -\frac{627200}{17929}, \frac{3763200}{17929}, \frac{2508800}{17929}, \frac{627200}{17929}, -\frac{2508800}{17929}\right].$$

We multiply $\mathbf{x}_6^*$ by $17929/39200$ to obtain a vector of relatively prime integers:

$$[-3, -16, 96, 64, 16, -64].$$

Since $t = 1$ and $X = 2$ we now form the polynomial

$$-3 - 16\left(\frac{x}{2}\right) + 96\left(\frac{x}{2}\right)^2 + 64\left(\frac{x}{2}\right)^3 + 16\left(\frac{x}{2}\right)^4 - 64\left(\frac{x}{2}\right)^5,$$

which simplifies to the integer polynomial

$$-3 - 8x + 24x^2 + 8x^3 + x^4 - 2x^5.$$

Using a floating-point algorithm to approximate the roots, we obtain

$$3, \quad 0.5, \quad -0.2294, \quad -1.386+1.563\,i, \quad -1.386-1.563\,i.$$

For our purposes, the only meaningful root is $x_0 = 3$, and this satisfies the original congruence:

$$p(3) = 3^2 + 14 \cdot 3 + 19 = 70 \equiv 0 \pmod{35}.$$

8.5 Projects

Project 8.1. Write a report and present a seminar talk on the applications of the LLL algorithm to cryptography. There are many excellent survey papers on this topic; some examples are Coupé et al. [34], Nguyen and Stern [108, 109], May [94], Hoffstein et al. [65], and Gentry [48].

Project 8.2. Write a report and present a seminar talk on algorithms for finding the roots of polynomials in one variable with integral coefficients. Start by reading Section 4.6.1 of Knuth [78].

Project 8.3. Study the paper by Howgrave-Graham [66] which presents an alternative technique for finding small roots of modular polynomial equations in one variable: "It will be proved, via a general result on dual lattices, that these two algorithms [Coppersmith's and Howgrave-Graham's] are in fact equivalent, though the present approach may be preferred for computational efficiency." Implement both algorithms in a computer algebra system.

8.6 Exercises

Exercise 8.1. Generalize Lemma 8.1 to an LLL-reduced lattice basis for an arbitrary reduction parameter $\frac{1}{4} < \alpha < 1$.

Exercise 8.2. Let $L \subset \mathbb{R}^n$ be a lattice of dimension n with determinant $\det(L)$, and let $\mathbf{x}_1, \mathbf{x}_2, \ldots, \mathbf{x}_n$ be an LLL-reduced basis of L. Prove that if an element $\mathbf{y} \in L$ satisfies $|\mathbf{y}| < |\mathbf{x}_j^*|$ for $j = i+1, \ldots, n$ then $\mathbf{y}$ belongs to the subspace spanned by $\mathbf{x}_1, \mathbf{x}_2, \ldots, \mathbf{x}_i$. (Coppersmith [32], Lemma 2, page 236.)

Exercise 8.3. Write the Coppersmith matrix for $d = 3$ and $h = 3$.

Exercise 8.4. Same as Exercise 8.3, but for $d = 3$ and $h = 4$.

Exercise 8.5. Same as Exercise 8.3, but for $d = 4$ and $h = 3$.

Exercise 8.6. Write a computer program to construct the matrix C of Definition 8.3 (see Figure 8.1) for general d and h.

Exercise 8.7. Prove Lemma 8.9.

Exercise 8.8. Prove Lemma 8.10.

Exercise 8.9. Generalize Example 8.7 to the case $d = 3$ and $h = 2$.

Exercise 8.10. Generalize Example 8.7 to the case $d = 2$ and $h = 3$.

Exercise 8.11. Generalize Example 8.7 to the case $d = 3$ and $h = 3$.

Exercise 8.12. For general d and h, write a computer program to transform the matrix C into the matrix C' and then extract the submatrix C_0.

Exercise 8.13. Apply the program of Exercise 8.12 to the Coppersmith matrix of Exercise 8.3.

Exercise 8.14. Apply the program of Exercise 8.12 to the Coppersmith matrix of Exercise 8.4.

Exercise 8.15. Apply the program of Exercise 8.12 to the Coppersmith matrix of Exercise 8.5.

Exercise 8.16. What happens to the computations in Example 8.14 if $X = 3$? $X = 4$? $X = 5$? Why does the algorithm break down for higher values of X?

Exercise 8.17. Redo the computations in Example 8.14 with $t = 1$ replaced by a better rational approximation to $1/\sqrt{dh} = 1/\sqrt{6}$, such as the partial convergents obtained from the continued fraction representation (see Section 9.1):
$$\frac{2}{5}, \quad \frac{5}{12}, \quad \frac{7}{17}, \quad \frac{9}{22}, \quad \cdots$$
Why do we always obtain the same polynomial at the end of the calculation?

Exercise 8.18. Make up your own example of a modular polynomial equation with a small solution similar to Example 8.14. Solve your equation using Coppersmith's algorithm.

9

Diophantine Approximation

CONTENTS

This chapter provides a brief introduction to one application of lattice basis reduction to number theory. We begin by reviewing the classical algorithm for computing the continued fraction expansion of a real number; the partial convergents give good rational approximations to an irrational number. We then consider the more general problem of computing simultaneous rational approximations (all with the same denominator) to a finite set of real numbers; this is where the LLL algorithm can be applied.

9.1 Continued fraction expansions

We first recall the Euclidean algorithm for the GCD of integers $r_0, r_1 \geq 1$:

$$(2) \qquad r_0 = q_2 r_1 + r_2 \qquad (0 < r_2 < r_1)$$
$$(3) \qquad r_1 = q_3 r_2 + r_3 \qquad (0 < r_3 < r_2)$$
$$\vdots$$
$$(i) \qquad r_{i-2} = q_i r_{i-1} + r_i \qquad (0 < r_i < r_{i-1})$$
$$\vdots$$
$$(n) \qquad r_{n-2} = q_n r_{n-1} + r_n \qquad (0 < r_n < r_{n-1})$$
$$(n+1) \qquad r_{n-1} = q_{n+1} r_n \qquad (r_{n+1} = 0)$$

The last nonzero remainder is $r_n = \gcd(r_0, r_1)$. For $i = 2, 3, \ldots, n+1$ we divide equation (i) by r_{i-1} and obtain the following equations:

$$(2)' \qquad \frac{r_0}{r_1} = q_2 + \frac{r_2}{r_1} \qquad \left(0 < \frac{r_2}{r_1} < 1\right)$$

$$(3)' \qquad \frac{r_1}{r_2} = q_3 + \frac{r_3}{r_2} \qquad \left(0 < \frac{r_3}{r_2} < 1\right)$$

$$\vdots$$

$$(i)' \qquad \frac{r_{i-2}}{r_{i-1}} = q_i + \frac{r_i}{r_{i-1}} \qquad \left(0 < \frac{r_i}{r_{i-1}} < 1\right)$$

$$\vdots$$

$$(n)' \qquad \frac{r_{n-2}}{r_{n-1}} = q_n + \frac{r_n}{r_{n-1}} \qquad \left(0 < \frac{r_n}{r_{n-1}} < 1\right)$$

$$(n+1)' \qquad \frac{r_{n-1}}{r_n} = q_{n+1} \qquad (r_{n+1} = 0)$$

Note that for $i = 2, 3, \ldots, n$, the second term on the right side of equation $(i)'$ is the reciprocal of the term on the left side of equation $(i+1)'$.

Definition 9.1. The (finite) **continued fraction expansion** of $r_0/r_1 \in \mathbb{Q}$ is obtained from equations $(2)'$ to $(n+1)'$ and denoted by $[q_1, q_2, \ldots, q_n]$:

$$\frac{r_0}{r_1} = [q_2, q_3, \ldots, q_{n+1}] = q_2 + \cfrac{1}{q_3 + \cfrac{1}{q_4 + \cfrac{1}{\ddots + \cfrac{1}{q_{n+1}}}}}$$

We want to generalize this so that the rational number r_0/r_1 can be replaced by any real number. We introduce auxiliary quantities γ_i ($i = 2, 3, \ldots$) and express the q_i as follows, using the fact that the second term on the right side of equation $(i)'$ is strictly less than 1:

$$\gamma_2 = \frac{r_0}{r_1}, \qquad q_2 = \lfloor \gamma_2 \rfloor, \qquad \gamma_3 = \frac{1}{\gamma_2 - q_2}, \qquad q_3 = \lfloor \gamma_3 \rfloor, \qquad \ldots$$

These equations make sense for any positive real number γ_2.

Definition 9.2. The (infinite) **continued fraction expansion** of $\gamma \in \mathbb{R}$ is given by the sequence of integers $q_2, q_3, \ldots$ and real numbers $\gamma_2, \gamma_3, \ldots$ defined by the following formulas:

$$\gamma_2 = \gamma, \qquad q_i = \lfloor \gamma_i \rfloor, \qquad \gamma_{i+1} = \frac{1}{\gamma_i - q_i} \qquad (i = 2, 3, \ldots)$$

Example 9.3. Applying these formulas to $\gamma = \pi$ for $i = 2, 3, \ldots, 10$ we obtain

$i = 2$	$\gamma_2 \approx 3.1415926535897932385\ldots$	$q_2 = 3$
$i = 3$	$\gamma_3 \approx 7.0625133059310457679\ldots$	$q_3 = 7$
$i = 4$	$\gamma_4 \approx 15.996594406685720373\ldots$	$q_4 = 15$
$i = 5$	$\gamma_5 \approx 1.0034172310133721161\ldots$	$q_5 = 1$

$$i = 6 \qquad \gamma_6 \approx 292.63459101443720781\ldots \qquad q_6 = 292$$
$$i = 7 \qquad \gamma_7 \approx 1.5758180895247281647\ldots \qquad q_7 = 1$$
$$i = 8 \qquad \gamma_8 \approx 1.7366595773769201739\ldots \qquad q_8 = 1$$
$$i = 9 \qquad \gamma_9 \approx 1.3574791270084020524\ldots \qquad q_9 = 1$$
$$i = 10 \qquad \gamma_{10} \approx 2.7973661241947040540\ldots \qquad q_{10} = 2$$

Taking the corresponding finite initial sequences of the continued fraction expansion, we obtain the following rational approximations to π:

[3]	$\dfrac{3}{1}$	3
[3, 7]	$\dfrac{22}{7}$	3.1428571428571428571...
[3, 7, 15]	$\dfrac{333}{106}$	3.14150943396226641509...
[3, 7, 15, 1]	$\dfrac{355}{113}$	3.1415929203539823009...
[3, 7, 15, 1, 292]	$\dfrac{103993}{33102}$	3.1415926530119026041...
[3, 7, 15, 1, 292, 1]	$\dfrac{104348}{33215}$	3.1415926539214210447...
[3, 7, 15, 1, 292, 1, 1]	$\dfrac{208341}{66317}$	3.1415926534674367055...
[3, 7, 15, 1, 292, 1, 1, 1]	$\dfrac{312689}{99532}$	3.1415926536189366234...
[3, 7, 15, 1, 292, 1, 1, 1, 2]	$\dfrac{833719}{265381}$	3.1415926535810777712...

We can use the continued fraction expansion to find good rational approximations to irrational numbers. One obvious but unsatisfactory way to find a rational approximation of a real number is to take the first n digits to the right of the decimal point; for example, for $\gamma = \pi$ and $n = 10$ we obtain

$$\gamma \approx 3.1415926535 = \frac{31415926535}{10000000000} = \frac{6283185307}{2000000000} = \frac{p}{q}.$$

Using this method, the original denominator is $q = 10^n$, and for this the error in the approximation satisfies the weak inequality

$$\left| \gamma - \frac{p}{q} \right| < \frac{1}{q}.$$

We can do much better; we can increase the denominator from q to q^2. The accuracy of a rational approximation p/q to an irrational number γ is then described by an inequality involving a parameter $c \in \mathbb{R}$, $c > 0$:

$$\left| \gamma - \frac{p}{q} \right| \le \frac{1}{cq^2}. \tag{9.1}$$

It can be shown that for every γ, every finite initial sequence $[q_2, q_3, \ldots, q_n]$ of the continued fraction expansion satisfies (9.1) with $c = 1$. Furthermore, for every γ, at least one of the three consecutive approximations

$$[q_2, q_3, \ldots, q_n], \qquad [q_2, q_3, \ldots, q_n, q_{n+1}], \qquad [q_2, q_3, \ldots, q_n, q_{n+1}, q_{n+2}],$$

satisfies (9.1) with $c = \sqrt{5} \approx 2.236$. In 1891, Hurwitz [67] (see also LeVeque [89], Chapter 9) showed that this is the best possible result: for any $c > \sqrt{5}$, there exists a real number γ which has only finitely many approximations that satisfy (9.1).

9.2 Simultaneous Diophantine approximation

Suppose we are given n distinct real numbers $x_1, x_2, \ldots, x_n$ for which we want to find rational approximations using the same denominator. That is, given a rational number ϵ for which $0 < \epsilon < 1$, we want to find integers $p_1, p_2, \ldots, p_n$ and an integer q such that

$$\left| x_i - \frac{p_i}{q} \right| < \epsilon \text{ for all } i.$$

In 1842, Dirichlet [39] (see also Cassels [22], §V.10) showed that there are infinitely many approximations satisfying the inequality

$$\left| x_i - \frac{p_i}{q} \right| < \frac{1}{q^{1+\frac{1}{n}}} \text{ for all } i. \tag{9.2}$$

Multiplying both sides by q we obtain the equivalent inequality

$$| qx_i - p_i | < \frac{1}{q^{1/n}} \text{ for all } i.$$

In this section we show how lattice basis reduction can be used to find simultaneous approximations satisfying a weaker inequality following Lenstra, Lenstra and Lovász [88], pages 524–525. Recall that we write α for the reduction parameter in the LLL algorithm, and set $\beta = 4/(4\alpha - 1)$.

We are given a positive integer n and real numbers $x_1, x_2, \ldots, x_n$. For $i = 1, 2, \ldots, n$, let y_i be a rational approximation to x_i; we do not assume that these rational approximations have the same denominator. Consider the lattice

L of dimension $n+1$ spanned by the rows $\mathbf{x}_1, \mathbf{x}_2, \ldots, \mathbf{x}_{n+1}$ of the $(n+1) \times (n+1)$ upper triangular matrix X:

$$X = \begin{bmatrix} \beta^{-n(n+1)/4}\epsilon^{n+1} & y_1 & y_2 & \cdots & y_n \\ 0 & -1 & 0 & \ldots & 0 \\ 0 & 0 & -1 & \ldots & 0 \\ \vdots & \vdots & \vdots & \ddots & \vdots \\ 0 & 0 & 0 & \ldots & -1 \end{bmatrix}$$

(Note that the entry in the upper left corner may not be rational.) Applying the LLL algorithm to the lattice L, we obtain in polynomial time a reduced basis $\mathbf{y}_1, \mathbf{y}_2, \ldots, \mathbf{y}_{n+1}$. In our analysis of the LLL algorithm we proved that

$$|\mathbf{y}_1| \leq \beta^{(n-1)/4}\big[\det(L)\big]^{1/n}.$$

(Note that here we have $n+1$ in place of n.) From this we obtain

$$|\mathbf{y}_1| \leq \beta^{n/4}\big[\det(L)\big]^{1/(n+1)} = \beta^{n/4}\big|\det(X)\big|^{1/(n+1)} = \beta^{n/4}\big(\beta^{-n/4}\epsilon\big) = \epsilon.$$

Since $\mathbf{y}_1 \in L$ there exist integers $q, p_1, p_2, \ldots, p_n$ such that

$$\mathbf{y}_1 = q\mathbf{x}_1 + \sum_{i=1}^{n} p_i\mathbf{x}_{i+1} = \big[\, q\beta^{-n(n+1)/4}\epsilon^{n+1}, \ qy_1 - p_1, \ \ldots, \ qy_n - p_n \,\big].$$

If $q = 0$ then at least one $p_i \neq 0$, and so $\mathbf{y}_1 = [0, -p_1, \ldots, -p_n]$ satisfies $|\mathbf{y}_1| \geq 1$, which contradicts $|\mathbf{y}_1| \leq \epsilon < 1$. Hence $q \neq 0$, and if $q < 0$ then we may replace $\mathbf{y}_1$ by $-\mathbf{y}_1$ to obtain $q > 0$. From the first component of $\mathbf{y}_1$ we obtain

$$q\beta^{-n(n+1)/4}\epsilon^{n+1} \leq |\mathbf{y}_1| < \epsilon,$$

and therefore

$$q < \beta^{n(n+1)/4}\epsilon^n.$$

From the other components of $\mathbf{y}_1$ we obtain

$$\big|qy_i - p_i\big| \leq |\mathbf{y}_1| < \epsilon \quad (1 \leq i \leq n).$$

This proves the following result from Lenstra, Lenstra and Lovász [88], Proposition 1.39, page 525.

Proposition 9.4. *There exists a polynomial-time algorithm that, on input consisting of a positive integer n and rational numbers $y_1, y_2, \ldots, y_n, \epsilon$ with $0 < \epsilon < 1$, computes integers $p_1, p_2, \ldots, p_n, q$ such that*

$$\left| y_i - \frac{p_i}{q} \right| < \frac{\epsilon}{q}, \qquad 0 \leq q < \beta^{n(n+1)/4}\epsilon^{-n}.$$

Example 9.5. Using the method of continued fractions to find rational approximations for the square roots of the first four prime numbers, we obtain

$$\sqrt{2} \approx \frac{1}{1}, \frac{3}{2}, \frac{7}{5}, \frac{17}{12}, \frac{41}{29}, \frac{99}{70}, \frac{239}{169}, \frac{577}{408}, \frac{1393}{985}, \cdots$$

$$\sqrt{3} \approx \frac{1}{1}, \frac{2}{1}, \frac{5}{3}, \frac{7}{4}, \frac{19}{11}, \frac{26}{15}, \frac{71}{41}, \frac{97}{56}, \frac{265}{153}, \cdots$$

$$\sqrt{5} \approx \frac{2}{1}, \frac{9}{4}, \frac{38}{17}, \frac{161}{72}, \frac{682}{305}, \frac{2889}{1292}, \frac{12238}{5473}, \frac{51841}{23184}, \frac{219602}{98209}, \cdots$$

$$\sqrt{7} \approx \frac{2}{1}, \frac{3}{1}, \frac{5}{2}, \frac{8}{3}, \frac{37}{14}, \frac{45}{17}, \frac{82}{31}, \frac{127}{48}, \frac{590}{223}, \cdots$$

Note that the only denominator that appears in every list of the first 9 approximations is the trivial denominator 1.

We now consider simultaneous approximation of these four square roots. We start with the first rational approximation in each list in which the denominator has three digits:

$$\sqrt{2} \approx \frac{239}{169}, \qquad \sqrt{3} \approx \frac{265}{153}, \qquad \sqrt{5} \approx \frac{682}{305}, \qquad \sqrt{7} \approx \frac{590}{223}.$$

The least common multiple of the denominators of these approximations is

$$1758663855 = 3^2 \cdot 5 \cdot 13^2 \cdot 17 \cdot 61 \cdot 223.$$

It is trivial to obtain a simultaneous approximation with this LCM as the denominator, but we want a much smaller denominator. We use lattice basis reduction with parameter $\alpha = 3/4$, so that $\beta = 2$.

For the first attempt, we choose $\epsilon = 1/10$, and obtain the matrix

$$X = \begin{bmatrix} \dfrac{1}{3200000} & \dfrac{239}{169} & \dfrac{265}{153} & \dfrac{682}{305} & \dfrac{590}{223} \\ 0 & -1 & 0 & 0 & 0 \\ 0 & 0 & -1 & 0 & 0 \\ 0 & 0 & 0 & -1 & 0 \\ 0 & 0 & 0 & 0 & -1 \end{bmatrix}$$

Applying the LLL algorithm to the rows of the matrix X, we obtain the reduced basis in the rows of the matrix Y:

$$
Y = \begin{bmatrix}
\dfrac{6189}{1600000} & \dfrac{-3}{169} & \dfrac{1}{51} & \dfrac{6}{305} & \dfrac{-7}{223} \\[2mm]
\dfrac{-15411}{800000} & \dfrac{-3}{169} & \dfrac{-1}{51} & \dfrac{-8}{305} & \dfrac{2}{223} \\[2mm]
\dfrac{169}{8000} & 0 & \dfrac{-5}{153} & \dfrac{2}{61} & \dfrac{4}{223} \\[2mm]
\dfrac{-33783}{3200000} & \dfrac{7}{169} & \dfrac{-2}{51} & \dfrac{-1}{305} & \dfrac{-7}{223} \\[2mm]
\dfrac{5129}{100000} & \dfrac{2}{169} & \dfrac{-2}{153} & \dfrac{-9}{305} & 0
\end{bmatrix}
$$

The Euclidean norms of the rows of Y are approximately

$$0.04568236, \quad 0.04288470, \quad 0.05395326, \quad 0.06603817, \quad 0.06174401.$$

The second row is the shortest basis vector. The matrix C with $Y = CX$ is

$$
C = \begin{bmatrix}
12378 & 17505 & 21439 & 27678 & 32749 \\
-61644 & -87177 & -106769 & -137840 & -163094 \\
67600 & 95600 & 117085 & 151158 & 178852 \\
-33783 & -47776 & -58513 & -75541 & -89381 \\
164128 & 232110 & 284274 & 367001 & 434240
\end{bmatrix}
$$

The first row has the smallest first entry, which is the denominator q in the simultaneous approximation; this is much smaller than the LCM of the original denominators. From the first row of C we obtain the following simultaneous approximations with denominator 12378; we have canceled common factors:

$$\sqrt{2} \approx \frac{5835}{4126} \approx 1.414202618, \qquad \sqrt{3} \approx \frac{21439}{12378} \approx 1.732024560,$$

$$\sqrt{5} \approx \frac{4613}{2063} \approx 2.236063984, \qquad \sqrt{7} \approx \frac{32749}{12378} \approx 2.645742446.$$

For the second attempt, we choose $\epsilon = 1/100$, and obtain the matrix

$$
X = \begin{bmatrix}
\dfrac{1}{320000000000} & \dfrac{239}{169} & \dfrac{265}{153} & \dfrac{682}{305} & \dfrac{590}{223} \\[2mm]
0 & -1 & 0 & 0 & 0 \\
0 & 0 & -1 & 0 & 0 \\
0 & 0 & 0 & -1 & 0 \\
0 & 0 & 0 & 0 & -1
\end{bmatrix}
$$

Applying the LLL algorithm to the rows of the matrix X, we obtain the reduced basis in the rows of the matrix Y:

$$Y = \begin{bmatrix} \dfrac{363264993}{320000000000} & 0 & 0 & \dfrac{1}{305} & 0 \\[2ex] \dfrac{74132019}{32000000000} & 0 & 0 & 0 & \dfrac{1}{223} \\[2ex] \dfrac{697699431}{160000000000} & 0 & 0 & \dfrac{-1}{305} & 0 \\[2ex] \dfrac{-147769389}{64000000000} & \dfrac{1}{169} & 0 & 0 & 0 \\[2ex] \dfrac{52874861}{64000000000} & 0 & \dfrac{1}{153} & 0 & 0 \end{bmatrix}$$

The Euclidean norms of the rows of Y are approximately

$$0.003469651, \quad 0.005047350, \quad 0.005455714, \quad 0.006351675, \quad 0.006587956.$$

The first row is the shortest basis vector. The change of basis matrix C is

$$\begin{bmatrix} 363264993 & 513729783 & 629184465 & 812284345 & 961104690 \\ 741320190 & 1048375890 & 1283985950 & 1657640556 & 1961340413 \\ 1395398862 & 1973374722 & 2416867310 & 3120203357 & 3691862460 \\ -738846945 & -1044878224 & -1279702225 & -1652110218 & -1954796850 \\ 264374305 & 373878455 & 457903208 & 591158282 & 699465650 \end{bmatrix}$$

The last row has in fact the smallest first entry, but the first entry in the first row is still smaller than the LCM of the original denominators. From the first row of C we obtain the following simultaneous approximations; only one of these differs from the original approximations:

$$\sqrt{2} \approx \frac{239}{169} \approx 1.414201183, \qquad \sqrt{3} \approx \frac{265}{153} \approx 1.732026144,$$

$$\sqrt{5} \approx \frac{812284345}{363264993} \approx 2.236065574, \qquad \sqrt{7} \approx \frac{590}{223} \approx 2.645739910.$$

For the third attempt, we choose $\epsilon = 1/1000$; but now the first row of C merely gives us back all of the four original approximations.

9.3 Projects

Project 9.1. This chapter has just scratched the surface of the problem of simultaneous Diophantine approximation. Write a report and present a seminar talk on the paper by Lagarias [83] which provides a detailed analysis of the computational complexity of this problem. See also the recent survey by Hanrot [53] on applications of LLL to Diophantine approximation.

Project 9.2. The famous Riemann Hypothesis in analytic number theory is closely related to the Mertens Conjecture, which states that for all real $x > 1$,

$$|M(x)| < \sqrt{x}.$$

The function $M(x)$ is defined by

$$M(x) = \sum_{n \leq x} \mu(n),$$

where $\mu(n)$ is the Möbius function defined by

$$\mu(n) = \begin{cases} 0 & \text{if } n \text{ is divisible by } p^2 \text{ for some prime } p \\ 1 & \text{if } n \text{ is the product of an even number of distinct primes} \\ -1 & \text{if } n \text{ is the product of an odd number of distinct primes} \end{cases}$$

Odlyzko and te Riele [115] used the LLL algorithm to provide a computational disproof of this conjecture. (Truth of the Mertens Conjecture would have implied the Riemann Hypothesis.) Their approach depends on numerical computation to 100 decimal digits of the first 2000 zeros of the Riemann ζ-function with real part $\frac{1}{2}$, combined with simultaneous Diophantine approximation. Write a report and present a seminar talk on this disproof of the Mertens Conjecture. Two further references are te Riele [136] and Pintz [117]. Numerous further applications of the LLL algorithm in number theory are discussed in a recent survey paper by Simon [130].

9.4 Exercises

Exercise 9.1. Use continued fractions to find the first ten rational approximations to the square roots of the first five prime numbers: 2, 3, 5, 7, 11. Then use lattice basis reduction to find simultaneous rational approximations.

Exercise 9.2. Same as Exercise 9.1 for 2, 3, 5, 7, 11, 13.

Exercise 9.3. Same as Exercise 9.1 for 2, 3, 5, 7, 11, 13, 17.

Exercise 9.4. Same as Exercise 9.1 for 2, 3, 5, 7, 11, 13, 17, 19.

Exercise 9.5. Same as Exercises 9.1–9.4 for the cube roots.

Exercise 9.6. Same as Exercises 9.1–9.4 for the binary logarithms (base 2).

Exercise 9.7. Same as Exercises 9.1–9.4 for the natural logarithms (base e).

10

The Fincke-Pohst Algorithm

CONTENTS

In this chapter we study an algorithm for finding not merely one short vector in a lattice, but for enumerating all vectors in a lattice with length less than a given upper bound. In particular, this allows us to determine a shortest nonzero lattice vector. This algorithm was introduced by Fincke and Pohst [43] in 1985; it originated in their earlier work on algebraic number theory [42]. Before the work of Fincke and Pohst, the standard methods for calculating short vectors in a lattice relied on enumerating all vectors in a suitable box. The contribution of Fincke and Pohst was to show that it suffices to consider only those vectors in a suitable ellipsoid of much smaller volume than the box; searching through this ellipsoid is in many cases much more efficient. The two papers of Fincke and Pohst also provide a detailed analysis of the complexity of their algorithm, which will not be discussed in this chapter. Of course, the decision version of the shortest vector problem is known to be NP-complete, and so we should expect that any algorithm to find a shortest vector in a lattice will have exponential time complexity with respect to the dimension.

10.1 The rational Cholesky decomposition

Suppose that $B = (b_{ij})$ is an $n \times m$ matrix $(m \le n)$ with real entries, and suppose that B has full rank; that is, $\text{rank}(B) = m$. We regard the m columns $\mathbf{b}_1, \mathbf{b}_2, \dots, \mathbf{b}_m$ of B as a basis for the lattice L in $\mathbb{R}^n$. (In this chapter we will put the lattice basis vectors into the columns of the matrix, to be consistent with the conventional notation for the Cholesky decomposition.)

Recall that the Gram matrix $G = B^t B$ is the $m \times m$ matrix in which the

(i, j) entry g_{ij} is the scalar product $\mathbf{b}_i \cdot \mathbf{b}_j$. Clearly G is symmetric, since

$$G^t = (B^t B)^t = B^t (B^t)^t = B^t B = G.$$

Furthermore, since B has full rank, we know that G is positive definite: for any column vector $\mathbf{x} \in \mathbb{R}^m$ we have

$$\mathbf{x}^t G \mathbf{x} = \mathbf{x}^t B^t B \mathbf{x} = (\mathbf{x}^t B^t)(B\mathbf{x}) = (B\mathbf{x})^t (B\mathbf{x}),$$

and so $\mathbf{x}^t G \mathbf{x} = \mathbf{0}$ if and only if $B\mathbf{x} = \mathbf{0}$; but $B\mathbf{x} = \mathbf{0}$ if and only if $\mathbf{x} = \mathbf{0}$. Any symmetric positive definite matrix G has a factorization $G = U^t U$ where U is an upper triangular matrix with nonzero entries on the diagonal; this is the **Cholesky decomposition** of G. Standard references for this topic are the textbooks by Golub and van Loan [49] and Trefethen and Bau [137].

We first give an algorithm for computing the Cholesky decomposition of a symmetric positive definite matrix G. This is a slightly modified version of Gaussian elimination: at each step, we use the diagonal entry in the current row to eliminate the entries below it in the same column. The assumptions on the matrix guarantee that the leading nonzero entry in each row will occur along the diagonal. We then write the resulting upper triangular matrix in the form DU where D is a diagonal matrix with positive entries $d_1, d_2, \ldots, d_m$ on the diagonal, and U is an upper triangular matrix with ones on the diagonal. (This algorithm is presented in Figure 10.1.) The output of this algorithm is the **rational Cholesky decomposition**,

$$G = U^t D U.$$

By the word "rational" we mean that if the entries of the input matrix G are rational numbers, then so are the entries of the output matrices D and U. From this we can obtain the ordinary Cholesky decomposition by the equation

$$G = \left(D^{1/2} U \right)^t \left(D^{1/2} U \right).$$

Example 10.1. Consider the lattice L in $\mathbb{R}^3$ with the rows of B^t as a basis:

$$B^t = \begin{bmatrix} 2 & 2 & -1 \\ 3 & -1 & -1 \\ 2 & -2 & -2 \end{bmatrix}$$

The corresponding Gram matrix G is

$$G = B^t B = \begin{bmatrix} 9 & 5 & 2 \\ 5 & 11 & 10 \\ 2 & 10 & 12 \end{bmatrix}$$

Apply these elementary row operations to G, where R_i denotes row i:

$$R_2 \leftarrow R_2 - \frac{5}{9} R_1, \quad R_3 \leftarrow R_3 - \frac{2}{9} R_1, \quad R_3 \leftarrow R_3 - \frac{40}{37} R_2.$$

- *Input:* A positive definite symmetric $m \times m$ matrix G.
- *Output:* A diagonal $m \times m$ matrix D and an upper triangular $m \times m$ matrix U with ones on the diagonal such that $G = U^t D U$.

(1) Create an $m \times m$ matrix U. Set $U \leftarrow G$.

(2) For $j = 1, 2, \ldots, m-1$ do:

    For $i = j+1, \ldots, m$ do:

        Add $-\dfrac{u_{ij}}{u_{jj}}$ times row j of U to row i.

(3) Create an $m \times m$ matrix D.

(4) For $i = 1, 2, \ldots, m$ do:

    For $j = 1, 2, \ldots, m$ do:

        If $i = j$ then set $d_{ij} \leftarrow u_{ij}$ else set $d_{ij} \leftarrow 0$.

(5) For $i = 1, 2, \ldots, m$ do:

    Multiply row i of U by d_{ii}^{-1}.

FIGURE 10.1
Algorithm for the rational Cholesky decomposition

We obtain the following upper triangular matrix and its factorization DU:

$$
\begin{bmatrix} 9 & 5 & 2 \\ 0 & \frac{74}{9} & \frac{80}{9} \\ 0 & 0 & \frac{72}{37} \end{bmatrix} = \begin{bmatrix} 9 & 0 & 0 \\ 0 & \frac{74}{9} & 0 \\ 0 & 0 & \frac{72}{37} \end{bmatrix} \begin{bmatrix} 1 & \frac{5}{9} & \frac{2}{9} \\ 0 & 1 & \frac{40}{37} \\ 0 & 0 & 1 \end{bmatrix}
$$

We now have the rational Cholesky decomposition $G = U^t D U$:

$$
\begin{bmatrix} 9 & 5 & 2 \\ 5 & 11 & 10 \\ 2 & 10 & 12 \end{bmatrix} = \begin{bmatrix} 1 & 0 & 0 \\ \frac{5}{9} & 1 & 0 \\ \frac{2}{9} & \frac{40}{37} & 1 \end{bmatrix} \begin{bmatrix} 9 & 0 & 0 \\ 0 & \frac{74}{9} & 0 \\ 0 & 0 & \frac{72}{37} \end{bmatrix} \begin{bmatrix} 1 & \frac{5}{9} & \frac{2}{9} \\ 0 & 1 & \frac{40}{37} \\ 0 & 0 & 1 \end{bmatrix}
$$

Example 10.2. Consider the lattice in $\mathbb{R}^4$ with the rows of B^t as a basis, and its Gram matrix $G = B^t B$:

$$
B^t = \begin{bmatrix} 1 & 3 & 4 & -4 \\ 2 & 2 & -3 & -3 \\ 1 & -1 & 0 & -4 \\ -1 & 3 & 3 & 2 \end{bmatrix}, \quad G = \begin{bmatrix} 42 & 8 & 14 & 12 \\ 8 & 26 & 12 & -11 \\ 14 & 12 & 18 & -12 \\ 12 & -11 & -12 & 23 \end{bmatrix}.
$$

We have the rational Cholesky decomposition $G = U^t D U$ where

$$D = \begin{bmatrix} 42 & 0 & 0 & 0 \\ 0 & \frac{514}{21} & 0 & 0 \\ 0 & 0 & \frac{2512}{257} & 0 \\ 0 & 0 & 0 & \frac{81}{628} \end{bmatrix}, \qquad U = \begin{bmatrix} 1 & \frac{4}{21} & \frac{1}{3} & \frac{2}{7} \\ 0 & 1 & \frac{98}{257} & -\frac{279}{514} \\ 0 & 0 & 1 & -\frac{1405}{1256} \\ 0 & 0 & 0 & 1 \end{bmatrix}.$$

Example 10.3. In this case, the lattice has dimension 3 but it is contained in a vector space of dimension 4:

$$B^t = \begin{bmatrix} -4 & -3 & -1 & -2 \\ 1 & 4 & 0 & -1 \\ 2 & 3 & -2 & -1 \end{bmatrix}, \qquad G = \begin{bmatrix} 30 & -14 & -13 \\ -14 & 18 & 15 \\ -13 & 15 & 18 \end{bmatrix}.$$

We have the rational Cholesky decomposition $G = U^t D U$ where

$$D = \begin{bmatrix} 30 & 0 & 0 \\ 0 & \frac{172}{15} & 0 \\ 0 & 0 & \frac{465}{86} \end{bmatrix}, \qquad U = \begin{bmatrix} 1 & -\frac{7}{15} & -\frac{13}{30} \\ 0 & 1 & \frac{67}{86} \\ 0 & 0 & 1 \end{bmatrix}.$$

10.2 Diagonalization of quadratic forms

A symmetric $m \times m$ matrix $G = (g_{ij})$ defines a symmetric bilinear form on column vectors $\mathbf{x}, \mathbf{y} \in \mathbb{R}^m$ by the formula

$$(\mathbf{x}, \mathbf{y}) = \mathbf{x}^t G \mathbf{y}.$$

More explicitly, if we write $\mathbf{x} = [x_1, \ldots, x_m]^t$ and $\mathbf{y} = [y_1, \ldots, y_m]^t$ then

$$(\mathbf{x}, \mathbf{y}) = \sum_{i=1}^{m} \sum_{j=1}^{m} g_{ij} x_i y_j.$$

The corresponding quadratic form is then defined by

$$(\mathbf{x}, \mathbf{x}) = \sum_{i=1}^{m} \sum_{j=1}^{m} g_{ij} x_i x_j.$$

If the matrix G is positive definite, then the corresponding quadratic form can be diagonalized; this means that, by a suitable change of basis, the quadratic form can be written as a linear combination of squares,

$$(\mathbf{x}, \mathbf{x}) = \sum_{i=1}^{m} d_i \left(x_i + \sum_{j=i+1}^{m} u_{ij} x_j \right)^2, \tag{10.1}$$

where $d_1, d_2, \ldots, d_m > 0$. This diagonalization process is essentially the same as computing the rational Cholesky decomposition of the matrix G. Indeed, if $G = U^t D U$ is the rational Cholesky decomposition of G then

$$(\mathbf{x}, \mathbf{x}) = \mathbf{x}^t G \mathbf{x} = \mathbf{x}^t U^t D U \mathbf{x} = (U\mathbf{x})^t D (U\mathbf{x}),$$

and this is clearly just the matrix form of equation (10.1).

Example 10.4. Recall the Gram matrix G from Example 10.1:

$$G = \begin{bmatrix} 9 & 5 & 2 \\ 5 & 11 & 10 \\ 2 & 10 & 12 \end{bmatrix}$$

Writing $\mathbf{x} = [x, y, z]^t$, the corresponding positive definite quadratic form is

$$(\mathbf{x}, \mathbf{x}) = 9x^2 + 10xy + 4xz + 11y^2 + 20yz + 12z^2.$$

It is easy to verify that

$$(\mathbf{x}, \mathbf{x}) = 9\left(x + \tfrac{5}{9}y + \tfrac{2}{9}z\right)^2 + \tfrac{74}{9}\left(y + \tfrac{40}{37}z\right)^2 + \tfrac{72}{37}z^2.$$

The coefficients come directly from the matrices D and U of Example 10.1.

Example 10.5. The positive definite quadratic form on $\mathbb{R}^4$ corresponding to the Gram matrix from Example 10.2 is

$$42w^2 + 16wx + 28wy + 24wz + 26x^2 + 24xy - 22xz + 18y^2 - 24yz + 23z^2.$$

It can be verified by direct calculation that this quadratic form can be written as the following linear combination of squares:

$$42\left(w + \tfrac{4}{21}x + \tfrac{1}{3}y + \tfrac{2}{7}z\right)^2 + \tfrac{514}{21}\left(x + \tfrac{98}{257}y - \tfrac{279}{514}z\right)^2 + \tfrac{2512}{257}\left(y - \tfrac{1405}{1256}z\right)^2 + \tfrac{81}{628}z^2.$$

The coefficients come directly from the matrices D and U of Example 10.2.

10.3 The original Fincke-Pohst algorithm

Any vector in the lattice $L \subset \mathbb{R}^n$ is an integral linear combination of the basis vectors $\mathbf{b}_1, \mathbf{b}_2, \ldots, \mathbf{b}_m$ which are the columns of the $n \times m$ matrix B. Hence any lattice vector can be written uniquely in the form

$$B\mathbf{x} = x_1\mathbf{b}_1 + x_2\mathbf{b}_2 + \cdots + x_m\mathbf{b}_m \ \text{ for some } \ \mathbf{x} = [x_1, x_2, \ldots, x_m]^t \in \mathbb{Z}^m.$$

The squared Euclidean length of $B\mathbf{x}$ is given by this quadratic form:

$$|B\mathbf{x}|^2 = (B\mathbf{x})^t(B\mathbf{x}) = \mathbf{x}^t(B^t B)\mathbf{x} = \mathbf{x}^t G\mathbf{x},$$

where G is the Gram matrix of the basis $\mathbf{b}_1, \mathbf{b}_2, \ldots, \mathbf{b}_m$. Computing the rational Cholesky decomposition $G = U^t D U$ corresponds to expressing this quadratic form as a positive linear combination of squares:

$$\mathbf{x}^t G\mathbf{x} = \mathbf{x}^t(U^t D U)\mathbf{x} = (U\mathbf{x})^t D(U\mathbf{x}).$$

Hence the squared Euclidean length of the lattice vector $B\mathbf{x}$ is equal to

$$|B\mathbf{x}|^2 = \sum_{i=1}^{m} d_i\Big(x_i + \sum_{j=i+1}^{m} u_{ij}x_j\Big)^2,$$

where $d_1, d_2, \ldots, d_m > 0$. Suppose we are given an upper bound $C > 0$, and we want to find all lattice vectors $B\mathbf{x} \in L$ satisfying the inequality

$$|B\mathbf{x}|^2 \le C.$$

Equivalently, we want to find all $\mathbf{x} = [x_1, \ldots, x_m]^t \in \mathbb{Z}^m$ for which

$$\sum_{i=1}^{m} d_i\Big(x_i + \sum_{j=i+1}^{m} u_{ij}x_j\Big)^2 \le C. \tag{10.2}$$

We consider the last term ($i = m$) in the summation (10.2) and obtain

$$d_m x_m^2 \le C \quad \text{or equivalently} \quad |x_m| \le \left\lfloor \sqrt{\frac{C}{d_m}} \right\rfloor.$$

It follows that the only possible values for x_m are the integers in the range

$$\left\lceil -\sqrt{\frac{C}{d_m}} \right\rceil \le x_m \le \left\lfloor \sqrt{\frac{C}{d_m}} \right\rfloor.$$

(As usual, we write $\lceil x \rceil$ for the smallest integer $\ge x$, and $\lfloor x \rfloor$ for the largest integer $\le x$.) For each of these values of x_m, we consider the last two terms ($i = m-1$ and $i = m$) in the summation (10.2) and obtain

$$d_{m-1}\big(x_{m-1} + u_{m-1,m}x_m\big)^2 + d_m x_m^2 \le C.$$

It follows that the only possible values for x_{m-1} are the integers in the range

$$\left\lceil -\sqrt{\frac{C - d_m x_m^2}{d_{m-1}}} - u_{m-1,m}x_m \right\rceil \le x_{m-1} \le \left\lfloor \sqrt{\frac{C - d_m x_m^2}{d_{m-1}}} - u_{m-1,m}x_m \right\rfloor.$$

Assuming that the $m-k$ values $x_{k+1}, \ldots, x_m$ have been determined, we consider the last $m-k+1$ terms in the summation (10.2) to obtain a range of possible values for x_k. To simplify the notation, we first define

$$S_k = \sum_{i=k+1}^{m} d_i \left(x_i + \sum_{j=i+1}^{m} u_{ij} x_j \right)^2, \qquad T_k = \sum_{j=k+1}^{m} u_{kj} x_j.$$

We then calculate as follows:

$$\sum_{i=k}^{m} d_i \left(x_i + \sum_{j=i+1}^{m} u_{ij} x_j \right)^2 \leq C$$

$$\implies d_k \left(x_k + \sum_{j=k+1}^{m} u_{kj} x_j \right)^2 + \sum_{i=k+1}^{m} d_i \left(x_i + \sum_{j=i+1}^{m} u_{ij} x_j \right)^2 \leq C$$

$$\implies d_k \left(x_k + T_k \right)^2 + S_k \leq C$$

$$\implies \left(x_k + T_k \right)^2 \leq \frac{C - S_k}{d_k}$$

$$\implies -\sqrt{\frac{C - S_k}{d_k}} \leq x_k + T_k \leq \sqrt{\frac{C - S_k}{d_k}}$$

$$\implies -\sqrt{\frac{C - S_k}{d_k}} - T_k \leq x_k \leq \sqrt{\frac{C - S_k}{d_k}} - T_k$$

$$\implies \left\lceil -\sqrt{\frac{C - S_k}{d_k}} - T_k \right\rceil \leq x_k \leq \left\lfloor \sqrt{\frac{C - S_k}{d_k}} - T_k \right\rfloor.$$

If we define $S_m = 0$ and $T_m = 0$ then we obtain the range of possible values for x_k as k decreases from m down to 1:

$$\left\lceil -\sqrt{\frac{C - S_k}{d_k}} - T_k \right\rceil \leq x_k \leq \left\lfloor \sqrt{\frac{C - S_k}{d_k}} - T_k \right\rfloor. \tag{10.3}$$

This is the essential inequality of the Fincke-Pohst algorithm.

Maple code implementing this algorithm is given in Figures 10.2 (initialization and procedures) and 10.3 (main loop). In addition to the **LinearAlgebra** package, this code uses two procedures: **NORM2** computes the squared length of a vector, and **PRINT** outputs a vector in the specified format. The program performs a specified number of trials; for each trial, a random matrix is generated whose columns contain the basis of the lattice. The Gram matrix and its rational Cholesky decomposition are then computed. The main loop on k then begins: for each k, the quantities S_k and T_k are calculated, and then the lower and upper bounds for x_k. The list of partial results from the previous step is extended by including each possible value of the new coefficient. A nonzero vector is included only if its last nonzero entry is positive. (The zero vector is also included.)

Two examples of the output of this Maple code are given in Figures 10.4

```
M,N := 5,5: # dimension of lattice, dimension of vector space
C    := 80:  # upper bound on square length of lattice vectors

with( LinearAlgebra ):

NORM2 := proc( v ) Norm( convert( v, Vector ), 2 )^2 end:
PRINT := proc( v, flag ) local x:
  printf( "   [" ):
  for x in v do printf( "%3d", x ) od:
  printf( " ]" ):
  if flag then printf( "\n" ) fi
end:
```

FIGURE 10.2
Maple code for the Fincke-Pohst algorithm: part 1

and 10.5. Each line contains an index number, the square length, the lattice vector, and the coordinate vector for the lattice vector with respect to the original lattice basis. The output vectors have been sorted by increasing square length.

Example 10.6. We trace the execution of the Maple code of Figures 10.2 and 10.3 on the lattice in $\mathbb{R}^5$ with basis given by the columns of the matrix

$$B = \begin{bmatrix} 1 & -4 & 9 & -7 & 9 \\ -3 & 2 & 0 & 5 & -5 \\ 9 & -8 & -1 & -4 & -5 \\ 6 & 6 & -3 & -8 & 8 \\ 4 & 3 & -6 & 5 & -8 \end{bmatrix}$$

The Gram matrix for this basis is

$$G = \begin{bmatrix} 143 & -34 & -42 & -86 & -5 \\ -34 & 129 & -64 & 37 & 18 \\ -42 & -64 & 127 & -65 & 110 \\ -86 & 37 & -65 & 179 & -172 \\ -5 & 18 & 110 & -172 & 259 \end{bmatrix}$$

The matrices D and U in the rational Cholesky decomposition are

$$D = \text{diagonal}\left[143, \frac{17291}{143}, \frac{1199889}{17291}, \frac{38978575}{1199889}, \frac{137545984}{38978575} \right]$$

```
for trial to 2 do
  printf("\n"): printf("   trial %d \n", trial):
  B := Matrix( N, M ):
  while Rank(B)<M do B := RandomMatrix(N,M,generator=-9..9) od:
  printf("\n"): printf("  lattice basis \n"): printf("\n"):
  for v in convert(Transpose(B),listlist) do PRINT(v,true) od:
  G := Transpose(B).B:  U := copy(G):
  for j to M-1 do for i from j+1 to M do
    U := RowOperation( U, [i,j], -U[i,j]/U[j,j] )
  od od:
  DD := Matrix( M, M, (i,j) -> if i=j then U[i,i] else 0 fi ):
  U := MatrixInverse(DD) . U:
  resultold := [ [] ]:
  for k from M to 1 by -1 do
    resultnew := []:
    for r in resultold do
      xvalue := [ seq(0,j=1..k), op(r) ]:
      S := add(DD[i,i]*add(U[i,j]*xvalue[j],j=i..M)^2,i=k+1..M):
      T := add(U[k,j]*xvalue[j],j=k+1..M):
      lowerbound := ceil( -sqrt( (C-S)/DD[k,k] ) - T ):
      upperbound := floor( sqrt( (C-S)/DD[k,k] ) - T ):
      for x from lowerbound to upperbound do
        xr := [ x, op(r) ]:
        mm := 0:
        for m to nops(xr) do if xr[m] <> 0 then mm := m fi od:
        if mm = 0 then ok := true else ok := evalb(xr[mm]>0) fi:
        if ok then resultnew := [ op(resultnew), xr ] fi
      od
    od:
    resultold := resultnew
  od:
  printf("\n"): printf("  short vectors \n"): printf("\n"):
  result :=
  [seq([v,convert(B.convert(v,Vector),list)], v in resultnew)]:
  result :=
  sort( result, proc(v1,v2) NORM2(v1[2]) < NORM2(v2[2]) end ):
  for i to nops(result) do
    v,vv := op(result[i]):
    printf( "  %2d:  %3d", i, NORM2(vv) ):
    PRINT( vv, false ): printf( "  from" ): PRINT( v, true )
  od
od:
```

FIGURE 10.3

Maple code for the Fincke-Pohst algorithm: part 2

```
trial 1

lattice basis

[ -6   5 -8 -5 -1 ]
[ -5  -4  1 -3  9 ]
[ -8  -6  6  6  2 ]
[ -4   9  2  0  0 ]
[ -1   4 -7 -7 -8 ]

short vectors

1:     0 [  0  0  0  0  0 ] from [  0  0  0  0  0 ]
2:    58 [  0 -5  2 -5  2 ] from [ -1  1  0  0  1 ]
3:    77 [ -4  4  4 -5  2 ] from [ -1  1  0  1  1 ]
4:    80 [  5 -1  1 -2 -7 ] from [ -1  0  0  0  1 ]
```

FIGURE 10.4
Output of Maple code in Figures 10.2 and 10.3: trial 1

$$U = \begin{bmatrix} 1 & -\dfrac{34}{143} & -\dfrac{42}{143} & -\dfrac{86}{143} & -\dfrac{5}{143} \\[2mm] 0 & 1 & -\dfrac{10580}{17291} & \dfrac{2367}{17291} & \dfrac{2404}{17291} \\[2mm] 0 & 0 & 1 & -\dfrac{1385539}{1199889} & \dfrac{2054480}{1199889} \\[2mm] 0 & 0 & 0 & 1 & -\dfrac{48123482}{38978575} \\[2mm] 0 & 0 & 0 & 0 & 1 \end{bmatrix}$$

We use $C = 80$ as the upper bound on the square lengths of the lattice vectors.
For $k = 5$ we have $S_5 = 0$ and $T_5 = 0$, and we obtain

$$\sqrt{\frac{C}{d_5}} \approx 4.76$$

Hence x_5 must satisfy the inequalities $-4 \leq x_5 \leq 4$. Since we exclude any coordinate vector which has a negative number as its last nonzero component, we assume $0 \leq x_5 \leq 4$. At this point, the list of partial coefficient vectors is

$$[x_5] = [0], \ [1], \ [2], \ [3], \ [4]$$

trial 2

lattice basis

```
[ -2 -5  5 -4 -5 ]
[ -3  8  3  3 -7 ]
[  2 -9  7  7 -8 ]
[  2 -5  2 -9  0 ]
[ -1 -3 -4 -6  5 ]
```

short vectors

```
 1:    0  [  0  0  0  0  0 ]  from  [  0  0  0  0  0 ]
 2:   18  [ -1 -2 -3  2  0 ]  from  [ -1  1  1  0  2 ]
 3:   30  [  4  1 -2  3  0 ]  from  [ -2  1  1  1  1 ]
 4:   36  [ -5 -3 -1 -1  0 ]  from  [  1  0  0 -1  1 ]
 5:   49  [  0  6 -3  0 -2 ]  from  [ -2  2  1  1  2 ]
 6:   55  [ -5  3 -4 -1 -2 ]  from  [ -1  2  1  0  3 ]
 7:   55  [  2 -4  3 -1 -5 ]  from  [ -1  1  1  1  1 ]
 8:   55  [ -4  5 -1 -3 -2 ]  from  [  0  1  0  0  1 ]
 9:   58  [  2  2  0 -1 -7 ]  from  [ -3  3  2  2  3 ]
10:   60  [  1  0 -3  1 -7 ]  from  [ -4  4  3  2  5 ]
11:   60  [  3 -1 -5  5  0 ]  from  [ -3  2  2  1  3 ]
12:   61  [ -1  4 -6  2 -2 ]  from  [ -3  3  2  1  4 ]
13:   63  [  1 -6  0  1 -5 ]  from  [ -2  2  2  1  3 ]
14:   64  [ -3 -1 -1 -2 -7 ]  from  [ -2  3  2  1  4 ]
15:   71  [  1 -1 -4 -7 -2 ]  from  [ -3  3  2  2  4 ]
16:   72  [ -2 -4 -6  4  0 ]  from  [ -2  2  2  0  4 ]
17:   73  [  1  8  0 -2 -2 ]  from  [ -1  1  0  1  0 ]
18:   74  [ -2  1  2 -4 -7 ]  from  [ -1  2  1  1  2 ]
19:   75  [  6 -3  1  2 -5 ]  from  [ -3  2  2  2  2 ]
20:   75  [  4  0 -3 -5  5 ]  from  [ -1  0  0  1  0 ]
21:   78  [ -6 -5 -4  1  0 ]  from  [  0  1  1 -1  3 ]
```

FIGURE 10.5
Output of Maple code in Figures 10.2 and 10.3: trial 2

We repeat the main loop with $k = 4$. For each partial vector we compute

$$S_4 = d_5 x_5^2 = \frac{137545984}{38978575} x_5^2, \qquad T_4 = u_{45} x_5 = -\frac{48123482}{38978575} x_5,$$

and then use inequality (10.3),

$$\left\lceil -\sqrt{\frac{C-S_4}{d_4}} - T_4 \right\rceil \leq x_4 \leq \left\lfloor \sqrt{\frac{C-S_4}{d_4}} - T_4 \right\rfloor, \qquad d_4 = \frac{38978575}{1199889},$$

to obtain the lower and upper bounds for x_4:

$$x_5 = 0: \quad S_4 = 0, \quad \sqrt{\frac{C-S_4}{d_4}} \approx 1.57, \quad T_4 = 0, \qquad -1 \leq x_4 \leq 1$$

$$x_5 = 1: \quad S_4 \approx 3.53, \quad \sqrt{\frac{C-S_4}{d_4}} \approx 1.54, \quad T_4 \approx -1.23, \qquad 0 \leq x_4 \leq 2$$

$$x_5 = 2: \quad S_4 \approx 14.1, \quad \sqrt{\frac{C-S_4}{d_4}} \approx 1.42, \quad T_4 \approx -2.47, \qquad 2 \leq x_4 \leq 3$$

$$x_5 = 3: \quad S_4 \approx 31.8, \quad \sqrt{\frac{C-S_4}{d_4}} \approx 1.22, \quad T_4 \approx -3.70, \qquad 3 \leq x_4 \leq 4$$

$$x_5 = 4: \quad S_4 \approx 56.5, \quad \sqrt{\frac{C-S_4}{d_4}} \approx 0.855, \quad T_4 \approx -4.94, \qquad 5 \leq x_4 \leq 5$$

At this point, the list of partial coefficient vectors is

$$\begin{bmatrix} x_4 \\ x_5 \end{bmatrix} = \begin{bmatrix} 0 \\ 0 \end{bmatrix}, \begin{bmatrix} 1 \\ 0 \end{bmatrix}, \begin{bmatrix} 0 \\ 1 \end{bmatrix}, \begin{bmatrix} 1 \\ 1 \end{bmatrix}, \begin{bmatrix} 2 \\ 1 \end{bmatrix}, \begin{bmatrix} 2 \\ 2 \end{bmatrix}, \begin{bmatrix} 3 \\ 2 \end{bmatrix}, \begin{bmatrix} 3 \\ 3 \end{bmatrix}, \begin{bmatrix} 4 \\ 3 \end{bmatrix}, \begin{bmatrix} 5 \\ 4 \end{bmatrix}$$

We repeat the main loop with $k = 3$. For each partial vector we compute

$$S_3 = d_4(x_4 + u_{45})^2 + d_5 x_5^2, \qquad T_3 = u_{34} x_4 + u_{35} x_5,$$

and then use inequality (10.3) to obtain the lower and upper bounds for x_3. We obtain the following table (in which some values are approximate):

x_4, x_5	S_3	$\sqrt{(C-S_3)/d_3}$	T_3	$\min x_3$	$\max x_3$
0, 0	0.0	1.07	0.0	-1	1
1, 0	32.5	0.825	-1.15	1	1
0, 1	53.0	0.624	1.71	-2	-2
1, 1	5.32	1.04	0.558	-1	0
2, 1	22.6	0.908	-0.597	0	1
2, 2	21.3	0.920	1.12	-2	-1
3, 2	23.3	0.906	-0.0397	0	0
3, 3	47.9	0.682	1.67	-2	-1
4, 3	34.6	0.809	0.518	-1	0
5, 4	56.6	0.583	1.08	-1	-1

At this point, the list of partial coefficient vectors is as follows:

$$
\begin{bmatrix} x_3 \\ x_4 \\ x_5 \end{bmatrix} = \begin{bmatrix} 0 \\ 0 \\ 0 \end{bmatrix} \begin{bmatrix} 1 \\ 0 \\ 0 \end{bmatrix} \begin{bmatrix} 1 \\ 1 \\ 0 \end{bmatrix} \begin{bmatrix} -2 \\ 0 \\ 1 \end{bmatrix} \begin{bmatrix} -1 \\ 1 \\ 1 \end{bmatrix} \begin{bmatrix} 0 \\ 1 \\ 1 \end{bmatrix} \begin{bmatrix} 0 \\ 2 \\ 1 \end{bmatrix} \begin{bmatrix} 1 \\ 2 \\ 1 \end{bmatrix}
$$
$$
\begin{bmatrix} -2 \\ 2 \\ 2 \end{bmatrix} \begin{bmatrix} -1 \\ 2 \\ 2 \end{bmatrix} \begin{bmatrix} 0 \\ 3 \\ 2 \end{bmatrix} \begin{bmatrix} -2 \\ 3 \\ 3 \end{bmatrix} \begin{bmatrix} -1 \\ 3 \\ 3 \end{bmatrix} \begin{bmatrix} -1 \\ 4 \\ 3 \end{bmatrix} \begin{bmatrix} 0 \\ 4 \\ 3 \end{bmatrix} \begin{bmatrix} -1 \\ 5 \\ 4 \end{bmatrix}
$$

We repeat the main loop with $k = 2$. We obtain the following table:

x_3, x_4, x_5	S_2	$\sqrt{(C-S_2)/d_2}$	T_2	min x_2	max x_2
$0, 0, 0$	0.0	0.813	0.0	0	0
$1, 0, 0$	69.4	0.296	-0.612	1	0
$1, 1, 0$	34.1	0.614	-0.475	0	1
$-2, 0, 1$	58.8	0.419	1.36	-1	-1
$-1, 1, 1$	18.9	0.711	0.888	-1	-1
$0, 1, 1$	26.9	0.661	0.276	0	0
$0, 2, 1$	47.3	0.520	0.413	0	0
$1, 2, 1$	33.8	0.618	-0.199	0	0
$-2, 2, 2$	75.6	0.191	1.78	-1	-2
$-1, 2, 2$	22.2	0.693	1.16	-1	-1
$0, 3, 2$	23.4	0.683	0.689	-1	-1
$-2, 3, 3$	55.3	0.451	2.05	-2	-2
$-1, 3, 3$	79.2	0.0795	1.44	-1	-2
$-1, 4, 3$	50.7	0.491	1.58	-2	-2
$0, 4, 3$	53.2	0.471	0.965	-1	-1
$-1, 5, 4$	57.0	0.435	1.85	-2	-2

This is the first time we have the situation that the lower bound on x_2 is greater than the upper bound, and so extending the partial vector is impossible. This causes the number of partial coefficient vectors to decrease; the previous list of 16 partial vectors becomes the following list of 14 partial vectors:

$$
\begin{bmatrix} x_2 \\ x_3 \\ x_4 \\ x_5 \end{bmatrix} = \begin{bmatrix} 0 \\ 0 \\ 0 \\ 0 \end{bmatrix} \begin{bmatrix} 0 \\ 1 \\ 1 \\ 0 \end{bmatrix} \begin{bmatrix} 1 \\ 1 \\ 1 \\ 0 \end{bmatrix} \begin{bmatrix} -1 \\ -2 \\ 0 \\ 1 \end{bmatrix} \begin{bmatrix} -1 \\ -1 \\ 1 \\ 1 \end{bmatrix} \begin{bmatrix} 0 \\ 0 \\ 1 \\ 1 \end{bmatrix} \begin{bmatrix} 0 \\ 0 \\ 2 \\ 1 \end{bmatrix}
$$
$$
\begin{bmatrix} 0 \\ 1 \\ 2 \\ 1 \end{bmatrix} \begin{bmatrix} -1 \\ -1 \\ 2 \\ 2 \end{bmatrix} \begin{bmatrix} -1 \\ 0 \\ 3 \\ 2 \end{bmatrix} \begin{bmatrix} -2 \\ -2 \\ 3 \\ 3 \end{bmatrix} \begin{bmatrix} -2 \\ -1 \\ 4 \\ 3 \end{bmatrix} \begin{bmatrix} -1 \\ 0 \\ 4 \\ 3 \end{bmatrix} \begin{bmatrix} -2 \\ -1 \\ 5 \\ 4 \end{bmatrix}
$$

We repeat the main loop with $k = 1$, and the number of partial vectors decreases further. We obtain the following final list of 13 coefficient vectors:

$$
\begin{bmatrix} x_1 \\ x_2 \\ x_3 \\ x_4 \\ x_5 \end{bmatrix} =
\begin{bmatrix} 0 \\ 0 \\ 0 \\ 0 \\ 0 \end{bmatrix}
\begin{bmatrix} 1 \\ 0 \\ 1 \\ 1 \\ 0 \end{bmatrix}
\begin{bmatrix} 1 \\ 1 \\ 1 \\ 1 \\ 0 \end{bmatrix}
\begin{bmatrix} 0 \\ -1 \\ -1 \\ 1 \\ 1 \end{bmatrix}
\begin{bmatrix} 1 \\ 0 \\ 0 \\ 1 \\ 1 \end{bmatrix}
\begin{bmatrix} 1 \\ 0 \\ 0 \\ 2 \\ 1 \end{bmatrix}
\begin{bmatrix} 1 \\ 0 \\ 1 \\ 2 \\ 1 \end{bmatrix}
$$

$$
\begin{bmatrix} 2 \\ 0 \\ 1 \\ 2 \\ 1 \end{bmatrix}
\begin{bmatrix} 1 \\ -1 \\ -1 \\ 2 \\ 2 \end{bmatrix}
\begin{bmatrix} 2 \\ -1 \\ 0 \\ 3 \\ 2 \end{bmatrix}
\begin{bmatrix} 1 \\ -2 \\ -2 \\ 3 \\ 3 \end{bmatrix}
\begin{bmatrix} 2 \\ -1 \\ 0 \\ 4 \\ 3 \end{bmatrix}
\begin{bmatrix} 2 \\ -2 \\ -1 \\ 5 \\ 4 \end{bmatrix}
$$

These coefficient vectors correspond to the following short lattice vectors with square length at most $C = 80$:

$$
\begin{bmatrix} 0 \\ 0 \\ 0 \\ 0 \\ 0 \end{bmatrix}
\begin{bmatrix} -3 \\ -2 \\ 0 \\ -3 \\ 0 \end{bmatrix}
\begin{bmatrix} 0 \\ -5 \\ 0 \\ 3 \\ 1 \end{bmatrix}
\begin{bmatrix} 3 \\ -3 \\ 4 \\ -2 \\ 4 \end{bmatrix}
\begin{bmatrix} 3 \\ -3 \\ 0 \\ 6 \\ 1 \end{bmatrix}
\begin{bmatrix} -3 \\ -7 \\ 0 \\ 0 \\ 1 \end{bmatrix}
\begin{bmatrix} 3 \\ 2 \\ 4 \\ -5 \\ 3 \end{bmatrix}
$$

$$
\begin{bmatrix} 5 \\ -3 \\ -5 \\ -2 \\ 1 \end{bmatrix}
\begin{bmatrix} 6 \\ -1 \\ 4 \\ 1 \\ 4 \end{bmatrix}
\begin{bmatrix} -1 \\ 4 \\ -4 \\ 1 \\ 6 \end{bmatrix}
\begin{bmatrix} -4 \\ 2 \\ -4 \\ -2 \\ 6 \end{bmatrix}
\begin{bmatrix} 5 \\ 2 \\ -5 \\ -5 \\ 0 \end{bmatrix}
\begin{bmatrix} 2 \\ -5 \\ -5 \\ -5 \\ 1 \end{bmatrix}
$$

10.4 The FP algorithm with LLL preprocessing

Fincke and Pohst [43] showed how to combine their algorithm with the LLL algorithm to produce a more efficient hybrid. This involves two insights:

(1) We can use the LLL algorithm to modify the quadratic form obtained from the Gram matrix of the lattice basis; this will diminish the ranges for the components of the partial coordinate vectors.

(2) We can reorder the vectors in the computation of the rational Cholesky decomposition of the Gram matrix; this will decrease the chance that a partial coordinate vector cannot be extended.

We start as before with an m-dimensional lattice L in n-dimensional Euclidean space defined by a basis $\mathbf{b}_1, \mathbf{b}_2, \ldots, \mathbf{b}_m$ stored in the columns of an $n \times m$ matrix B. The Gram matrix is the $m \times m$ matrix G defined by the equation

$$ G = B^t B. $$

The rational Cholesky decomposition of the Gram matrix has the form

$$G = U^t DU,$$

where U is an upper triangular matrix and D is a positive diagonal matrix. We define $D^{1/2}$ to be the (unique) positive diagonal matrix for which

$$(D^{1/2})^2 = D,$$

and we define the upper triangular matrix R by the equation

$$R = D^{1/2}U.$$

We then have the classical Cholesky decomposition of the Gram matrix:

$$G = R^t R.$$

We consider the inverse matrix R^{-1} and the m-dimensional lattice in $\mathbb{R}^m$ with basis consisting of the rows (not columns) of R^{-1}. We apply the LLL algorithm to this basis, and obtain a reduced basis which we regard as the rows of the matrix S^{-1}. The change of basis matrix from the rows of R^{-1} to the rows of S^{-1} will be denoted by X^{-1}. This application of lattice basis reduction can be expressed by the matrix equation

$$S^{-1} = X^{-1} R^{-1}.$$

We have the following formulas for the matrices X and S:

$$X = (S^{-1}R)^{-1}, \qquad S = RX.$$

Let P^{-1} be the permutation matrix for which the matrix $P^{-1}S^{-1} = (SP)^{-1}$ has rows of decreasing (that is, non-increasing) Euclidean norm. We apply the same permutation to the columns of S, and obtain the matrix SP. We now consider the new Gram matrix H, which is the symmetric positive definite matrix defined by the equation

$$H = (SP)^t(SP).$$

We compute the rational Cholesky decomposition of H:

$$H = V^t EV.$$

We apply the original Fincke-Pohst algorithm to the quadratic form corresponding to the new Gram matrix H. If $\mathbf{z}$ is one of the coefficient vectors, we first permute the components by calculating

$$\mathbf{y} = P\mathbf{z}.$$

We then apply the change of basis matrix and obtain

$$\mathbf{x} = X\mathbf{y}.$$

A final step gives the short lattice vectors as elements of $\mathbb{R}^n$:

$$\mathbf{w} = B\mathbf{x}.$$

This procedure will be made clearer by an example.

Example 10.7. The rows of this matrix form a basis for the lattice L in $\mathbb{R}^5$:

$$B^t = \begin{bmatrix} 7 & 7 & -1 & -9 & 3 \\ 8 & -2 & -7 & 7 & 6 \\ -5 & -5 & 0 & -1 & 6 \\ -6 & -8 & -2 & -2 & 1 \\ 7 & 8 & -1 & 4 & -9 \end{bmatrix}$$

We find all vectors in L with squared length at most $C = 100$. The Gram matrix of this basis is

$$G = \begin{bmatrix} 189 & 4 & -43 & -75 & 43 \\ 4 & 202 & -1 & -26 & 21 \\ -43 & -1 & 87 & 78 & -133 \\ -75 & -26 & 78 & 109 & -121 \\ 43 & 21 & -133 & -121 & 211 \end{bmatrix}$$

The rational Cholesky decomposition of $G = U^t D U$ is given by the following diagonal matrix D and upper triangular matrix U:

$$D = \text{diagonal} \left[189, \frac{38162}{189}, \frac{2946751}{38162}, \frac{83142434}{2946751}, \frac{164019249}{83142434} \right]$$

$$U = \begin{bmatrix} 1 & \dfrac{4}{189} & -\dfrac{43}{189} & -\dfrac{25}{63} & \dfrac{43}{189} \\ 0 & 1 & -\dfrac{17}{38162} & -\dfrac{2307}{19081} & \dfrac{3797}{38162} \\ 0 & 0 & 1 & \dfrac{2325044}{2946751} & -\dfrac{4701863}{2946751} \\ 0 & 0 & 0 & 1 & -\dfrac{12653409}{83142434} \\ 0 & 0 & 0 & 0 & 1 \end{bmatrix}$$

We now use floating point arithmetic; we somewhat arbitrarily choose to use six significant decimal digits. The classical Cholesky decomposition $G = R^t R$ where $R = D^{1/2} U$ is given by the following upper triangular matrix R:

$$R \approx \begin{bmatrix} 13.7477 & 0.290957 & -3.12778 & -5.45544 & 3.12778 \\ 0 & 14.2097 & -0.00632999 & -1.71803 & 1.41382 \\ 0 & 0 & 8.78734 & 6.93338 & -14.0212 \\ 0 & 0 & 0 & 5.31179 & -0.808399 \\ 0 & 0 & 0 & 0 & 1.40455 \end{bmatrix}$$

We compute the inverse matrix R^{-1}:

$$
\begin{bmatrix}
0.0727394 & -0.00148941 & 0.0258899 & 0.0404312 & 0.121238 \\
0 & 0.0703745 & 0.0000506945 & 0.0226955 & -0.0572703 \\
0 & 0 & 0.113800 & -0.148541 & 1.05054 \\
0 & 0 & 0 & 0.188260 & 0.108355 \\
0 & 0 & 0 & 0 & 0.711972
\end{bmatrix}
$$

We apply the LLL algorithm to the rows of R^{-1}, and obtain a new lattice basis given by the rows of S^{-1}:

$$
\begin{bmatrix}
0 & 0.0703745 & 0.0000506945 & 0.0226955 & -0.0572703 \\
0.0727394 & 0.0688851 & 0.0259406 & 0.0631267 & 0.0639677 \\
-0.0727394 & 0.00148941 & -0.0258899 & 0.147829 & -0.0128830 \\
0.145479 & -0.0733533 & -0.0620709 & 0.0184479 & -0.147177 \\
-0.0727394 & 0.0718639 & -0.139639 & -0.0574547 & 0.0865407
\end{bmatrix}
$$

This application of the LLL algorithm was performed by the Maple command

```
convert(IntegerRelations[LLL](convert(RI,listlist)),Matrix);
```

where RI represents R^{-1}. The change of basis matrix corresponding to this application of the LLL algorithm is

$$
X = \begin{bmatrix}
-1 & 1 & 0 & 0 & 0 \\
1 & 0 & 0 & 0 & 0 \\
-7 & 4 & -1 & -2 & 1 \\
-1 & 1 & 1 & 0 & 0 \\
-5 & 3 & 0 & -1 & 1
\end{bmatrix}
$$

The matrix S is the inverse of S^{-1}:

$$
S \approx \begin{bmatrix}
-1.74574 & 5.16448 & -2.32766 & 3.12778 & 0 \\
8.90294 & 2.49811 & -1.71170 & -1.40116 & 1.40749 \\
1.66124 & 0.0191400 & -1.85396 & -3.55348 & -5.23386 \\
-1.26980 & 2.88659 & 5.31179 & 0.808399 & -0.808399 \\
-7.02275 & 4.21365 & 0 & -1.40455 & 1.40455
\end{bmatrix}
$$

We sort the rows of S^{-1} by decreasing norm and obtain the matrix $P^{-1}S^{-1}$:

$$
\begin{bmatrix}
0.145479 & -0.0733533 & -0.0620709 & 0.0184479 & -0.147177 \\
-0.0727394 & 0.0718639 & -0.139639 & -0.0574547 & 0.0865407 \\
-0.0727394 & 0.00148941 & -0.0258899 & 0.147829 & -0.0128830 \\
0.0727394 & 0.0688851 & 0.0259406 & 0.0631267 & 0.0639677 \\
0 & 0.0703745 & 0.0000506945 & 0.0226955 & -0.0572703
\end{bmatrix}
$$

The permutation $\pi^{-1} = 45321$ is represented by the matrix

$$
P^{-1} = \begin{bmatrix}
0 & 0 & 0 & 1 & 0 \\
0 & 0 & 0 & 0 & 1 \\
0 & 0 & 1 & 0 & 0 \\
0 & 1 & 0 & 0 & 0 \\
1 & 0 & 0 & 0 & 0
\end{bmatrix}
$$

We apply this permutation to the columns of S and obtain the matrix SP:

$$SP = \begin{bmatrix} 3.12778 & 0 & -2.32766 & 5.16448 & -1.74574 \\ -1.40116 & 1.40749 & -1.71170 & 2.49811 & 8.90294 \\ -3.55348 & -5.23386 & -1.85396 & 0.0191400 & 1.66124 \\ 0.808399 & -0.808399 & 5.31179 & 2.88659 & -1.26980 \\ -1.40455 & 1.40455 & 0 & 4.21365 & -7.02275 \end{bmatrix}$$

From this we obtain the new Gram matrix $H = (SP)^t(SP)$:

$$H = \begin{bmatrix} 27 & 14 & 6 & 9 & -15 \\ 14 & 32 & 3 & 7 & -5 \\ 6 & 3 & 40 & -1 & -21 \\ 9 & 7 & -1 & 59 & -20 \\ -15 & -5 & -21 & -20 & 136 \end{bmatrix}$$

(The floating-point entries of H are nearly integers; we have rounded them off.) It is important to observe that the entries of H are significantly smaller than the entries of the original Gram matrix G: we have $|G|_\infty = 1888$ and $|H|_\infty = 496$, where $|\ |_\infty$ denotes the sum of the absolute values of the entries.

The rational Cholesky decomposition of the new Gram matrix $H = V^t E V$ is given by the diagonal matrix E and the upper triangular matrix V:

$$E = \text{diagonal} \left[27, \ \frac{668}{27}, \ \frac{25829}{668}, \ \frac{1434770}{25829}, \ \frac{164019249}{1434770} \right]$$

$$V = \begin{bmatrix} 1 & \dfrac{14}{27} & \dfrac{2}{9} & \dfrac{1}{3} & -\dfrac{5}{9} \\ 0 & 1 & -\dfrac{3}{668} & \dfrac{63}{668} & \dfrac{75}{668} \\ 0 & 0 & 1 & -\dfrac{1997}{25829} & -\dfrac{11793}{25829} \\ 0 & 0 & 0 & 1 & -\dfrac{429457}{1434770} \\ 0 & 0 & 0 & 0 & 1 \end{bmatrix}$$

We now apply the original Fincke-Pohst algorithm to this decomposition of the corresponding quadratic form. We obtain the results in Table 10.6, in which the vectors are sorted by increasing Euclidean norm:

- Column $\mathbf{z}$ gives the coefficient vectors from the application of the original Fincke-Pohst algorithm to the reduced Gram matrix H.

- Column $\mathbf{y}$ gives the permuted coefficient vectors $\mathbf{y} = P\mathbf{z}$.

- Column $\mathbf{x}$ gives the coefficient vectors $\mathbf{x} = X\mathbf{y}$ before the LLL algorithm.

- Column $\mathbf{w}$ gives the short lattice vectors $\mathbf{w} = B\mathbf{x}$.

The Fincke-Pohst algorithm with LLL preprocessing is presented in pseudocode in Figure 10.7.

#	$|w|^2$	w_1	w_2	w_3	w_4	w_5	x_1	x_2	x_3	x_4	x_5	y_1	y_2	y_3	y_4	y_5	z_1	z_2	z_3	z_4	z_5
1	0	0	0	0	0	0	0	0	0	0	0	0	0	0	0	0	0	0	0	0	0
2	27	3	2	1	-2	-3	0	0	-2	0	-1	0	0	1	-1	-1	0	-1	0	1	0
3	31	-1	1	-2	5	0	0	0	3	0	2	0	0	1	-1	-1	0	-1	1	-1	0
4	32	2	3	-1	3	-3	0	0	-1	0	-1	0	0	0	0	1	0	0	1	0	0
5	40	-1	-3	-2	-1	-5	0	0	-1	1	0	0	1	0	0	0	0	1	1	0	0
6	55	-4	-5	-3	1	-2	0	0	4	1	1	0	1	1	-1	-1	0	1	0	-1	0
7	59	2	3	-6	-3	1	-1	0	2	1	3	0	0	0	0	0	0	0	0	0	0
8	65	-2	-2	-4	4	5	0	0	-2	1	2	0	1	1	-1	-1	0	1	1	-1	0
9	66	-3	-6	-1	-4	-2	0	0	6	1	-1	0	1	0	0	1	0	1	-1	-1	0
10	68	-1	1	-7	-1	4	-1	0	3	1	4	0	0	1	-1	0	0	0	0	0	0
11	77	0	0	-5	-6	4	1	0	-4	1	2	0	1	0	0	1	0	1	1	-1	0
12	77	0	-4	0	-6	-5	0	0	0	1	-2	0	1	1	-1	0	0	1	0	-1	0
13	78	1	0	-3	2	-8	0	0	-3	1	-1	2	0	0	0	-1	0	0	1	0	0
14	79	2	-1	-1	-3	-8	0	0	5	1	-1	0	0	1	-1	1	0	1	1	-1	0
15	84	-4	-1	-3	7	3	-1	0	7	0	3	0	0	-2	-2	-2	0	-2	0	-1	0
16	86	1	4	-8	2	1	0	0	-1	1	5	0	1	1	-1	-1	0	1	1	-1	0
17	87	5	5	0	1	-6	-1	0	5	0	0	0	0	1	-1	-1	0	1	0	-1	0
18	94	-2	-2	-9	-2	-1	-1	0	1	2	4	0	1	1	-1	-1	0	1	1	-1	0
19	94	3	2	-4	-8	1	1	0	3	1	1	0	0	1	-1	-1	0	0	0	0	0
20	97	1	0	-8	-4	-4	0	0	4	2	3	0	1	0	0	0	0	1	1	-1	0
21	99	1	4	-3	8	-3	0	0	4	0	3	0	0	1	-1	-1	0	-1	2	2	0

FIGURE 10.6
Output for Example 10.7

- *Input:* An $n \times m$ matrix B of rank $m \le n$, and an upper bound C.
- *Output:* All vectors of squared length $\le C$ in the lattice L spanned by the columns of B.

(1)   Set $G \leftarrow B^t B$ and set $Q \leftarrow G$.

(2)   For $j = 1, 2, \ldots, m-1$ do:

> For $i = j+1, \ldots, m$ do: Add $-\frac{Q_{ij}}{Q_{jj}}(\text{row } j)$ of Q to row i.

(3)   Create the $m \times m$ matrix D. For $i = 1, 2, \ldots, m$ do: Set $D_{ii} \leftarrow Q_{ii}$.

(4)   Set $U \leftarrow D^{-1}Q$ and set $R \leftarrow D^{1/2}U$. Calculate R^{-1}.

(5)   Apply LLL to the rows of R^{-1} to obtain $S^{-1} = X^{-1}R^{-1}$.

(6)   Sort the rows of S^{-1} by decreasing norm to obtain $P^{-1}S^{-1}$.

(7)   Set $S \leftarrow RX$ and $P \leftarrow (P^{-1})^{-1}$.

(8)   Set $H \leftarrow (SP)^t(SP)$ and $Q \leftarrow H$.

(9)   For $j = 1, 2, \ldots, m-1$ do:

> For $i = j+1, \ldots, m$ do: Add $-\frac{Q_{ij}}{Q_{jj}}(\text{row } j)$ of Q to row i.

(10) Create the $m \times m$ matrix E. For $i = 1, 2, \ldots, m$ do: Set $E_{ii} \leftarrow Q_{ii}$.

(11) Set $V \leftarrow E^{-1}Q$.

(12) Set old $\leftarrow [[\,]]$ *(a list containing one element, the empty list).*

(13) For $k = m, m-1, \ldots, 1$ do:

(a) Set new $\leftarrow [\,]$ *(the empty list).*

(b) For r in old do:

- For $j = 1, 2, \ldots, k$ do: Set $x_j \leftarrow 0$. For $j = k+1, \ldots, m$ do: Set $x_j \leftarrow r_{j-k}$.
- Set $S \leftarrow \sum_{i=k+1}^{m} E_{ii}\left(\sum_{j=i}^{m} V_{ij}x_j\right)^2$ and $T \leftarrow \sum_{j=k+1}^{m} V_{kj}x_j$.
- Set $\ell \leftarrow \left\lceil -\sqrt{\frac{C-S}{E_{kk}}} - T \right\rceil$ and $u \leftarrow \left\lfloor \sqrt{\frac{C-S}{E_{kk}}} - T \right\rfloor$.
- For $x_k = \ell, \ldots, u$ do:
  - If all $x_j = 0$, or $x_j > 0$ where j is maximal for $x_j \ne 0$, then: Append $[x_k, x_{k+1}, \ldots, x_m]$ to new.

(c) Set old $\leftarrow$ new.

(14) For z in old do: Set $w \leftarrow BXPz$ and output w.

FIGURE 10.7
The Fincke-Pohst algorithm with LLL preprocessing

10.5 Projects

Project 10.1. Implement a recursive version of the Fincke-Pohst algorithm: replace the main loop by a recursive procedure call. Compare the time and space usage of the two algorithms on a number of examples.

Project 10.2. Implement on a computer the two versions of the Fincke-Pohst algorithm: the first (original) version, and the second version with LLL preprocessing. Run both versions on the same sample of lattices of dimensions up to 20 (or higher, if sufficient computer time and memory are available). Keep track of the execution times; for the second algorithm, keep track of both the time for LLL preprocessing and the time for the application of the original Fincke-Pohst algorithm to the reduced Gram matrix. Estimate the dimension at which the second algorithm becomes preferable to the first algorithm.

Project 10.3. Following the original papers of Fincke and Pohst [42, 43], write a report giving a detailed analysis of the complexity of the original Fincke-Pohst algorithm. Extend this to a detailed analysis of the complexity of the Fincke-Pohst algorithm with LLL preprocessing.

Project 10.4. Before the LLL algorithm and the Fincke-Pohst algorithm, other authors had developed methods for finding short vectors in lattices. Two of the most important contributions are by Deiter [38] and Knuth [78]. Study these references, implement these algorithms on a computer, and write a report comparing these algorithms with the Fincke-Pohst algorithm.

10.6 Exercises

Exercise 10.1. Consider the lattice L in $\mathbb{R}^3$ with the rows of B^t as a basis:

$$B^t = \begin{bmatrix} -3 & -2 & 1 \\ 2 & 3 & -1 \\ -2 & -3 & 2 \end{bmatrix}$$

Calculate the rational Cholesky decomposition $G = U^t D U$ of the Gram matrix $G = B B^t$. Diagonalize the quadratic form on $\mathbb{R}^3$ defined by G.

Exercise 10.2. Same as Exercise 10.1, but for the matrix

$$B^t = \begin{bmatrix} -2 & 0 & 3 \\ 1 & -1 & -2 \\ 4 & 2 & -1 \end{bmatrix}$$

Exercise 10.3. Same as Exercise 10.1, but for the matrix

$$B^t = \begin{bmatrix} -2 & 5 & -3 \\ 1 & 5 & 5 \\ -1 & -2 & -5 \end{bmatrix}$$

Exercise 10.4. Consider the lattice in $\mathbb{R}^4$ with the rows of B^t as a basis:

$$B^t = \begin{bmatrix} -2 & -4 & 1 & 2 \\ 4 & -3 & 4 & 3 \\ 3 & -1 & 0 & -2 \\ -2 & -2 & -1 & -1 \end{bmatrix}$$

Calculate the rational Cholesky decomposition $G = U^t D U$ of the Gram matrix $G = BB^t$. Diagonalize the quadratic form on $\mathbb{R}^3$ defined by G.

Exercise 10.5. Same as Exercise 10.4, but for the matrix

$$B^t = \begin{bmatrix} 1 & 1 & 5 & -5 \\ 1 & 3 & 3 & 1 \\ -4 & -2 & -1 & -5 \\ -5 & 0 & -2 & -1 \end{bmatrix}$$

Exercise 10.6. Same as Exercise 10.4, but for the matrix

$$B^t = \begin{bmatrix} 1 & 2 & -1 & 5 \\ 0 & -4 & -3 & 3 \\ 3 & 2 & 5 & 0 \\ -1 & 1 & 3 & -5 \end{bmatrix}$$

Exercise 10.7. Consider the lattice in $\mathbb{R}^4$ spanned by the three rows of B^t:

$$B^t = \begin{bmatrix} 4 & -4 & 0 & 2 \\ 1 & -5 & -1 & 4 \\ 5 & -3 & -2 & -1 \end{bmatrix}$$

Calculate the rational Cholesky decomposition $G = U^t D U$ of the Gram matrix $G = BB^t$. Diagonalize the quadratic form on $\mathbb{R}^3$ defined by G.

Exercise 10.8. Same as Exercise 10.7, but for the matrix

$$B^t = \begin{bmatrix} 2 & 5 & 0 & -2 \\ 3 & -3 & 4 & 1 \\ -4 & -4 & -3 & 5 \end{bmatrix}$$

Exercise 10.9. Same as Exercise 10.7, but for the matrix

$$B^t = \begin{bmatrix} -2 & -4 & 0 & 3 \\ -2 & 3 & 4 & -1 \\ -1 & 3 & -1 & -3 \end{bmatrix}$$

Exercise 10.10. Diagonalize the positive definite quadratic forms corresponding to the Gram matrices from Exercises 10.1 to 10.9.

Exercise 10.11. In each case, consider the lattice in $\mathbb{R}^3$ with basis consisting of the columns of the matrix. Trace the execution of the Fincke-Pohst algorithm as it finds all lattice vectors of square length at most 80:

$$
\begin{bmatrix} 9 & 0 & -7 \\ 2 & -1 & -7 \\ 0 & 4 & -8 \end{bmatrix}, \quad
\begin{bmatrix} 1 & -8 & 6 \\ -3 & -6 & 2 \\ 9 & 6 & -4 \end{bmatrix}, \quad
\begin{bmatrix} -6 & 5 & -1 \\ 5 & -8 & -5 \\ -6 & -5 & -4 \end{bmatrix},
$$

$$
\begin{bmatrix} -8 & 2 & -1 \\ 2 & -9 & -3 \\ -5 & 0 & -4 \end{bmatrix}, \quad
\begin{bmatrix} -3 & 3 & -9 \\ 8 & -7 & 7 \\ 3 & 2 & 7 \end{bmatrix}, \quad
\begin{bmatrix} 8 & -9 & 5 \\ -1 & -2 & -4 \\ 4 & -5 & -5 \end{bmatrix},
$$

$$
\begin{bmatrix} 0 & -6 & -2 \\ -1 & -8 & 1 \\ 6 & -2 & 7 \end{bmatrix}, \quad
\begin{bmatrix} -9 & -2 & 6 \\ 3 & -7 & -5 \\ 8 & 7 & -5 \end{bmatrix}.
$$

Exercise 10.12. In each case, consider the lattice in $\mathbb{R}^4$ with basis consisting of the columns of the matrix. Trace the execution of the Fincke-Pohst algorithm as it finds all lattice vectors of square length at most 60:

$$
\begin{bmatrix} 9 & 6 & 2 & 4 \\ -8 & 2 & 0 & -7 \\ -6 & -4 & 0 & -7 \\ 6 & 9 & -1 & -8 \end{bmatrix}, \quad
\begin{bmatrix} -9 & -4 & 5 & -5 \\ 0 & -6 & -8 & -4 \\ -1 & 5 & -5 & 1 \\ -3 & -6 & -1 & -3 \end{bmatrix}, \quad
\begin{bmatrix} 5 & 8 & 2 & -8 \\ -4 & 3 & -9 & 2 \\ -5 & 3 & 7 & -5 \\ -3 & -7 & 7 & 2 \end{bmatrix},
$$

$$
\begin{bmatrix} -5 & -6 & 1 & 4 \\ 0 & -8 & 7 & -9 \\ -1 & -2 & 8 & -2 \\ 6 & -2 & -1 & -5 \end{bmatrix}, \quad
\begin{bmatrix} 7 & -1 & -9 & -7 \\ 6 & 7 & 3 & 7 \\ 9 & 7 & 8 & 6 \\ -5 & -1 & -2 & -5 \end{bmatrix}, \quad
\begin{bmatrix} 8 & 0 & -8 & 3 \\ -1 & -3 & -8 & 6 \\ 5 & -9 & 3 & 2 \\ -9 & -7 & 1 & -9 \end{bmatrix}.
$$

Exercise 10.13. In each case, consider the lattice in $\mathbb{R}^5$ with basis consisting of the columns of the matrix. Trace the execution of the Fincke-Pohst algorithm as it finds all lattice vectors of square length at most 100:

$$
\begin{bmatrix} 9 & 6 & 4 & -6 & -1 \\ 8 & 6 & -4 & -1 & 8 \\ 7 & -7 & -1 & -3 & 9 \\ -2 & -6 & 8 & -4 & -3 \\ 5 & -8 & -5 & 5 & -3 \end{bmatrix}, \quad
\begin{bmatrix} -1 & -5 & 1 & 1 & -2 \\ 7 & 8 & -1 & 7 & 0 \\ -1 & 6 & -3 & 8 & 6 \\ 7 & -4 & -8 & 2 & -2 \\ -4 & -2 & 2 & -2 & -5 \end{bmatrix},
$$

$$
\begin{bmatrix} -3 & -4 & -2 & 0 & 8 \\ -1 & 3 & 4 & 6 & 6 \\ -6 & -4 & -2 & 3 & -1 \\ 2 & -1 & -9 & -5 & -5 \\ -2 & 1 & -2 & -6 & 1 \end{bmatrix}, \quad
\begin{bmatrix} 1 & -9 & 8 & 6 & -1 \\ -9 & -3 & -4 & 2 & -2 \\ 5 & -9 & -6 & 8 & -3 \\ -6 & -2 & -9 & 4 & -6 \\ -4 & 2 & 7 & 3 & 0 \end{bmatrix}.
$$

Exercise 10.14. In each case, consider the lattice in $\mathbb{R}^6$ with basis consisting of the columns of the matrix. Trace the execution of the Fincke-Pohst algorithm as it finds all lattice vectors of square length at most 50:

$$
\begin{bmatrix}
-4 & 8 & 8 & -9 & 3 & 9 \\
-8 & -8 & -1 & -7 & 6 & -5 \\
5 & -5 & 5 & -8 & 2 & -1 \\
9 & -2 & -9 & -8 & -9 & 7 \\
-5 & 3 & 0 & 3 & 7 & 7 \\
-5 & -5 & -3 & 1 & 6 & -1
\end{bmatrix},
\begin{bmatrix}
7 & 8 & 3 & 6 & -4 & 0 \\
-9 & -9 & 7 & 1 & 2 & -1 \\
1 & -7 & 5 & -3 & -8 & -3 \\
1 & 9 & 1 & 9 & 6 & -6 \\
4 & 7 & -3 & 6 & 3 & -7 \\
-7 & -3 & -8 & 4 & 9 & 5
\end{bmatrix},
$$

$$
\begin{bmatrix}
8 & -7 & 8 & 5 & -9 & 2 \\
7 & -6 & -5 & -1 & 2 & -3 \\
-2 & -8 & -6 & 8 & 4 & -3 \\
5 & 4 & -1 & 9 & -1 & -2 \\
6 & -4 & -3 & -3 & -5 & -1 \\
6 & -1 & -4 & -3 & -4 & -2
\end{bmatrix},
\begin{bmatrix}
0 & 6 & -1 & -4 & 2 & -2 \\
6 & -1 & 7 & -2 & 1 & 0 \\
3 & -5 & -4 & 1 & 7 & 6 \\
-5 & 1 & -5 & -1 & 8 & -2 \\
-6 & -1 & 8 & -3 & 2 & -5 \\
8 & 7 & 6 & -8 & -2 & 9
\end{bmatrix}.
$$

Exercise 10.15. Use the Fincke-Pohst algorithm with LLL preprocessing as in Example 10.7 to redo Exercises 10.12, 10.13 and 10.14.

11

Kannan's Algorithm

CONTENTS

In this chapter we present the algorithm of Kannan [70, 71] which finds a lattice basis which is reduced in the strong sense of Definition 11.9 below; in particular, the first vector in the basis is a shortest (nonzero) lattice vector.

11.1 Basic definitions

We recall the Gram-Schmidt orthogonalization process , and introduce some further notation. Suppose that the vectors $\mathbf{b}_1, \mathbf{b}_2, \ldots, \mathbf{b}_n \in \mathbb{R}^n$ form a basis of the lattice L. The orthogonal vectors $\mathbf{b}_1^*, \mathbf{b}_2^*, \ldots, \mathbf{b}_n^*$ are defined by

$$\mathbf{b}_1^* = \mathbf{b}_1, \qquad \mathbf{b}_i^* = \mathbf{b}_i - \sum_{j=1}^{i-1} \mu_{ij}\mathbf{b}_j^* \quad (i = 2, 3, \ldots, n), \qquad \mu_{ij} = \frac{\mathbf{b}_i \cdot \mathbf{b}_j^*}{|\mathbf{b}_j^*|^2}.$$

We define $\mu_{ii} = 1$ for all i, and rewrite these equations in the form

$$\mathbf{b}_i = \sum_{j=1}^{i} \mu_{ij}\mathbf{b}_j^* \quad (i = 1, 2, \ldots, n).$$

We introduce the following normalized vectors (that is, vectors of length 1):

$$\mathbf{u}_i = \frac{1}{|\mathbf{b}_i^*|}\mathbf{b}_i^* \quad (i = 1, 2, \ldots, n).$$

179

Definition 11.1. We define real numbers $b_i(j)$ for $1 \leq j \leq i \leq n$ by

$$\mathbf{b}_i = \sum_{j=1}^{i} b_i(j) \mathbf{u}_j.$$

That is, $b_i(j) = \mu_{ij} |\mathbf{b}_j^*|$, and in particular $b_i(i) = |\mathbf{b}_i^*|$.

Lemma 11.2. *If* $\mathbf{b}_1, \mathbf{b}_2, \ldots, \mathbf{b}_n \in \mathbb{Q}^n$ *then* $b_i(j)^2 \in \mathbb{Q}$ *for* $1 \leq j \leq i \leq n$.

Proof. Exercise 11.1. $\square$

Definition 11.3. We define vectors $\mathbf{b}(i, j)$ for $1 \leq j \leq i \leq n$ by the equations

$$\mathbf{b}(i, j) = \sum_{k=j}^{i} b_i(k) \mathbf{u}_k.$$

Geometrically speaking, $\mathbf{b}(i, j)$ is the component of $\mathbf{b}_i$ orthogonal to the subspace spanned by $\mathbf{b}_1, \mathbf{b}_2, \ldots, \mathbf{b}_{j-1}$.

Consider a vector $\mathbf{v}$ in $\mathbb{R}^n$ and a basis $\mathbf{b}_1, \mathbf{b}_2, \ldots, \mathbf{b}_n$ of $\mathbb{R}^n$. We express $\mathbf{v}$ with respect to the Gram-Schmidt orthonormalization $\mathbf{u}_1, \mathbf{u}_2, \ldots, \mathbf{b}_n$:

$$\mathbf{v} = \sum_{j=1}^{n} (\mathbf{v} \cdot \mathbf{u}_j) \mathbf{u}_j. \tag{11.1}$$

Definition 11.4. For any $j = 1, 2, \ldots, n$ we obtain from equation (11.1) an expression for $\mathbf{v}$ as the sum of two orthogonal components:

$$\mathbf{v} = \sum_{k=1}^{j-1} (\mathbf{b} \cdot \mathbf{u}_k) \mathbf{u}_k + \sum_{k=j}^{n} (\mathbf{v} \cdot \mathbf{u}_k) \mathbf{u}_k.$$

The first sum is the **projection** of $\mathbf{v}$ to the subspace spanned by $\mathbf{b}_1, \mathbf{b}_2, \ldots,$ $\mathbf{b}_{j-1}$, and the second sum is the **projection** of $\mathbf{v}$ **orthogonal** to this subspace. The projection of a set of vectors to a subspace is the set of projections of its elements. The projection of a lattice to a subspace is a lattice in that subspace; but in general the projected vectors do not belong to the original lattice.

Definition 11.5. Suppose that $L \subset \mathbb{R}^n$ is a lattice with basis $\mathbf{b}_1, \mathbf{b}_2, \ldots,$ $\mathbf{b}_n$. To avoid ambiguity, we sometimes write $L = L(\mathbf{b}_1, \mathbf{b}_2, \ldots, \mathbf{b}_n)$. For $j = 1, 2, \ldots, n$ we write $L_j = L_j(\mathbf{b}_1, \mathbf{b}_2, \ldots, \mathbf{b}_n)$ for the projection of L orthogonal to the subspace spanned by $\mathbf{b}_1, \mathbf{b}_2, \ldots, \mathbf{b}_{j-1}$. In particular, $L_1 = L$ since the subspace spanned by the empty set is $\{\mathbf{0}\}$.

Lemma 11.6. *The change of basis matrix from the basis* $\mathbf{b}_1$, $\mathbf{b}_2$, ..., $\mathbf{b}_n$ *to the corresponding orthonormal basis* $\mathbf{u}_1, \mathbf{u}_2, \ldots, \mathbf{u}_n$ *has the form*

$$
\begin{bmatrix}
b_1(1) & 0 & 0 & \cdots & 0 & 0 & \cdots & 0 \\
b_2(1) & b_2(2) & 0 & \cdots & 0 & 0 & \cdots & 0 \\
\vdots & \vdots & \vdots & \ddots & \vdots & \vdots & \ddots & \vdots \\
b_i(1) & b_i(2) & b_i(3) & \cdots & b_i(i) & 0 & \cdots & 0 \\
b_{i+1}(1) & b_{i+1}(2) & b_{i+1}(3) & \cdots & b_{i+1}(i) & b_{i+1}(i+1) & \cdots & 0 \\
\vdots & \vdots & \vdots & \ddots & \vdots & \vdots & \ddots & \vdots \\
b_n(1) & b_n(2) & b_n(3) & \cdots & b_n(i) & b_n(i+1) & \cdots & b_n(n)
\end{bmatrix}
$$

In this matrix, the (i, j) *entry is the coefficient of* $\mathbf{u}_j$ *in the expression for* $\mathbf{b}_i$.

Proof. Exercise 11.2. $\qquad\square$

Let L be a lattice in $\mathbb{R}^n$ and let $\mathbf{v}$ be a nonzero vector in L. Let L' be the projection of L orthogonal to $\mathbf{v}$; that is, for all $\mathbf{u} \in L$, the lattice L' contains the component of $\mathbf{u}$ orthogonal to $\mathbf{v}$:

$$
L' = \left\{ \mathbf{u} - \frac{\mathbf{u} \cdot \mathbf{v}}{|\mathbf{v}|^2} \mathbf{v} \,\middle|\, \mathbf{u} \in L \right\}.
$$

If $\mathbf{w} \in L'$ then there are infinitely many $\mathbf{u} \in L$ such that $\mathbf{u}$ projects to $\mathbf{w}$: given any such $\mathbf{u}$, the vectors $\mathbf{u} + n\mathbf{v}$ for $n \in \mathbb{Z}$ also project to $\mathbf{w}$.

Definition 11.7. With the previous notation, the **lifting** of $\mathbf{w} \in L'$ is the unique vector $\mathbf{u} \in L$ which projects to $\mathbf{w}$ and satisfies

$$
-\frac{1}{2} < \frac{\mathbf{u} \cdot \mathbf{v}}{|\mathbf{v}|^2} \leq \frac{1}{2}.
$$

Geometrically speaking, the lifting of $\mathbf{w}$ is the vector $\mathbf{u} \in L$ which projects to $\mathbf{w}$ and is closest to the hyperplane orthogonal to $\mathbf{v}$.

Definition 11.8. If L is an n-dimensional lattice in $\mathbb{R}^n$, then $\Lambda_1(L)$ denotes the **first minimum** of L; that is, the length of a shortest nonzero vector in L.

We now give the definition of reduced lattice basis used in this chapter.

Definition 11.9. We say that the basis $\mathbf{v}_1, \mathbf{v}_2, \ldots, \mathbf{v}_n$ of the lattice $L \subset \mathbb{R}^n$ is **Hermite-reduced** if it satisfies the following two conditions:

(a) $v_j(j) = \Lambda_1(L_j)$ for $1 \leq j \leq n$; that is, the projection of $\mathbf{v}_j$ orthogonal to $\mathbf{v}_1, \mathbf{v}_2, \ldots, \mathbf{v}_{j-1}$ is the shortest vector in the projection of the lattice L orthogonal to $\mathbf{v}_1, \mathbf{v}_2, \ldots, \mathbf{v}_{j-1}$.

(b) $|v_i(j)| \leq \frac{1}{2} v_j(j)$ for $1 \leq j < i$, or equivalently $|\mu_{ij}| \leq \frac{1}{2}$; that is, $\mathbf{v}_i$ is nearly orthogonal to $\mathbf{v}_1, \mathbf{v}_2, \ldots, \mathbf{v}_{i-1}$.

Some authors call such a basis **Korkine-Zolotareff-reduced**. These two conditions were introduced by Hermite [60] and Korkine and Zolotareff [79].

Lemma 11.6, which gives a lower triangular representation of a lattice basis, shows that condition (a) in Definition 11.9 means that for all $j = 1, 2, \ldots, n$, the j-th diagonal entry of the matrix is the length of the shortest vector in the $(n-j+1)$-dimensional lattice generated by the rows of the lower right $(n-j+1) \times (n-j+1)$ block. In contrast, the definition of reduced basis used by the LLL algorithm includes the much weaker condition that the j-th diagonal entry of the matrix is the length of the shortest vector in the 2-dimensional lattice generated by the 2×2 block in rows $j, j+1$ and columns $j, j+1$.

11.2 Results from the geometry of numbers

The analysis of Kannan's algorithm depends on some results from the geometry of numbers, which we now recall. The first result tells us when a set of linearly independent lattice vectors can be extended to a basis of the lattice (see also Corollary 1.25).

Proposition 11.10. (Cassels [22], page 14) *Let L be an n-dimensional lattice and let $\mathbf{a}_1, \mathbf{a}_2, \ldots, \mathbf{a}_m \in L$ ($m < n$) be linearly independent. The following two conditions are equivalent:*

> *(a) There exist $\mathbf{a}_{m+1}, \ldots, \mathbf{a}_n \in L$ such that $\mathbf{a}_1, \mathbf{a}_2, \ldots, \mathbf{a}_n$ is a basis of L.*
>
> *(b) If $\mathbf{c} = x_1\mathbf{a}_1 + x_2\mathbf{a}_2 + \cdots + x_m\mathbf{a}_m \in L$ with $x_1, x_2, \ldots, x_m \in \mathbb{R}$ then $x_1, x_2, \ldots, x_m \in \mathbb{Z}$.*

If $m = 1$ then we obtain the following special case.

Proposition 11.11. *Let L be a lattice and let $\mathbf{v} \in L$ be nonzero. There exists a basis of L containing $\mathbf{v}$ if and only if $x\mathbf{v} \in L$ for $x \in \mathbb{R}$ implies $x \in \mathbb{Z}$.*

This result can be restated as follows.

Proposition 11.12. (Kannan [71], page 419) *Suppose that $\mathbf{v}$ is a nonzero element of the lattice L such that $x\mathbf{v}$ does not belong to L for $0 < x < 1$. Then there is a basis of L containing $\mathbf{v}$.*

The second result is Minkowski's famous Convex Body Theorem.

Theorem 11.13. (Cassels [22], page 71) *Let S be a subset of $\mathbb{R}^n$ which is convex and symmetric about the origin, and let $\mathrm{vol}(S)$ be its volume. Let L be a lattice in $\mathbb{R}^n$ with determinant $\det(L)$. Let m be a positive integer. If*

$$\mathrm{vol}(S) > m2^n \det(L),$$

then S contains at least m pairs $\pm\mathbf{x}_1, \pm\mathbf{x}_2, \ldots, \pm\mathbf{x}_m \in L$ which are distinct from each other and from $\mathbf{0}$.

If $m = 1$ then we obtain the following special case (see also Theorem 1.32).

Proposition 11.14. *Let S be a subset of $\mathbb{R}^n$ which is convex and symmetric about the origin, and let* $\mathrm{vol}(S)$ *be its volume. Let L be a lattice in $\mathbb{R}^n$ with determinant* $\det(L)$. *If* $\mathrm{vol}(S) > 2^n \det(L)$ *then S contains a nonzero* $\mathbf{x} \in L$.

Lemma 11.15. *Let S_n be the n-dimensional sphere of radius $r > 0$ in $\mathbb{R}^n$:*

$$S_n = \{\, \mathbf{x} \in \mathbb{R}^n \mid |\mathbf{x}| \le r \,\}.$$

The n-dimensional volume of S_n equals

$$\mathrm{vol}(S_n) = \frac{\pi^{n/2}}{\Gamma(\frac{n}{2}+1)}\, r^n,$$

where Γ is the Gamma function.

Proof. Exercise 11.3. □

Corollary 11.16. (Kannan [71], page 420) *If $L \subset \mathbb{R}^n$ is an n-dimensional lattice with determinant* $\det(L)$ *then there is a nonzero* $\mathbf{x} \in L$ *such that*

$$|\mathbf{x}| \le n^{1/2} \big(\det(L) \big)^{1/n}.$$

Proof. Let S be the n-dimensional sphere in $\mathbb{R}^n$ with center $\mathbf{0}$ and radius

$$r = n^{1/2} \big(\det(L) \big)^{1/n}.$$

By Lemma 11.15 and Exercise 11.4 we have

$$\mathrm{vol}(S) = \frac{\pi^{n/2}}{\Gamma(\frac{n}{2}+1)} n^{n/2} \det(L) \ge 2^n \det(L).$$

Since S is clearly convex and symmetric about the origin, the result follows from Proposition 11.14. □

11.3 Kannan's algorithm

Let $L \subset \mathbb{R}^n$ be a lattice with ordered basis $\mathbf{b}_1, \mathbf{b}_2, \dots, \mathbf{b}_n \in \mathbb{Z}^n$ (the vector components are integers). Kannan's algorithm recursively finds a shortest vector in L by computing shortest vectors in lattices of lower dimension.

Definition 11.17. The first stage of the algorithm finds a new lattice basis $\mathbf{a}_1, \mathbf{a}_2, \dots, \mathbf{a}_n \in \mathbb{Z}^n$ satisfying the following conditions:

(1) For $2 \leq j \leq n$ we have $a_j(j) = \Lambda_1(L_j(\mathbf{a}_1, \mathbf{a}_2, \ldots, \mathbf{a}_n))$. In terms of the lower triangular representation of $\mathbf{a}_1, \mathbf{a}_2, \ldots, \mathbf{a}_n$ as in Lemma 11.6, this means that for $j \geq 2$, the j-th diagonal entry is the length of the shortest vector in the lattice generated by the rows of the lower right $(n-j+1) \times (n-j+1)$ block.

(2) $|\mathbf{a}_1|^2 \leq \frac{4}{3}|\mathbf{a}_2|^2$.

(3) $|a_2(1)|^2 \leq \frac{1}{4}|\mathbf{a}_1|^2$; recall that $a_2(1)$ is the component of $\mathbf{a}_2$ parallel to $\mathbf{a}_1$.

It follows from these conditions that

$$|\mathbf{a}_1|^2 \leq 2n \left(\det(L)^{2/n} \right),$$

where $\det(L)$ is the determinant of the lattice L. If we compare this with the inequality achieved by the LLL algorithm in polynomial time, namely

$$|\mathbf{a}_1|^2 \leq 2^n \left(\det(L)^{2/n} \right),$$

we see that Kannan's algorithm finds a much shorter vector, but uses superexponential time. (We will see that the complexity is $O(n^n s)$ arithmetic operations where s is the length of the input.)

The second stage of the algorithm uses the fact that, given a basis $\mathbf{a}_1, \mathbf{a}_2, \ldots, \mathbf{a}_n$ satisfying conditions (1) to (3) of Definition 11.17, a shortest lattice vector must have the form

$$\mathbf{x} = \sum_{i=1}^{n} x_i \mathbf{a}_i, \qquad [x_1, x_2, \ldots, x_n] \in T \subset \mathbb{Z}^n,$$

where the number of elements in the set T satisfies the inequality

$$|T| \leq \frac{3^n |\mathbf{a}_1|^n}{\det(L)}.$$

The algorithm then enumerates all the elements of T and finds the shortest corresponding lattice vector $\mathbf{x}$.

The main procedure of Kannan's algorithm is REDUCEDBASIS; this calls itself recursively as well as two other procedures, COMPUTEBASIS and SHORTESTVECTOR. We discuss these last two procedures first.

11.3.1   Procedure COMPUTEBASIS

In Kannan [70, 71] this procedure is called SELECTBASIS. This algorithm is similar to the special case of the MLLL algorithm which takes as input $n+1$ linearly dependent vectors spanning a lattice L of dimension n, and produces as output n linearly independent vectors forming a basis of L. However, COMPUTEBASIS preserves the direction of the first vector: the first vector in

- *Input:* A list of $n+1$ linearly dependent vectors $\mathbf{b}_1, \mathbf{b}_2, \ldots, \mathbf{b}_{n+1} \in \mathbb{Q}^n$ which generate the n-dimensional lattice $L \subset \mathbb{R}^n$.

- *Output:* A basis of L in which the first vector is a scalar multiple of $\mathbf{b}_1$.

- If $\mathbf{b}_1 = \mathbf{0}$ then

  Set **basis** $\leftarrow [\,]$ *(the empty list)*

  else

  (1) If $\mathbf{b}_1$ is not a linear combination of $\mathbf{b}_2, \ldots, \mathbf{b}_{n+1}$ then
    Set $\mathbf{a}_1 \leftarrow \mathbf{b}_1$
    else
    (a) Compute $x_2, \ldots, x_{n+1} \in \mathbb{Q}$ such that

    $$\mathbf{b}_1 = x_2\mathbf{b}_2 + \cdots + x_{n+1}\mathbf{b}_{n+1}.$$

    (b) For $i = 2, \ldots, n+1$ do:
        Write $x_i = y_i/z_i$ with $y_i, z_i \in \mathbb{Z}$ and $\mathrm{GCD}(y_i, z_i) = 1$.
    (c) Set $m \leftarrow \mathrm{LCM}(z_2, \ldots, z_{n+1})$.
    (d) Set $d \leftarrow \mathrm{GCD}(mx_2, \ldots, mx_{n+1})$.
    (e) Write $m/d = p/q$ with $p, q \in \mathbb{Z}$ and $\mathrm{GCD}(p, q) = 1$.
    (f) Set $\mathbf{a}_1 \leftarrow (1/q)\,\mathbf{b}_1$.
  (2) For $i = 2, \ldots, n+1$ do:
      Set $\overline{\mathbf{b}}_i$ equal to the projection of $\mathbf{b}_i$ orthogonal to $\mathbf{a}_1$.
  (3) Set $[\mathbf{c}_2, \ldots, \mathbf{c}_n] \leftarrow \mathrm{COMPUTEBASIS}(\overline{\mathbf{b}}_2, \ldots, \overline{\mathbf{b}}_{n+1})$.
  (4) For $i = 2, \ldots, n$ do: Set $\mathbf{a}_i$ equal to the unique minimal lifting of $\mathbf{c}_i$ (see Definition 11.7) to the lattice generated by $\mathbf{b}_1, \mathbf{b}_2, \ldots, \mathbf{b}_{n+1}$.
  (5) Set **basis** $\leftarrow [\mathbf{a}_1, \mathbf{a}_2, \ldots, \mathbf{a}_n]$.

- Return **basis**.

FIGURE 11.1
Kannan's procedure COMPUTEBASIS

the output is always a scalar multiple of the first vector in the input. This procedure is presented in Figure 11.1.

To explain and justify this procedure, consider the computation of $\mathbf{a}_1$ in the case that $\mathbf{b}_1$ is a rational linear combination of $\mathbf{b}_2, \ldots, \mathbf{b}_{n+1}$:

$$\mathbf{b}_1 = x_2 \mathbf{b}_2 + \cdots + x_{n+1} \mathbf{b}_{n+1} \quad \left(x_i = \frac{y_i}{z_i}; \quad y_i, z_i \in \mathbb{Z}; \quad \mathrm{GCD}(y_i, z_i) = 1 \right).$$

Multiplying both sides of this equation by the least common multiple m of the denominators z_i expresses $m\mathbf{b}_1$ as an integral linear combination of $\mathbf{b}_2, \ldots, \mathbf{b}_{n+1}$:

$$m\mathbf{b}_1 = mx_2 \mathbf{b}_2 + \cdots + mx_{n+1} \mathbf{b}_{n+1}.$$

Dividing both sides of this equation by the greatest common divisor d of the integral coefficients $mx_2, \ldots, mx_{n+1}$ expresses a rational multiple of $\mathbf{b}_1$ as an integral linear combination of $\mathbf{b}_2, \ldots, \mathbf{b}_{n+1}$:

$$\frac{m}{d} \mathbf{b}_1 = \frac{mx_2}{d} \mathbf{b}_2 + \cdots + \frac{mx_{n+1}}{d} \mathbf{b}_{n+1}.$$

Cancelling any common factors in m and d gives the equation

$$\frac{p}{q} \mathbf{b}_1 = \frac{px_2}{q} \mathbf{b}_2 + \cdots + \frac{px_{n+1}}{q} \mathbf{b}_{n+1},$$

where $m/d = p/q$ with $\mathrm{GCD}(p, q) = 1$. Since p and q are relatively prime, we have $rp + sq = 1$ for some $r, s \in \mathbb{Z}$. By assumption $\mathbf{b}_1 \in L$, and the last displayed equation shows that $(p/q)\mathbf{b}_1 \in L$. But we have

$$s\mathbf{b}_1 + r\left(\frac{p}{q}\mathbf{b}_1\right) = \frac{1}{q}\left(sq\mathbf{b}_1 + rp\mathbf{b}_1\right) = \frac{1}{q}(sq + rp)\mathbf{b}_1 = \frac{1}{q}\mathbf{b}_1.$$

Thus $(1/q)\mathbf{b}_1$ is an integral linear combination of $\mathbf{b}_1, \mathbf{b}_2, \ldots, \mathbf{b}_{n+1}$ and hence

$$\mathbf{a}_1 = \frac{1}{q}\mathbf{b}_1 \in L.$$

In fact, since $\mathrm{GCD}(p, q) = 1$, it follows that this is the shortest (nonzero) multiple of $\mathbf{b}_1$ which belongs to the lattice L (see Exercise 11.5).

The projection $\overline{\mathbf{b}}_i$ of $\mathbf{b}_i$ orthogonal to $\mathbf{a}_1$ is easily computed by the formula

$$\overline{\mathbf{b}}_i = \mathbf{b}_i - \frac{\mathbf{b}_i \cdot \mathbf{a}_i}{|\mathbf{a}_i|^2}.$$

Computation of the minimal lifting of a vector $\mathbf{c}_i$ to the lattice generated by $\mathbf{b}_1, \mathbf{b}_2, \ldots, \mathbf{b}_{n+1}$ is left as an exercise for the reader (Exercise 11.6).

11.3.2 Procedure SHORTESTVECTOR

In Kannan [70, 71] this procedure is called ENUMERATE. This procedure is similar to the Fincke-Pohst algorithm, in the sense that it uses a complete enumeration to find a shortest vector in a lattice with a given basis. However, the Fincke-Pohst algorithm is more sophisticated, since it uses the Cholesky decomposition of the Gram matrix to limit the search to the integer points in a certain ellipsoid; on the other hand, SHORTESTVECTOR relies on a naive search over a rectangular parallelipiped. The bounds on each coordinate in the Fincke-Pohst algorithm depend dynamically on the coordinates already computed, whereas the bounds on the coordinates in SHORTESTVECTOR are computed once and for all at the start of the procedure.

Let $\mathbf{b}_1, \mathbf{b}_2, \ldots, \mathbf{b}_m \in \mathbb{Q}^n$ be a basis for the m-dimensional lattice $L \subset \mathbb{Q}^n$, and let $\mathbf{b}_1^*, \mathbf{b}_2^*, \ldots, \mathbf{b}_m^* \in \mathbb{Q}^n$ be its Gram-Schmidt orthogonalization. Suppose that a shortest nonzero vector $\mathbf{y} \in L$ has the form

$$\mathbf{y} = \sum_{i=1}^{m} y_i \mathbf{b}_i \quad (y_i \in \mathbb{Z}).$$

If we are given an index $i \leq m-1$ and values $y_{i+1}, \ldots, y_m \in \mathbb{Z}$ then we want to find a range of possible values of y_i. We write $\mathbf{y}$ in the form

$$\mathbf{y} = \sum_{j=1}^{i-1} y_j \mathbf{b}_j + y_i \mathbf{b}_i + \mathbf{y}', \qquad \mathbf{y}' = \sum_{j=i+1}^{m} y_j \mathbf{b}_j.$$

Since $\mathbf{y}$ is a shortest vector in the lattice, we have $|\mathbf{y}| \leq |\mathbf{b}_1|$, and the same inequality is satisfied by the component of $\mathbf{y}$ in the direction of $\mathbf{b}_i^*$, namely

$$y_i \mathbf{b}_i^* + t \mathbf{b}_i^* \quad (t \in \mathbb{Q}),$$

where $t\mathbf{b}_i^*$ is the component of $\mathbf{y}'$ in the direction of $\mathbf{b}_i^*$. We have

$$|(y_i + t)\mathbf{b}_i^*| \leq |\mathbf{b}_1| \implies |y_i + t| \leq \frac{|\mathbf{b}_1|}{|\mathbf{b}_i^*|} \implies -t - \frac{|\mathbf{b}_1|}{|\mathbf{b}_i^*|} \leq y_i \leq -t + \frac{|\mathbf{b}_1|}{|\mathbf{b}_i^*|}.$$

Using these inequalities, it is not difficult to write an algorithm to enumerate all possible coefficient vectors $[y_1, y_2, \ldots, y_m]$ which provide candidates for the shortest nonzero vector in the lattice L. (This algorithm is similar to, but simpler than, the Fincke-Pohst algorithm; see Exercise 11.8.)

Let T denote the set of all coefficient vectors $[y_1, y_2, \ldots, y_m] \in \mathbb{Z}^m$ satisfying the above inequalities. Since $|\mathbf{b}_i^*| = |\mathbf{b}_i(i)|$ it is clear that

$$|T| \leq \prod_{i=1}^{m} \left(1 + 2 \frac{|\mathbf{b}_1|}{|\mathbf{b}_i(i)|} \right).$$

Proposition 11.18. (Kannan [71], page 423) *We have* $|T| \leq (18n)^{n/2}$.

Proof. When we call SHORTESTVECTOR from the main procedure REDUCEDBASIS (see below), the index $m = k-1$ is chosen so that $|\mathbf{b}_i(i)| \le |\mathbf{b}_1|$ for all $i = 1, 2, \ldots, m$. Hence

$$|T| \le \prod_{i=1}^{m} \frac{|\mathbf{b}_i(i)| + 2|\mathbf{b}_1|}{|\mathbf{b}_i(i)|} \le \prod_{i=1}^{m} \frac{3|\mathbf{b}_1|}{|\mathbf{b}_i(i)|} = \frac{3^m |\mathbf{b}_1|^m}{\prod_{i=1}^{m} |\mathbf{b}_i(i)|}.$$

The denominator $\prod_{i=1}^{m} |\mathbf{b}_i(i)|$ is the determinant $\det(L)$ of the lattice L generated by $\mathbf{b}_1, \mathbf{b}_2, \ldots, \mathbf{b}_m$ and so we have

$$|T| \le \frac{3^m |\mathbf{b}_1|^m}{\det(L)}.$$

Consider the lattice L_2: the projection of L orthogonal to $\mathbf{b}_1$. Apply Corollary 11.16 to the $(m-1)$-dimensional lattice L_2 to obtain

$$|\mathbf{x}| \le (m-1)^{1/2} \big(\det(L_2) \big)^{1/(m-1)},$$

where $\mathbf{x}$ is a shortest nonzero element of L_2. At this point in the procedure REDUCEDBASIS, we have performed a recursive call to REDUCEDBASIS, and so $|\mathbf{b}_2(2)| = \Lambda_1(L_2)$ which gives

$$|\mathbf{b}_2(2)| \le (m-1)^{1/2} \big(\det(L_2) \big)^{1/(m-1)}.$$

Conditions (2) and (3) in Definition 11.17 imply that

$$|\mathbf{b}_2|^2 \ge \frac{3}{4}|\mathbf{b}_1|^2, \qquad |\mathbf{b}_2(1)|^2 \le \frac{1}{4}|\mathbf{b}_1|^2.$$

Considering the 2-dimensional lattice spanned by $\mathbf{b}_1$ and $\mathbf{b}_2$ shows that

$$|\mathbf{b}_2|^2 = |\mathbf{b}_2(1)|^2 + |\mathbf{b}_2(2)|^2,$$

and hence

$$|\mathbf{b}_1|^2 \le 2|\mathbf{b}_2(2)|^2.$$

We have

$$\det(L) = |\mathbf{b}_1^*| \det(L_2) = |\mathbf{b}_1| \det(L_2).$$

Combining this with the previous inequalities gives

$$|\mathbf{b}_1| \le 2^{1/2}|\mathbf{b}_2(2)| \le 2^{1/2}(m-1)^{1/2}\big(\det(L_2)\big)^{1/(m-1)}$$
$$= \big(2(m-1)\big)^{1/2} \left(\frac{\det(L)}{|\mathbf{b}_1|} \right)^{1/(m-1)}.$$

Hence

$$|\mathbf{b}_1|^{m-1} \le \big(2(m-1)\big)^{(m-1)/2} \frac{\det(L)}{|\mathbf{b}_1|},$$

which gives

$$|\mathbf{b}_1|^m \leq \left(2(m-1)\right)^{(m-1)/2} \det(L) \leq (2m)^{m/2} \det(L).$$

Returning to our upper bound for $|T|$ we obtain

$$|T| \leq \frac{3^m |\mathbf{b}_1|^m}{\det(L)} \leq 3^m (2m)^{m/2} = (18m)^{m/2}.$$

Since $m \leq n$ this completes the proof. □

11.3.3 Procedure REDUCEDBASIS

In Kannan [70, 71] this procedure is called SHORTEST. The preceding discussion provides a theoretical justification of the algorithm; some additional explanatory comments follow. This procedure is presented in Figure 11.2.

The algorithm begins in step (1) by reducing the input basis using the LLL algorithm. (The reduction parameter can be given its standard value $\alpha = \frac{3}{4}$ or any value in the range $\frac{1}{4} < \alpha < 1$.) The importance of this initial LLL reduction will be clear from the proof of Proposition 11.20 below.

The loop in steps (2) and (3) first projects the n-dimensional lattice L spanned by $\mathbf{b}_1, \mathbf{b}_2, \ldots, \mathbf{b}_n$ to the $(n-1)$-dimensional lattice L_1 in the hyperplane orthogonal to $\mathbf{b}_1$, then calls REDUCEDBASIS recursively on this lattice of lower dimension, and finally lifts the reduced basis of L_1 back to L. Once the execution of this loop is complete, the lattice L satisfies conditions (1), (2) and (3) in Definition 11.17.

Steps (4) and (5) determine the value of k needed for Proposition 11.18; in the proof of that result, m corresponds to $k-1$ in REDUCEDBASIS. Step (6) finds a shortest vector $\mathbf{v}_1$ in the lattice generated by $\mathbf{b}_1, \mathbf{b}_2, \ldots, \mathbf{b}_{k-1}$; by Proposition 11.19 (see below), this is also a shortest vector in the lattice generated by $\mathbf{b}_1, \mathbf{b}_2, \ldots, \mathbf{b}_n$. Step (7) replaces the basis $\mathbf{b}_1, \mathbf{b}_2, \ldots, \mathbf{b}_n$ by a new basis in which the first vector is a multiple of the shortest vector $\mathbf{v}_1$.

Finally, steps (8), (9) and (10) repeat the computation inside the loop of step (3); see Exercise 11.11.

Proposition 11.19. (Kannan [71], page 422) *If k is defined as in steps (4) and (5) of* REDUCEDBASIS *then a shortest vector of the lattice generated by* $\mathbf{b}_1, \mathbf{b}_2, \ldots, \mathbf{b}_{k-1}$ *is a shortest vector of the lattice generated by* $\mathbf{b}_1, \mathbf{b}_2, \ldots, \mathbf{b}_n$.

Proof. Consider a shortest nonzero vector $\mathbf{x}$ of the lattice $L(\mathbf{b}_1, \mathbf{b}_2, \ldots, \mathbf{b}_n)$:

$$\mathbf{x} = \sum_{i=1}^{n} x_i \mathbf{b}_i \quad (x_i \in \mathbb{Z}).$$

If this vector does not belong to the lattice $L(\mathbf{b}_1, \mathbf{b}_2, \ldots, \mathbf{b}_{k-1})$ then at least one of the coefficients $x_k, x_{k+1}, \ldots, x_n$ is not zero. Let V_k be the subspace of

- *Input:* A basis $\mathbf{b}_1, \mathbf{b}_2, \ldots, \mathbf{b}_n$ of the lattice L.

- *Output:* A Hermite-reduced basis of L in the sense of Definition 11.9.

- If $n = 1$ then

    Set reduced $\leftarrow [\mathbf{b}_1]$

  else

    (1) Set $[\mathbf{b}_1, \mathbf{b}_2, \ldots, \mathbf{b}_n] \leftarrow$ LLL$(\mathbf{b}_1, \mathbf{b}_2, \ldots, \mathbf{b}_n)$.
    (2) Set done $\leftarrow$ false.
    (3) While not done do:
        (a) For $i = 2, 3, \ldots, n$ do:
              Set $\overline{\mathbf{b}}_i'$ to the projection of $\mathbf{b}_i$ orthogonal to $\mathbf{b}_1$.
        (b) Set $[\overline{\mathbf{b}}_2, \overline{\mathbf{b}}_3, \ldots, \overline{\mathbf{b}}_n] \leftarrow$ REDUCEDBASIS$(\overline{\mathbf{b}}_2', \overline{\mathbf{b}}_3', \ldots, \overline{\mathbf{b}}_n')$.
        (c) For $i = 2, 3, \ldots, n$ do:
              Set $\mathbf{b}_i$ to the unique lifting of $\overline{\mathbf{b}}_i$ (see Definition 11.7).
        (d) If $|\mathbf{b}_2|^2 \geq \frac{3}{4}|\mathbf{b}_1|^2$
              then set done $\leftarrow$ true
              else interchange $\mathbf{b}_1 \leftrightarrow \mathbf{b}_2$.
    (4) Set indexset $\leftarrow \{\, j \mid j \geq 2,\ b_j(j) \geq |\mathbf{b}_1| \,\}$.
    (5) If indexset $= \emptyset$ then set $k \leftarrow n{+}1$ else set $k \leftarrow \min(\text{indexset})$.
    (6) Set $\mathbf{v}_1 \leftarrow$ SHORTESTVECTOR$([\mathbf{b}_1, \mathbf{b}_2, \ldots, \mathbf{b}_{k-1}])$.
    (7) Set $[\mathbf{b}_1, \mathbf{b}_2, \ldots, \mathbf{b}_n] \leftarrow$ COMPUTEBASIS$(\mathbf{v}_1, \mathbf{b}_1, \mathbf{b}_2, \ldots, \mathbf{b}_n)$.
    (8) For $i = 2, 3, \ldots, n$ do:
          Set $\overline{\mathbf{b}}_i'$ to the projection of $\mathbf{b}_i$ orthogonal to $\mathbf{b}_1$.
    (9) Set $[\overline{\mathbf{b}}_2, \overline{\mathbf{b}}_3, \ldots, \overline{\mathbf{b}}_n] \leftarrow$ REDUCEDBASIS$(\overline{\mathbf{b}}_2', \overline{\mathbf{b}}_3', \ldots, \overline{\mathbf{b}}_n')$.
    (10) For $i = 2, 3, \ldots, n$ do:
          Set $\mathbf{b}_i$ to the unique lifting of $\overline{\mathbf{b}}_i$ (see Definition 11.7).
    (11) Set reduced $\leftarrow [\mathbf{b}_1, \mathbf{b}_2, \ldots, \mathbf{b}_n]$.

- Return reduced.

FIGURE 11.2
Kannan's procedure REDUCEDBASIS

$\mathbb{R}^n$ spanned by $\mathbf{b}_1, \mathbf{b}_2, \ldots, \mathbf{b}_{k-1}$ and let $V_k^\perp$ be its orthogonal complement. Since $x_\ell \neq 0$ for some $\ell \geq k$, it follows that the projection $\mathbf{x}'$ of $\mathbf{x}$ to $V_k^\perp$ is not $\mathbf{0}$. Hence $\mathbf{x}'$ is no shorter than the shortest nonzero vector in the lattice $L_k(\mathbf{b}_1, \mathbf{b}_2, \ldots, \mathbf{b}_n)$ obtained by projecting $L(\mathbf{b}_1, \mathbf{b}_2, \ldots, \mathbf{b}_n)$ to $V_k^\perp$:

$$|\mathbf{x}'| \geq \Lambda_1\big(L_k(\mathbf{b}_1, \mathbf{b}_2, \ldots, \mathbf{b}_n)\big).$$

Condition (1) in Definition 11.17 implies that at this stage we have

$$\Lambda_1\big(L_k(\mathbf{b}_1, \mathbf{b}_2, \ldots, \mathbf{b}_n)\big) = b_k(k).$$

By the choice of k we have $|\mathbf{b}_1| \leq b_k(k)$, and therefore

$$|\mathbf{x}| \geq |\mathbf{x}'| \geq b_k(k) \geq \mathbf{b}_1.$$

Since $\mathbf{b}_1 \neq \mathbf{0}$, it follows that $\mathbf{b}_1$ is a shortest nonzero vector of the lattice $L(\mathbf{b}_1, \mathbf{b}_2, \ldots, \mathbf{b}_n)$, and $\mathbf{b}_1$ clearly belongs to $L(\mathbf{b}_1, \mathbf{b}_2, \ldots, \mathbf{b}_{k-1})$. $\qquad \square$

11.4 Complexity of Kannan's algorithm

In this section we establish an upper bound on the number of arithmetic operations performed by Kannan's algorithm. We assume without proof some basic results about the size of the rational numbers appearing during the execution of the algorithm expressed in terms of the size of the input. Some of these results can be proved in a manner similar to the complexity analysis of the LLL algorithm. The remaining details may be found in Kannan's paper [71] (see especially §3, pages 425–428).

Proposition 11.20. (Kannan [71], page 423) *The recursive call in step (3)(b) of procedure* REDUCEBASIS *is performed at most $5n/2$ times where n is the dimension of the input lattice L.*

Proof. After the LLL reduction in step (1) of REDUCEBASIS, the basis $\mathbf{b}_1, \mathbf{b}_2, \ldots, \mathbf{b}_n$ of the input lattice L satisfies the inequality

$$|\mathbf{b}_1| \leq 2^{n/2} \Lambda_1(L),$$

that is, $\mathbf{b}_1$ is no longer than $2^{n/2}$ times the length of a shortest nonzero vector in L. Each execution (except the first) of the loop in step (3) of REDUCEBASIS multiplies the length of the first basis vector $\mathbf{b}_1$ by a factor no greater than $\sqrt{3}/2$. Hence every 5 executions of the loop multiply $|\mathbf{b}_1|$ by a factor no greater than $9\sqrt{3}/32 < 1/2$. Thus once the loop has been executed $5n/2$ times, the length of $\mathbf{b}_1$ will have become no greater than $\Lambda_1(L)$, at which point the length of $\mathbf{b}_1$ cannot be reduced any further. $\qquad \square$

In our analysis of the LLL algorithm, we saw that the number of arithmetic operations performed depends on the sizes of the input and of the intermediate results. However, it is not hard to see that the number of arithmetic operations performed by procedure REDUCEDBASIS, excluding the operations performed by the calls to the LLL algorithm in step (1), depends only on the dimension n of the input lattice and not on the size of the components of the lattice basis vectors $\mathbf{b}_1, \mathbf{b}_2, \ldots, \mathbf{b}_n$.

Theorem 11.21. (Kannan [71], page 424) *If $X(n)$ is the number of arithmetic operations performed by procedure* REDUCEDBASIS, *excluding the calls to the LLL algorithm in step (1), then $X(n) \leq cn^n$ for some constant $c \geq 1$.*

Proof. Excluding the calls to the LLL algorithm, we consider the recursive calls to REDUCEDBASIS (Proposition 11.20), the calls to SHORTESTVECTOR (Proposition 11.18), and the calls to COMPUTEBASIS. Therefore

$$X(n) \leq (18n)^{n/2}\, p(n) + \frac{5n}{2}\, X(n-1), \tag{11.2}$$

where $p(n)$ is a polynomial in n alone. Recall the well-known limit

$$\lim_{n\to\infty} \frac{(n-1)^{n-1}}{n^{n-1}} = \lim_{n\to\infty} \left(1 - \frac{1}{n}\right)^{n-1} = \frac{1}{e}. \tag{11.3}$$

Hence

$$\lim_{n\to\infty} \frac{5n}{2} \cdot \frac{(n-1)^{n-1}}{n^n} = \frac{5}{2e} \approx 0.9196986.$$

Since the sequence in (11.3) is strictly decreasing, there exists an integer P such that

$$n \geq P \quad \Longrightarrow \quad \frac{5n}{2} \cdot \frac{(n-1)^{n-1}}{n^n} \leq 0.95.$$

There is also an integer Q such that

$$n \geq Q \quad \Longrightarrow \quad \frac{(18n)^{n/2} p(n)}{n^n} \leq 0.05.$$

Set $N = \max(P, Q)$ and choose $c \geq 1$ such that

$$n \leq N \quad \Longrightarrow \quad T(n) \leq c\,n^n.$$

This is the basis of our proof by induction on n that $X(n) \leq cn^n$ for all $n \geq 1$. Assume that $n > N$ and that the claim holds for $n-1$. Inequality (11.2) gives

$$\frac{X(n)}{c\,n^n} \leq \frac{(18n)^{n/2} p(n)}{c\,n^n} + \frac{\frac{5n}{2} X(n-1)}{c\,n^n} \leq \frac{0.05}{c} + \frac{\frac{5n}{2} c(n-1)^{n-1}}{c\,n^n}$$
$$\leq 0.05 + 0.95 = 1,$$

and this completes the proof. $\square$

Proposition 11.22. *The total number of arithmetic operations performed by procedure* REDUCEDBASIS *during the calls to the LLL algorithm in step (1) is bounded by a constant multiple of $n^n \log B$, where B is an upper bound on the squared lengths of the input basis vectors $b_1, b_2, \ldots, b_n$.*

Proof. Kannan [71], Proposition 3.8, page 427. □

Corollary 11.23. *The procedure* REDUCEDBASIS *computes a Hermite-reduced basis of the input lattice using $O(n^n s)$ arithmetic operations, where n is the dimension of the lattice and s is the size of the input basis in binary digits.*

11.5 Improvements to Kannan's algorithm

Helfrich [58] and Schnorr [123] have suggested two improvements to Kannan's original algorithm:

(1) Replace the interchange of the basis vectors $\mathbf{b}_1$ and $\mathbf{b}_2$ in step (3)(d) of procedure REDUCEDBASIS by an application of the Gaussian algorithm to the 2-dimensional lattice spanned by $\mathbf{b}_1$ and $\mathbf{b}_2$.

(2) Reduce the size of the set T of coefficient vectors tested by procedure SHORTESTVECTOR from $(18n)^{n/2}$ to $n^{n/2}$.

More precisely, Helfrich proves the following result.

Lemma 11.24. (Helfrich [58], page 129) *Let $\mathbf{b}_1, \mathbf{b}_2, \ldots, \mathbf{b}_n$ be a basis of the lattice L in $\mathbb{R}^d$ which satisfies the following conditions:*

(1) $\mu_{ij} \leq \frac{1}{2}$ for all $1 \leq j < i \leq n$,

(2) $|\mathbf{b}_1|^2 \leq \frac{4}{3}|\mathbf{b}_2|^2$,

(3) $\mathbf{b}(2, 2), \ldots, \mathbf{b}(n, 2)$ is a Hermite-reduced basis of the lattice L_2.

Let $\mathbf{x} = x_1\mathbf{b}_1 + x_2\mathbf{b}_2 + \cdots + x_n\mathbf{b}_n$ be a shortest nonzero vector of L. There exists a subset $T \subset \mathbb{Z}^n$ such that

(4) either $|\mathbf{x}| = |\mathbf{b}_1|$ or $[x_1, x_2, \ldots, x_n] \in T$, and

(5) $|T| \leq n^{n/2+o(1)}$ and T can be computed in $n^{n/2+O(1)}$ arithmetic operations.

Helfrich then proves the following bounds on the improved algorithm.

Proposition 11.25. (Helfrich [58], page 132) *Let $\mathbf{b}_1, \mathbf{b}_2, \ldots, \mathbf{b}_n \in \mathbb{Z}^d$ be a basis of the lattice $L \in \mathbb{R}^d$ and let $B = \max(2, |\mathbf{b}_1|^2, |\mathbf{b}_2|^2, \ldots, |\mathbf{b}_n|^2)$. Helfrich's improved version of Kannan's algorithm computes a Hermite-reduced basis of L using $d(n^{n/2+O(1)} + n^{o(n)} \log B)$ arithmetic operations on integers with binary length $O(n^2(\log n + \log B))$.*

A few years ago, Hanrot and Stehlé [54] considered a different improvement to Kannan's algorithm:

(1) Replace the enumeration procedures of Kannan and Helfrich, which search over all integer points in an n-dimensional paralleliped, by a call to the Fincke-Pohst algorithm, which searches over all integer points in an n-dimensional ellipsoid of smaller volume. (The ratio between the two volumes tends to 0 as n increases.)

(2) Use a more sophisticated analysis to obtain a smaller upper bound on the number of integer points in the n-dimensional ellipsoid.

This allows Hanrot and Stehlé to improve the complexity factor $n^{n/2} = n^{0.5n}$ to $n^{n/2e} \approx n^{0.184n}$.

11.6   Projects

Project 11.1. Write a report and give a seminar talk on the basic results in the geometry of numbers which underlie Kannan's algorithm, especially Minkowski's convex body theorem and the theory of Hermite's lattice constants. A very useful reference is Kannan's survey paper [72].

Project 11.2. Implement Kannan's algorithm (and/or the improved versions by Helfrich/Schnorr and Hanrot/Stehlé). Test your implementation on a large family of pseudorandom low-dimensional lattices.

Project 11.3. Work out all the details of a complete analysis of the complexity of Kannan's original algorithm.

Project 11.4. Work out all the details of a complete analysis of the complexity of (one or both of) the improved versions of Kannan's algorithm.

Project 11.5. Study Schnorr's algorithm [123] (Appendix D) to compute a Hermite-reduced basis of a lattice of dimension ≤ 5. Attempt to extend this approach to dimensions ≥ 6.

Project 11.6. Helfrich [57, 58] gives an algorithm which takes as input an arbitrary basis of a lattice and computes a basis $\mathbf{b}_1, \mathbf{b}_2, \dots, \mathbf{b}_n$ which is **Minkowski-reduced** in the following sense: For all $i = 1, 2, \dots, n$ the vector $\mathbf{b}_i$ is a shortest lattice vector which is linearly independent of $\mathbf{b}_1, \mathbf{b}_2, \dots \mathbf{b}_{i-1}$. (In particular, $\mathbf{b}_1$ is a shortest lattice vector). Write a report and give a seminar talk explaining this algorithm.

Project 11.7. The Fincke-Pohst algorithm and Kannan's algorithm (and its improved versions) find a shortest nonzero lattice vector given a basis of the lattice. An alternative approach, using a randomized sieve algorithm, to this

Shortest Vector Problem (SVP) has been proposed by Ajtai et al. [7, 8]; see also Nguyen and Vidick [111]. Write a report and present a seminar talk on this algorithm.

11.7 Exercises

Exercise 11.1. Prove Lemma 11.2.

Exercise 11.2. Prove Lemma 11.6.

Exercise 11.3. Prove Lemma 11.15.

Exercise 11.4. Prove the following inequality for $n \geq 1$:

$$\frac{\pi^{n/2}}{\Gamma(\frac{n}{2}+1)} \, n^{n/2} \geq 2^n.$$

Exercise 11.5. Suppose that $\mathbf{b}_1, \mathbf{b}_2, \ldots, \mathbf{b}_{n+1} \in \mathbb{Q}^n$ generate an n-dimensional lattice L. Suppose that $\mathbf{b}_1$ is a rational linear combination of $\mathbf{b}_2, \ldots, \mathbf{b}_{n+1}$ and that p and q are relatively prime integers such that $(p/q)\mathbf{b}_1$ is an integral linear combination of $\mathbf{b}_2, \ldots, \mathbf{b}_{n+1}$. Prove that $(1/q)\mathbf{b}_1$ is the shortest (nonzero) multiple of $\mathbf{b}_1$ in L.

Exercise 11.6. Suppose that $\mathbf{b}_1, \mathbf{b}_2, \ldots, \mathbf{b}_{n+1} \in \mathbb{Q}^n$, with $\mathbf{b}_1 \neq \mathbf{0}$, are linearly dependent vectors which generate an n-dimensional lattice. Let $\overline{\mathbf{b}}_2, \ldots, \overline{\mathbf{b}}_{n+1}$ be the projections of $\mathbf{b}_2, \ldots, \mathbf{b}_{n+1}$ orthogonal to $\mathbf{b}_1$, and let $\mathbf{c}$ be any vector in the lattice generated by $\overline{\mathbf{b}}_2, \ldots, \overline{\mathbf{b}}_{n+1}$. Write down an algorithm to compute the unique minimal lifting (in the sense of Definition 11.7) of $\mathbf{c}$ to the lattice generated by $\mathbf{b}_1, \mathbf{b}_2, \ldots, \mathbf{b}_{n+1}$. Hint: See Helfrich [58] (page 128) and Kannan [71] (page 428).

Exercise 11.7. Write a computer program to implement the procedure COMPUTEBASIS based on the description given in subsection 11.3.1. Compare the performance of this algorithm with the corresponding special case of the MLLL algorithm in which the input consists of $n+1$ vectors generating a lattice of dimension n.

Exercise 11.8. Write a computer program to implement the enumeration procedure SHORTESTVECTOR based on the description given in subsection 11.3.2. Compare the performance of this algorithm with that of the Fincke-Pohst algorithm.

Exercise 11.9. Work out precisely the complexity (the number of arithmetic operations performed as a function of the size of the input) for both the Fincke-Pohst algorithm (without LLL preprocessing) and the procedure SHORTESTVECTOR described in subsection 11.3.2.

Exercise 11.10. Using computer implementations of both the Fincke-Pohst algorithm (without LLL preprocessing) and the procedure SHORTESTVECTOR, compare the running times of these two methods on a set of pseudorandom lattice bases. In particular, estimate the dimension at which the Fincke-Pohst algorithm starts to perform better than SHORTESTVECTOR on the lattice bases in your sample.

Exercise 11.11. Why is it necessary to perform steps (8), (9) and (10) in the procedure REDUCEDBASIS of Figure 11.2? What would go wrong if these steps were omitted?

Exercise 11.12. Write down the details of the proof of Lemma 11.24.

12

Schnorr's Algorithm

CONTENTS

This chapter discusses Schnorr's paper [123] which introduced a one-parameter family of lattice basis reduction algorithms interpolating between the LLL algorithm [88] and Kannan's algorithm [70, 71] for a Hermite-reduced basis. For an n-dimensional lattice $L \subset \mathbb{Z}^n$ and a fixed integer $k \geq 2$, Schnorr's algorithm repeatedly applies Hermite reduction to blocks of length k in the lattice basis $\mathbf{b}_1, \mathbf{b}_2, \ldots, \mathbf{b}_n \in \mathbb{Z}^n$. For $k = 2$, the algorithm is equivalent to LLL reduction; for $k = n$, the algorithm is equivalent to Hermite reduction. Schnorr's algorithm outputs a (nonzero) vector $\mathbf{x} \in L$ for which

$$|\mathbf{x}|^2 \leq (6k^2)^{n/k} \Lambda_1(L)^2,$$

where $\Lambda_1(L)$ is the first minimum of L. If B is an upper bound for $|\mathbf{b}_1|^2, |\mathbf{b}_2|^2, \ldots, |\mathbf{b}_n|^2$ then the complexity of Schnorr's algorithm is given by

$$O\big(n^2(k^{k/2+o(k)} + n^2)\log B\big),$$

arithmetic operations on integers with $O(n \log B)$ binary digits.

The basic idea of the LLL algorithm is to repeatedly perform the Gaussian algorithm on sublattices of dimension 2. The output of the Gaussian algorithm is a Hermite-reduced basis of a 2-dimensional lattice (Exercise 12.1). The basic idea of Schnorr's algorithm is that this strategy can be generalized, using Kannan's algorithm for Hermite reduction, to repeatedly compute a Hermite-reduced basis for sublattices of dimension k, where k is any integer ≥ 2.

12.1 Basic definitions and theorems

Let $\mathbf{b}_1, \mathbf{b}_2, \ldots, \mathbf{b}_n \in \mathbb{R}^d$ be a sequence of linearly independent vectors (hence $n \leq d$) forming a basis of the lattice L.

Definition 12.1. For $1 \leq j \leq i \neq n$ we denote by $\mathbf{b}_i(j)$ the component of $\mathbf{b}_i$ orthogonal to the subspace spanned by $\mathbf{b}_1, \mathbf{b}_2, \ldots, \mathbf{b}_{j-1}$. (Note the slight difference between this notation and that of the previous chapter.) We also write $\mathbf{b}_j^* = \mathbf{b}_j(j)$ so that $\mathbf{b}_1^*, \mathbf{b}_2^*, \ldots, \mathbf{b}_n^*$ is the Gram-Schmidt orthogonalization of $\mathbf{b}_1, \mathbf{b}_2, \ldots, \mathbf{b}_n$. We write L_i for the projection of L orthogonal to the subspace spanned by $\mathbf{b}_1, \mathbf{b}_2, \ldots, \mathbf{b}_{i-1}$; this can also be expressed as

$$L_i = \mathbb{Z}\,\mathbf{b}_i(i) + \mathbb{Z}\,\mathbf{b}_{i+1}(i) + \cdots + \mathbb{Z}\,\mathbf{b}_n(i).$$

Thus L_i is a lattice of dimension $n-i+1$.

Definition 12.2. A lattice basis $\mathbf{b}_1, \mathbf{b}_2, \ldots, \mathbf{b}_n$ is **size-reduced** if its Gram-Schmidt coefficients satisfy

$$|\mu_{ij}| \leq \frac{1}{2} \quad (1 \leq j < i \leq n).$$

The lattice basis is **Hermite-reduced** if it is size-reduced and also satisfies

$$|\mathbf{b}_i^*| = \Lambda_1(L_i) \quad (1 \leq i \leq n).$$

That is, $\mathbf{b}_i^*$ is a shortest nonzero vector in the lattice L_i.

The relation between LLL-reduced bases and Hermite-reduced bases can be clarified by recalling the definition of the successive minima of a lattice.

Definition 12.3. Let L be a lattice of dimension n in $\mathbb{R}^d$. We write $\Lambda_i(L)$ for the i-**th minimum** of the lattice and define this to be the least real number c for which there exist i linearly independent vectors $\mathbf{x}_1, \mathbf{x}_2, \ldots, \mathbf{x}_i \in L$ such that $|\mathbf{x}_j| \leq c$ for $j = 1, 2, \ldots, i$. In other words, $\Lambda_i(L)$ is the radius of the smallest ball in $\mathbb{R}^d$ containing i linearly independent lattice vectors.

The next two results express the relative quality of LLL-reduced and Hermite-reduced bases in terms of the sucessive minima of the lattice.

Theorem 12.4. (Lenstra, Lenstra, Lovász [88]) *If* $\mathbf{b}_1, \mathbf{b}_2, \ldots, \mathbf{b}_n$ *is an LLL-reduced basis (with parameter* $\alpha = \frac{3}{4}$*) of an* n*-dimensional lattice* L *then*

$$\frac{1}{2^{i-1}} \leq \frac{|\mathbf{b}_i|^2}{\Lambda_i(L)^2} \leq 2^{i-1} \quad (1 \leq i \leq n).$$

Theorem 12.5. (Lagarias et al. [84], page 336) *If* $\mathbf{b}_1, \mathbf{b}_2, \ldots, \mathbf{b}_n$ *is a Hermite-reduced basis of an* n*-dimensional lattice* L *then*

$$\frac{4}{i+3} \leq \frac{|\mathbf{b}_i|^2}{\Lambda_i(L)^2} \leq \frac{i+3}{4} \quad (1 \leq i \leq n).$$

The basic idea of Schnorr [123] is to introduce lattice bases that are locally Hermite-reduced for arbitrary block size $k \geq 2$ in the same sense that an LLL-reduced lattice basis is locally Hermite-reduced for block size $k = 2$.

Definition 12.6. Assume $k \geq 2$. A lattice basis $\mathbf{b}_1, \mathbf{b}_2, \ldots, \mathbf{b}_n \in \mathbb{R}^d$ is called **k-reduced** if it is size-reduced and for each $i = 1, 2, \ldots, n-k+1$ the vectors

$$\mathbf{b}_i(i), \quad \mathbf{b}_{i+1}(i), \quad \ldots, \quad \mathbf{b}_{i+k-1}(i)$$

form a Hermite-reduced basis of a sublattice of L_i. That is, the projections of the k consecutive basis vectors $\mathbf{b}_i, \mathbf{b}_{i+1}, \ldots, \mathbf{b}_{i+k-1}$ to the orthogonal complement of the span of the first $i-1$ basis vectors $\mathbf{b}_1, \mathbf{b}_2, \ldots, \mathbf{b}_{i-1}$ form a Hermite-reduced basis of the k-dimensional lattice that they generate.

Definition 12.7. Consider a lattice basis $\mathbf{b}_1, \mathbf{b}_2, \ldots, \mathbf{b}_n \in \mathbb{R}^d$. If $k \geq 2$ then by a **k-block** we mean a sequence of k consecutive basis vectors,

$$\mathbf{b}_i, \quad \mathbf{b}_{i+1}, \quad \ldots, \quad \mathbf{b}_{i+k-1}.$$

We use the same term for the corresponding sequence of projected vectors,

$$\mathbf{b}_i(i), \quad \mathbf{b}_{i+1}(i), \quad \ldots, \quad \mathbf{b}_{i+k-1}(i).$$

Using this terminology, we can say that the LLL algorithm is based on the idea of repeatedly applying the Gaussian algorithm to reduce 2-blocks in the basis of an n-dimensional lattice. The 2-blocks $\mathbf{b}_i(i), \mathbf{b}_{i+1}(i)$ of an LLL-reduced basis are almost Hermite-reduced; they would be Hermite-reduced if we used the limiting value $\alpha = 1$ of the reduction parameter.

Definition 12.8. Assume $k \geq 2$ and $n = mk$ for $m \geq 1$. Let $\mathbf{b}_1, \mathbf{b}_2, \ldots, \mathbf{b}_n$ be a basis for the n-dimensional lattice $L \subset \mathbb{R}^d$. This basis is **block $2k$-reduced** if it is size-reduced and for each $i = 0, 1, \ldots, m-2$ the vectors

$$\mathbf{b}_{ik+1}(ik+1), \quad \mathbf{b}_{ik+2}(ik+1), \quad \ldots, \quad \mathbf{b}_{(i+2)k}(ik+1)$$

form a Hermite-reduced basis of a sublattice of L_{ik}. (Note that the consecutive $2k$-blocks overlap in k consecutive basis vectors.) That is, the projections of the vectors in the i-th $2k$-block, to the orthogonal complement of the span of the preceding ik basis vectors $\mathbf{b}_1, \mathbf{b}_2, \ldots, \mathbf{b}_{ik}$, form a Hermite-reduced basis of the $2k$-dimensional lattice that they generate.

We will see below that every block $2k$-reduced basis contains a vector of length at most $(4k^2)^{n/k}$ times the length of a shortest nonzero lattice vector.

In order to describe more precisely the quality of a k-reduced lattice basis, Schnorr [123] introduced the following lattice constants. (To justify the terminology, recall that a Hermite-reduced basis is also called a Korkine-Zolotareff reduced basis.)

Definition 12.9. (Schnorr [123], page 205) For an integer $k \geq 1$ we define the **Korkine-Zolotareff constant** (or **KZ-constant** for short) as follows:

$$\alpha_k = \max \frac{|\mathbf{b}_1|^2}{|\mathbf{b}_k^*|^2},$$

where the maximum is over all Hermite-reduced bases $\mathbf{b}_1, \mathbf{b}_2, \ldots, \mathbf{b}_k$ of lattices of dimension k.

Lemma 12.10. *We have $\alpha_k \le \alpha_{k+1}$ for all $k \ge 1$.*

Proof. Suppose that the vectors $\mathbf{b}_2, \ldots, \mathbf{b}_{k+1}$ form a Hermite-reduced basis of a k-dimensional lattice in $\mathbb{R}^d$ where $d \ge k+1$. Let $\mathbf{b}_1 \in \mathbb{R}^d$ be such that $|\mathbf{b}_1| = |\mathbf{b}_2|$ and $\mathbf{b}_1 \cdot \mathbf{b}_i = 0$ for $i = 2, \ldots, k+1$. Then $\mathbf{b}_1, \mathbf{b}_2, \ldots, \mathbf{b}_{k+1}$ form a Hermite-reduced basis of a $(k+1)$-dimensional lattice in $\mathbb{R}^d$ (Exercise 12.2). $\qquad\square$

We now quote a sequence of results from Schnorr [123]; the proofs will be left as exercises, but details may be found in the original paper. The first result relates the Korkine-Zolotareff constant to the length of the first vector in a k-reduced lattice basis.

Theorem 12.11. (Schnorr [123], page 205) *If $k-1$ is a divisor of $n-1$ then every k-reduced basis $\mathbf{b}_1, \mathbf{b}_2, \ldots, \mathbf{b}_n$ of an n-dimensional lattice L satisfies*

$$|\mathbf{b}_1|^2 \le \alpha_k^{\frac{n-1}{k-1}} \Lambda_1(L)^2.$$

Proof. Exercise 12.3. $\qquad\square$

Definition 12.12. For $n \ge 1$ the **Hermite constant** is defined by

$$\gamma_n = \max \frac{\Lambda_1(L)^2}{\det(L)^{2/n}},$$

where the maximum is over all lattices L of dimension n.

Lemma 12.13. (Schnorr [123], page 206) *If $\mathbf{b}_1, \mathbf{b}_2, \ldots, \mathbf{b}_n$ is a Hermite-reduced basis of an n-dimensional lattice then for $j = 1, 2, \ldots, n-1$ we have*

$$|\mathbf{b}_1^*|^2 \le \gamma_n^{n/(n-1)} \prod_{i=1}^{j-1} \gamma_{n-i}^{1/(n-i-1)} \left(\prod_{i=j+1}^{n} |\mathbf{b}_i^*|^2 \right)^{1/(n-j)}.$$

Proof. Exercise 12.4. $\qquad\square$

Corollary 12.14. (Schnorr [123], page 207) *For $k \ge 1$ we have*

$$\alpha_k \le k^{1+\ln k},$$

where $\ln k$ is the natural logarithm.

Proof. Exercise 12.5. $\qquad\square$

Remark 12.15. Ajtai [9] has shown that, for some $\epsilon > 0$ and all $k \ge 1$,

$$\alpha_k \ge k^{\epsilon \ln k}.$$

Corollary 12.16. *If $\mathbf{b}_1, \mathbf{b}_2, \ldots, \mathbf{b}_n$ is a k-reduced basis of the lattice L then*

$$|\mathbf{b}_1|^2 \le (1 + \epsilon_k)^{n-1} \Lambda_1(L),$$

where ϵ_k is a constant depending only on k which satisfies

$$\lim_{k \to \infty} \epsilon_k = 0.$$

Proof. It follows from Corollary 12.14 that

$$\lim_{k \to \infty} \alpha_k^{1/k} = 1.$$

Combining this with Theorem 12.11 gives the required inequality. □

Definition 12.17. (Schnorr [123], page 207) For an integer $k \geq 1$ we define

$$\beta_k = \max \left(\frac{|\mathbf{b}_1^*|^2 |\mathbf{b}_2^*|^2 \cdots |\mathbf{b}_k^*|^2}{|\mathbf{b}_{k+1}^*|^2 |\mathbf{b}_{k+2}^*|^2 \cdots |\mathbf{b}_{2k}^*|^2} \right)^{1/k},$$

where the maximum is over all Hermite-reduced bases $\mathbf{b}_1, \mathbf{b}_2, \ldots, \mathbf{b}_{2k}$ of lattices of dimension $2k$. (We call β_k the **Schnorr constant**.)

Theorem 12.18. (Schnorr [123], page 207) *If $\mathbf{b}_1, \mathbf{b}_2, \ldots, \mathbf{b}_n$ for $n = mk$ is a block $2k$-reduced basis of an n-dimensional lattice L then*

$$|\mathbf{b}_1|^2 \leq \gamma_k \beta_k^{m-1} \Lambda_1(L)^2.$$

Proof. Exercise 12.6. □

Theorem 12.19. (Schnorr [123], page 208) *For all $k \geq 1$ we have*

$$\beta_k \leq 4k^2.$$

Proof. Exercise 12.7. □

Remark 12.20. Ajtai [9] has shown that, for some $\epsilon > 0$ and all $k \geq 1$,

$$\beta_k \geq k^\epsilon.$$

Gama et al. [44] have shown that, for all $k \geq 1$,

$$\beta_k \geq \frac{1}{12} k.$$

Corollary 12.21. *For $k \geq 1$ we have the following upper bound for the Schnorr constants β_k in terms of the Hermite constants γ_k:*

$$\beta_k \leq \prod_{i=1}^{k} \gamma_{2k-i+1}^{2/(2k-i)}.$$

Proof. Combine the inequality of Lemma 12.13 with the formula in Definition 12.17. The details are left to the reader as Exercise 12.8. □

The exact value of γ_k is known only for $k \leq 8$ and $k = 24$; in particular, we have the following table, taken from Nguyen [105], page 33:

k	1	2	3	4	5	6	7	8	$\cdots$	24
γ_k	1	$\left(\frac{4}{3}\right)^{1/2}$	$2^{1/3}$	$2^{1/2}$	$8^{1/5}$	$\left(\frac{64}{3}\right)^{1/6}$	$64^{1/7}$	2	$\cdots$	4

Using these values and Corollary 12.21 we obtain

$$\beta_1 \leq \gamma_2^2 = \frac{4}{3} \approx 1.333333333,$$

$$\beta_2 \leq \gamma_4^{2/3}\gamma_3 = 2^{2/3} \approx 1.587401052,$$

$$\beta_3 \leq \gamma_6^{2/5}\gamma_5^{1/2}\gamma_4^{2/3} = \frac{2}{3}\, 2^{1/30}\, 3^{14/15} \approx 1.902200296,$$

$$\beta_4 \leq \gamma_8^{2/7}\gamma_7^{1/3}\gamma_6^{2/5}\gamma_5^{1/2} = \frac{2}{3}\, 2^{19/70}\, 2^{14/15} \approx 2.243520558.$$

12.2 A hierarchy of polynomial-time algorithms

To prove that the algorithms in Schnorr's hierarchy run in polynomial time, we introduce the weakened concepts of semi-k-reduction and semi-block-$2k$-reduction. We have already seen this weakening process before in the case $k = 2$: the Gaussian algorithm for 2-dimensional lattices is modified by the introduction of a reduction parameter. For simplicity we will follow Schnorr [123] and base the discussion on the original version of the LLL algorithm with parameter $\alpha = \frac{3}{4}$. Similar algorithms can be developed for any value of the reduction parameter: $\frac{1}{4} < \alpha < 1$ (Project 12.2).

The complexity analysis of the LLL algorithm depends on the fact that a reduction step changes only one of the Gram determinants

$$d_i = \prod_{j=1}^{i} |\mathbf{b}_j^*|^2. \tag{12.1}$$

This determinant decreases by a scalar factor depending on the reduction parameter; for $\alpha = \frac{3}{4}$ the determinant decreases by a factor of at least $\frac{4}{3}$. Extending this analysis to Hermite reduction is not straightforward, since Hermite reduction of a k-block

$$\mathbf{b}_{s+1}(s+1), \quad \mathbf{b}_{s+2}(s+1), \quad \cdots, \quad \mathbf{b}_{s+k}(s+1),$$

may change the Gram determinants d_i for $i = s+1, s+2, \ldots, s+k-1$ and some of these determinants may increase. We make the following definitions.

Definition 12.22. Let $k \geq 1$, let $n = mk$ for some $m \geq 1$, and let $\mathbf{b}_1, \mathbf{b}_2, \ldots, \mathbf{b}_n$ be a basis of the lattice $L \subset \mathbb{R}^d$. We generalize the Gram determinants to the m pairwise disjoint k-blocks by defining

$$C_i = \prod_{j=1}^{k} |\mathbf{b}_{ik+j}^*|^2, \qquad D_i = \prod_{j=0}^{i-1} C_j \qquad (i = 0, 1, \ldots, m-1).$$

Definition 12.23. Let $\mathbf{b}_1, \mathbf{b}_2, \ldots, \mathbf{b}_n$ where $n = mk$ be a basis of the lattice $L \subset \mathbb{R}^d$. This basis is **semi-k-reduced** if it is size-reduced and satisfies

(1) We have $|\mathbf{b}_{ik}^*|^2 \leq 2|\mathbf{b}_{ik+1}^*|^2$ for $i = 0, 1, \ldots, m-1$.

(2) For $i = 0, 1, \ldots, m-1$ the i-th k-block is Hermite-reduced:

$$\mathbf{b}_{ik+1}(ik+1), \quad \mathbf{b}_{ik+2}(ik+1), \quad \ldots, \quad \mathbf{b}_{(i+1)k}(ik+1).$$

This basis is **semi-block-$2k$-reduced** if it is semi-k-reduced and also satisfies

(3) We have $C_i \leq \frac{4}{3}\beta_k^k C_{i+1}$ for $i = 0, 1, \ldots, m-1$.

Here β_k is the Schnorr constant of Definition 12.17.

Theorem 12.24. (Schnorr [123], page 209) *If $\mathbf{b}_1, \mathbf{b}_2, \ldots, \mathbf{b}_n$ for $n = mk$ is a semi-block-$2k$-reduced basis of an n-dimensional lattice L then*

$$|\mathbf{b}_1|^2 \leq 2\gamma_k \alpha_k \left(\tfrac{4}{3}\beta_k\right)^{m-2} \Lambda_1(L)^2.$$

Proof. Consider a shortest nonzero vector in L:

$$\mathbf{x} = x_1 \mathbf{b}_1 + x_2 \mathbf{b}_2 + \cdots + x_n \mathbf{b}_n.$$

Let ℓ be the largest index for which $x_\ell \neq 0$:

$$\ell = \max\{j \mid x_j \neq 0\}.$$

Let i be the index of the k-block containing $\mathbf{b}_\ell$: that is, the unique i for which

$$ik < \ell \leq (i+1)k.$$

Using Definition 12.12, equation (12.1), and Definition 12.22, we have

$$|\mathbf{b}_1|^2 \leq \gamma_k D_1^{1/k} = \gamma_k C_1^{1/k}.$$

Condition (3) of Definition 12.23 and induction on i imply that

$$|\mathbf{b}_1|^2 \leq \gamma_k \left(\tfrac{4}{3}\beta_k\right)^{i-2} C_{i-1}^{1/k} \qquad (i \geq 2). \tag{12.2}$$

Using Definition 12.9, condition (1) of Definition 12.23, and Lemma 12.10, we obtain

$$|\mathbf{b}_{(i-1)k+j}^*|^2 \leq \alpha_k |\mathbf{b}_{ik}^*|^2 \qquad (1 \leq j \leq k).$$

Therefore

$$C_{i-1}^{1/k} = \prod_{j=1}^{k} |\mathbf{b}_{(i-1)k+j}^*|^{2/k} \leq \alpha_k |\mathbf{b}_{ik}^*|^2. \tag{12.3}$$

Conditions (1) and (2) of Definition 12.23 imply that

$$|\mathbf{b}_{ik}^*|^2 \leq 2|\mathbf{b}_{ik+1}^*|^2 \leq 2|\mathbf{x}|^2 = 2\Lambda_1(L)^2. \tag{12.4}$$

- *Input:* A basis $\mathbf{b}_1, \mathbf{b}_2, \ldots, \mathbf{b}_n$ $(n = mk)$ for a lattice $L \subset \mathbb{Z}^d$.
- *Output:* A semi-k-reduced basis of L.

(1) For $i = 0, 1, \ldots, m-1$ do:

Perform Hermite reduction on the i-th k-block:
$$\mathbf{b}_{ik+1}(ik+1), \quad \mathbf{b}_{ik+2}(ik+1), \quad \ldots, \quad \mathbf{b}_{(i+1)k}(ik+1).$$

(2) Set $i \leftarrow 1$.

(3) While $i \leq m-1$ do:

(Check condition (1) of Definition 12.23)
If $|\mathbf{b}_{ik}^*|^2 > 2\,|\mathbf{b}_{ik+1}^*|^2$ then

    (a) *(Size-reduce $\mathbf{b}_{ik+1}$ as in the LLL algorithm)*
        For $j = ik, \ldots, 2, 1$ do:
           Set $\mathbf{b}_{ik+1} \leftarrow \mathbf{b}_{ik+1} - \lceil \mu_{ik+1,j} \rfloor \mathbf{b}_j$.
           For $\ell = 1, \ldots, j$ do: Set $\mu_{ik+1,\ell} \leftarrow \mu_{ik+1,\ell} - \lceil \mu_{ik+1,j} \rfloor \mu_{j,\ell}$.
    (b) Exchange $\mathbf{b}_{ik} \leftrightarrow \mathbf{b}_{ik+1}$.
    (c) Perform Hermite reduction on k-blocks $s = i-1$ and $s = i$:
        $\mathbf{b}_{sk+1}(sk+1), \quad \mathbf{b}_{sk+2}(sk+1), \quad \ldots, \quad \mathbf{b}_{(s+1)k}(sk+1).$

    else

    (d) Set $i \leftarrow i + 1$.

(4) Return $\mathbf{b}_1, \mathbf{b}_2, \ldots, \mathbf{b}_n$ $(n = mk)$.

FIGURE 12.1
Schnorr's algorithm for semi-k-reduction

Combining inequalities (12.2), (12.3) and (12.4) we see that for $2 \leq i \leq m$,

$$|\mathbf{b}_1|^2 \leq 2\gamma_k \alpha_k \left(\tfrac{4}{3}\beta_k\right)^{m-2} \Lambda_1(L)^2.$$

If $i = 1$ then this argument shows that

$$|\mathbf{b}_1|^2 \leq 2\gamma_k \alpha_k \Lambda_1(L)^2.$$

If $i = 0$ then $|\mathbf{b}_1|^2 = \Lambda_1(L)^2$, and this completes the proof. $\qquad\square$

Schnorr's algorithms for semi-k-reduction and semi-block-$2k$-reduction are presented in Figures 12.1 and 12.2. Both call a subroutine for Hermite reduction; for this we can use either Kannan's algorithm or one of its improved versions. The algorithm for semi-block-$2k$-reduction tests inequalities involving the unknown constants β_k of Definition 12.17. However, the algorithm of Figure 12.2 still works well if we replace these occurrences of β_k by the upper bound $4k^2$ of Theorem 12.19.

- *Input:* A basis $\mathbf{b}_1, \mathbf{b}_2, \ldots, \mathbf{b}_n$ $(n = mk)$ for a lattice $L \subset \mathbb{Z}^d$.
- *Output:* A semi-block-$2k$-reduced basis of L.

(1) For $i = 0, 1, \ldots, m-1$ do:

Perform Hermite reduction on the i-th k-block:
$$\mathbf{b}_{ik+1}(ik+1), \quad \mathbf{b}_{ik+2}(ik+1), \quad \ldots, \quad \mathbf{b}_{(i+1)k}(ik+1).$$

(2) Set $i \leftarrow 1$.

(3) While $i \leq m-1$ do:

(Check conditions (1) and (3) of Definition 12.23)
If $|\mathbf{b}_{ik}^*|^2 > 2\,|\mathbf{b}_{ik+1}^*|^2$ then

(a) If $|\mathbf{b}_{ik}^*|^2 > 2\,|\mathbf{b}_{ik+1}^*|^2$ or $C_i > \frac{4}{3}\beta_k^k C_{i+1}$ then
   (i) *(Size-reduce $\mathbf{b}_{ik+1}$ as in the LLL algorithm)*
       For $j = ik, \ldots, 2, 1$ do:
           Set $\mathbf{b}_{ik+1} \leftarrow \mathbf{b}_{ik+1} - \lceil \mu_{ik+1,j} \rfloor \mathbf{b}_j$.
           For $\ell = 1, 2, \ldots, j$ do:
               Set $\mu_{ik+1,\ell} \leftarrow \mu_{ik+1,\ell} - \lceil \mu_{ik+1,j} \rfloor \mu_{j,\ell}$.
   (ii) Exchange $\mathbf{b}_{ik} \leftrightarrow \mathbf{b}_{ik+1}$.
   (iii) For $s = i-1$ and $s = i$, perform Hermite reduction on the k-block:
       $$\mathbf{b}_{sk+1}(sk+1), \quad \mathbf{b}_{sk+2}(sk+1), \quad \ldots, \quad \mathbf{b}_{(s+1)k}(sk+1).$$
(b) If $C_i > \frac{4}{3}\beta_k^k C_{i+1}$ then
   (i) For $s = i-1$, perform Hermite reduction on the $2k$-block:
       $$\mathbf{b}_{sk+1}(sk+1), \quad \mathbf{b}_{sk+2}(sk+1), \quad \ldots, \quad \mathbf{b}_{(s+2)k}(sk+1).$$

else

(c) Set $i \leftarrow i + 1$.

(4) Return $\mathbf{b}_1, \mathbf{b}_2, \ldots, \mathbf{b}_n$ $(n = mk)$.

FIGURE 12.2
Schnorr's algorithm for semi-block-$2k$-reduction

A detailed analysis of the complexity of these algorithms is presented in Schnorr's original paper [123]. We assume that the lattice has dimension n and consists of vectors of dimension d where $d = O(n)$. During the algorithm for semi-block-$2k$-reduction (Figure 12.2), the number of calls to the subroutine for Hermite reduction (of k-blocks or $2k$-blocks) is

$$O\left(\frac{n^2}{k} \log B\right),$$

where $n = mk$ and B is an upper bound for $|\mathbf{b}_1|^2, |\mathbf{b}_2|^2, \ldots, |\mathbf{b}_n|^2$ (the square-lengths of the input basis vectors). The number of arithmetical operations performed by this algorithm is

$$O\left(n^2(k^{k/2+o(k)} + n^2) \log B\right),$$

and the number of bits in the integer operands is $O(n \log B)$.

The worst-case behavior of Schnorr's algorithm for semi-block-$2k$-reduction has been analyzed by Ajtai [9]; see also the related work of Schnorr [124]. Gama et al. [44] have reconsidered Schnorr's algorithm for semi-block-$2k$-reduction and have reformulated it as a generalization of the LLL algorithm. They have also shown that the output of Schnorr's algorithm is better than was previously realized: they decrease Schnorr's original complexity factors by raising them to the power $\ln 2 \approx 0.6931471806$.

12.3 Projects

Project 12.1. As we have seen in the last two chapters, the analysis of algorithms for computing lattice bases of very good quality (such as Hermite-reduced bases) depends on classical results in the geometry of numbers, and in particular on the theory of Hermite's constants γ_n. Write a report and present a seminar talk on these constants, based on standard texts: Cassels [22], Rogers [121], Lekkerkerker [87], Gruber and Lekkerkerker [52], Conway and Sloane [27], and Martinet [93]. For a recent survey paper relating Hermite's constant to algorithms for lattice basis reduction, see Nguyen [105].

Project 12.2. Schnorr's paper [123] is based on the original version of the LLL algorithm with reduction parameter $\alpha = \frac{3}{4}$. Generalize Schnorr's hierarchy of algorithms and his complexity analysis by using the general version of the LLL algorithm with $\frac{1}{4} < \alpha < 1$.

Project 12.3. Section 4 of Schnorr's paper [123] presents an improvement of Kannan's algorithm for Hermite reduction, which is different from the improved algorithms of Helfrich and Hanrot-Stehlé mentioned in the previous chapter. Compare these three modifications of Kannan's algorithm.

Project 12.4. Work out the details of a complexity analysis of Schnorr's algorithm for semi-block-$2k$-reduction based on his original paper [123] (see especially Sections 4 and 5). Note that Schnorr's analysis is based on his improved version of Kannan's algorithm for Hermite reduction.

Project 12.5. An algorithm for lattice basis reduction similar to the algorithms in Schnorr's hierarchy is given by Schnorr and Euchner [127]. Compare this algorithm (block Hermite reduction) with the algorithms in Schnorr [123].

Project 12.6. Koy and Schnorr [80, 81] (see also Schnorr [125]) introduced the concept of segment LLL-reduced bases which is closely related to the block reduction discussed in this chapter. Write a report and present a seminar talk explaining the Koy-Schnorr algorithm.

12.4 Exercises

Exercise 12.1. Show that the output of the Gaussian algorithm applied to a basis of a 2-dimensional lattice is a Hermite-reduced basis of the lattice.

The remaining exercises are based directly on results in Schnorr [123].

Exercise 12.2. Complete the proof of Lemma 12.10.

Exercise 12.3. Prove Theorem 12.11.

Exercise 12.4. Prove Lemma 12.13.

Exercise 12.5. Prove Corollary 12.14.

Exercise 12.6. Prove Theorem 12.18.

Exercise 12.7. Prove Theorem 12.19.

Exercise 12.8. Complete the proof of Corollary 12.21.

13

NP-Completeness

CONTENTS

In this chapter we present the proof by van Emde Boas [144] that the problem of finding a shortest lattice vector with respect to the max norm on $\mathbb{R}^n$ is NP-complete. By the max norm we mean as usual

$$|\mathbf{x}|_\infty = \max \big(|x_1|, |x_2|, \ldots, |x_n| \big) \quad \text{for} \quad \mathbf{x} = [x_1, \dot{x}_2, \ldots, x_n] \in \mathbb{R}^n \ .$$

Proving NP-completeness of the shortest vector problem for the Euclidean norm is much more difficult (Project 13.2).

13.1 Combinatorial problems for lattices

We consider a family of closely related combinatorial problems. The first of these, POSITIVE PARTITION, is famous for being among the first problems shown to be NP-complete: see Karp [74], Garey and Johnson [46].

> **POSITIVE PARTITION**
> INSTANCE: A vector $[a_1, a_2, \ldots, a_n]$ of positive integers.
> QUESTION: Does there exist a subset $I \subset N = \{1, 2, \ldots, n\}$ such that
> $$\sum_{i \in I} a_i = \sum_{j \in N \setminus I} a_j?$$

This problem is called PARTITION in Garey and Johnson [46]. We follow van Emde Boas [144] and use the name PARTITION for the following variant, in which the vector components are not required to be positive.

PARTITION
INSTANCE: A vector $[a_1, a_2, \ldots, a_n]$ of integers.
QUESTION: Does there exist a subset $I \subset N = \{1, 2, \ldots, n\}$
such that
$$\sum_{i \in I} a_i = \sum_{j \in N \setminus I} a_j \ ?$$

It is not difficult to show that NP-completeness of PARTITION follows from the known NP-completeness of POSITIVE PARTITION (Exercise 13.1). We rewrite the equation which must be satisfied in PARTITION this way:

$$\sum_{i \in I} a_i - \sum_{j \in N \setminus I} a_j = 0 \ .$$

The question asked by PARTITION is then equivalent to the following: Do there exist $x_1, x_2, \ldots, x_n \in \{\pm 1\}$ such that

$$\sum_{i=1}^{n} x_i a_i = 0 \ ?$$

If we now weaken the condition on the coefficients $x_1, x_2, \ldots, x_n$ to allow some but not all of them to be zero, then we obtain our third problem:

WEAK PARTITION
INSTANCE: A vector $[a_1, a_2, \ldots, a_n]$ of integers.
QUESTION: Do there exist $x_1, x_2, \ldots, x_n \in \{-1, 0, 1\}$, not all zero,
such that
$$\sum_{i=1}^{n} x_i a_i = 0 \ ?$$

In PARTITION, we must partition the vector components into two complementary subsets with equal sum, but in WEAK PARTITION, it suffices to produces two disjoint nonempty subsets with equal sum. The question asked by WEAK PARTITION is equivalent to the following: Do there exist integers $x_1, x_2, \ldots, x_n$, not all zero, with $|x_i| \leq 1$ for all i, such that

$$\sum_{i=1}^{n} x_i a_i = 0 \ ?$$

If we weaken the condition on the coefficients $x_1, x_2, \ldots, x_n$ to allow an arbitrary but fixed upper bound K, then we obtain our fourth problem:

BOUNDED HOMOGENEOUS LINEAR EQUATION
INSTANCE: A vector $[a_1, a_2, \ldots, a_n]$ of integers; an integer $K \geq 1$.
QUESTION: Do there exist integers $x_1, x_2, \ldots, x_n$, not all zero,
with $|x_i| \leq K$ for all i, such that

$$\sum_{i=1}^{n} x_i a_i = 0 \ ?$$

We abbreviate the name of this problem by BHLE.

We now introduce our fifth problem, the main topic of this chapter:

SHORTEST VECTOR PROBLEM IN THE MAX-NORM
INSTANCE: Vectors $\mathbf{v}_1, \mathbf{v}_2, \ldots, \mathbf{v}_n$ in $\mathbb{Z}^n$, linearly independent in $\mathbb{R}^n$,
which form a basis for a lattice L, together with an integer $K \geq 1$.
QUESTION: Do there exist $x_1, x_2, \ldots, x_n \in \mathbb{Z}$, not all zero, such that
the lattice vector
$$\mathbf{y} = x_1 \mathbf{v}_1 + x_2 \mathbf{v}_2 + \cdots + x_n \mathbf{v}_n = [y_1, y_2, \ldots, y_n] \in \mathbb{Z}^n$$
satisfies $|y_i| \leq K$ for all i?

We write SVP for the shortest vector problem (with respect to any norm), and SVPM for the shortest vector problem in the max norm.

In problem SVPM we consider only vectors with integer components. Moreover, we have assumed that the dimension n of the lattice L is equal to the dimension of the ambient vector space $\mathbb{R}^n$. This is not an essential restriction on the generality of the problem, for the following reasons. Suppose that we have m vectors $\mathbf{v}_1, \mathbf{v}_2, \ldots, \mathbf{v}_m$ in $\mathbb{Z}^n$:

Case 1: Suppose that $m > n$. Hence the vectors are linearly dependent, and so the zero vector can be written as a linear combination

$$x_1 \mathbf{v}_1 + x_2 \mathbf{v}_2 + \cdots + x_m \mathbf{v}_m = \mathbf{0},$$

where the x_i are integers, not all zero. (Here $\mathbf{y} = \mathbf{0}$ and so $y_i = 0$ for all i, clearly satisfying $|y_i| \leq K$ for all i.) This can be verified in polynomial time, and the answer to the question in SVPM is "yes".

Case 2: Suppose that $m < n$, and that the vectors are linearly independent. We can find vectors $\mathbf{v}_{m+1}, \ldots, \mathbf{v}_n$ in $\mathbb{Z}^n$ such that

  (i)   The vectors $\mathbf{v}_1, \mathbf{v}_2, \ldots, \mathbf{v}_m, \mathbf{v}_{m+1}, \ldots, \mathbf{v}_n$ form a basis of $\mathbb{R}^n$.

  (ii) For $m+1 \leq i \leq n$, the vector $\mathbf{v}_i$ is orthogonal to the other $n-1$ vectors $\mathbf{v}_1, \ldots, \mathbf{v}_{i-1}, \mathbf{v}_{i+1}, \ldots, \mathbf{v}_n$.

  (iii) For $m+1 \leq i \leq n$, the vector $\mathbf{v}_i$ has Euclidean norm at least nK.

These conditions imply that none of the vectors $\mathbf{v}_{m+1}, \ldots, \mathbf{v}_n$ can occur in a linear combination

$$\mathbf{y} = x_1\mathbf{v}_1 + x_2\mathbf{v}_2 + \cdots + x_n\mathbf{v}_n = [y_1, y_2, \ldots, y_n] \in \mathbb{Z}^n,$$

satisfying $|y_i| \leq K$ for all i. Vectors $\mathbf{v}_{m+1}, \ldots, \mathbf{v}_n$ satisfying conditions (i)–(iii) can be computed in polynomial time, and this reduces Case 2 to SVPM.

13.2   A brief introduction to NP-completeness

The complexity classes P and NP, and the concept of NP-completeness, were introduced by Cook [28] in 1971. One year later, Karp [74] showed that a large number of classical combinatorial problems were NP-complete. This theory developed very rapidly during the 1970s, culminating in the survey monograph of Garey and Johnson [46]. The P = NP problem is now regarded as one of the most famous unsolved problems in mathematics. The Clay Mathematics Institute [23] has included it among their seven Millenium Prize Problems [24], and is offering US $1 million to anyone who can resolve this question. Cook [29] has written the official statement of the problem, which also provides a very readable introduction to the theory of NP-completeness. There is a great deal of information about the theory of NP-completness on the internet; see, for example, the Wikipedia article [148], and its links to other articles.

We briefly summarize the theory of NP-completeness. Roughly speaking, a **decision problem** is a precise mathematical question, depending on some input parameters called the **instance**, which has a yes-or-no answer. The class P consists of all decision problems which can be **solved** in polynomial time (as a function of the size of the problem instance) by a deterministic Turing machine. The class NP consists of all decision problems for which a "yes" instance can be **verified** in polynomial time by a deterministic Turing machine. It is clear that P $\subseteq$ NP, and it is widely believed that this containment is strict; that is, P $\neq$ NP. A decision problem X is **NP-hard** if any instance of any problem in NP can be converted into an instance of X by a transformation of the input data which is computable in polynomial time by a deterministic Turing machine. A problem is **NP-complete** if it is NP-hard and in NP. In order to prove that a decision problem X is NP-hard, it suffices to give an algorithm for converting any instance of a particular NP-complete decision problem Y into an instance of X by a polynomial-time transformation of the input data. In order to prove that a decision problem X is NP-complete, it suffices to prove that it is NP-hard and that a "yes" instance can be verified in polynomial time.

13.3   NP-completeness of SVP in the max norm

NP-completeness for SVPM will be proved in two steps:

(1) We first show that any instance of PARTITION, which is known to be NP-complete, can be reduced by a polynomial-time computation to an instance of WEAK PARTITION, thus establishing the NP-hardness of WEAK PARTITION. A modification of this construction reduces PARTITION to BHLE, thus establishing the NP-hardness of BHLE.

(2) We next show that any instance of BHLE can be reduced by a polynomial-time computation to an instance of SVPM. Combining this with step (1) shows that any instance of the NP-complete problem PARTITION can be reduced by a polynomial-time computation to an instance of SVPM, thus establishing the NP-hardness of SVPM. Since it is not difficult to show that SVPM is in NP (Exercise 13.10), it follows SVPM is NP-complete.

Lemma 13.1. (van Emde Boas [144], page 3) *Let $[a_1, a_2, \ldots, a_n]$ be a vector of integers, let $K, M \geq 1$ be integers, and write $a_i = b_i + Mc_i$ for all i. If*

$$M > K \sum_{i=1}^{n} |b_i|,$$

then the equation

$$\sum_{i=1}^{n} x_i a_i = 0 \ \text{ with } \ |x_i| \leq K \ \text{ for all } i,$$

is equivalent to the pair of equations

$$\left. \begin{array}{l} \displaystyle\sum_{i=1}^{n} x_i b_i = 0 \\[2mm] \displaystyle\sum_{i=1}^{n} x_i c_i = 0 \end{array} \right\} \ \text{ with } |x_i| \leq K \text{ for all } i.$$

Proof. We have

$$0 = \sum_{i=1}^{n} x_i a_i = \sum_{i=1}^{n} x_i \big(b_i + Mc_i\big) = \sum_{i=1}^{n} x_i b_i + M \sum_{i=1}^{n} x_i c_i,$$

and also

$$\left| \sum_{i=1}^{n} x_i b_i \right| \leq \sum_{i=1}^{n} |x_i|\, |b_i| \leq K \sum_{i=1}^{n} |b_i| < M.$$

Since all quantities are integers, the claim follows.   □

Example 13.2. Consider the integer vector

$$[a_1, a_2, \ldots, a_{10}] = [-81, 80, -2, -71, 60, -68, -78, 89, -20, -60],$$

and the coefficient vector

$$[x_1, x_2, \ldots, x_{10}] = [1, 0, -1, 0, 1, -1, 0, -1, 1, -1],$$

which satisfy

$$x_1 a_1 + x_2 a_2 + \cdots x_{10} a_{10} = 0. \tag{13.1}$$

Since $|x_i| \le 1$ for all i, we set $K = 1$. If $M = 10$ then $a_i = b_i + M c_i$ for

$$[b_1, b_2, \ldots, b_{10}] = [-1, 0, -2, -1, 0, 2, 2, -1, 0, 0],$$
$$[c_1, c_2, \ldots, c_{10}] = [-8, 8, 0, -7, 6, -7, -8, 9, -2, -6].$$

We have

$$K \sum_{i=1}^{n} |b_i| = 9 < 10 = M,$$

and also

$$x_1 b_1 + x_2 b_2 + \cdots x_{10} b_{10} = 0, \qquad x_1 c_1 + x_2 c_2 + \cdots x_{10} c_{10} = 0. \tag{13.2}$$

In fact, for any vector $[x_1, x_2, \ldots, x_{10}]$ satisfying $|x_i| \le 1$ for all i, Lemma 13.1 guarantees that equation (13.1) holds if and only equations (13.2) hold.

Theorem 13.3. (van Emde Boas [144], pages 3–5) *The decision problem* WEAK PARTITION *is* NP-*hard*.

Proof. We need to show that any instance of PARTITION, which is known to be NP-complete, can be reduced in polynomial time to an instance of WEAK PARTITION. Let $[a_1, a_2, \ldots, a_n]$ be an instance of PARTITION, and let the integers M and d satisfy

$$M = 2 \sum_{i=1}^{n} |a_i| + 1, \qquad d > 4.$$

For each a_i we define new integers b_{ij} for $j = 1, 2, \ldots, 5$ as follows. For $i = 1, 2, \ldots, n-1$ we set

$$
\begin{aligned}
b_{i1} &= a_i + M\left(d^{4i-4} + d^{4i-3} + 0 \phantom{{}+ d^{4i-2}} + d^{4i-1} + 0 \phantom{{}+ d^{4i}} \right) \\
b_{i2} &= 0 \phantom{{}+ a_i} + M\left(0 \phantom{{}+ d^{4i-4}} + d^{4i-3} + 0 \phantom{{}+ d^{4i-2}} + 0 \phantom{{}+ d^{4i-1}} + d^{4i} \right) \\
b_{i3} &= 0 \phantom{{}+ a_i} + M\left(d^{4i-4} + 0 \phantom{{}+ d^{4i-3}} + d^{4i-2} + 0 \phantom{{}+ d^{4i-1}} + 0 \phantom{{}+ d^{4i}} \right) \\
b_{i4} &= a_i + M\left(0 \phantom{{}+ d^{4i-4}} + 0 \phantom{{}+ d^{4i-3}} + d^{4i-2} + d^{4i-1} + d^{4i} \right) \\
b_{i5} &= 0 \phantom{{}+ a_i} + M\left(0 \phantom{{}+ d^{4i-4}} + 0 \phantom{{}+ d^{4i-3}} + 0 \phantom{{}+ d^{4i-2}} + d^{4i-1} + 0 \phantom{{}+ d^{4i}} \right)
\end{aligned}
$$

For $i = n$ we replace the term Md^{4n} by $Md^0 = M$, and so we set

$$
\begin{aligned}
b_{n1} &= a_n + M\left(d^{4n-4} + d^{4n-3} + 0 && + d^{4n-1} + 0 \right)\\
b_{n2} &= 0 + M\left(0 && + d^{4n-3} + 0 && + 0 && + 1 \right)\\
b_{n3} &= 0 + M\left(d^{4n-4} + 0 && + d^{4n-2} + 0 && + 0 \right)\\
b_{n4} &= a_n + M\left(0 && + 0 && + d^{4n-2} + d^{4n-1} + 1 \right)\\
b_{n5} &= 0 + M\left(0 && + 0 && + 0 && + d^{4n-1} + 0 \right)
\end{aligned}
$$

(That is, in the d-ary representation of the quotient of b_{n2} and b_{n4} by M, we wrap around the $4n$-th digit from the $4n$-th position to the 0-th position.)

We now consider the following instance of WEAK PARTITION:

$$
\sum_{i=1}^{n} \sum_{j=1}^{5} x_{ij} b_{ij} = 0 \text{ where } x_{ij} \in \{-1, 0, 1\} \text{ for all } i, j.
$$

Lemma 13.1 (with $K = 1$) shows that this single equation is equivalent to two equations: the first involves only the original integers a_i,

$$
\sum_{i=1}^{n} (x_{i1} + x_{i4}) a_i = 0, \tag{13.3}
$$

and the second contains only terms of the form $x_{ij} d^\ell$. In this second equation, for each ℓ there exist at most four corresponding terms. Since $d > 4$, we can apply Lemma 13.1 again (with M replaced by a suitable power of d) to convert the second equation into the following $4n$ equations:

$$
\begin{aligned}
x_{11} + x_{13} + x_{n2} + x_{n4} &= 0\\
x_{i1} + x_{i3} + x_{i-1,2} + x_{i-1,4} &= 0 && (i = 2, 3, \ldots, n)\\
x_{i1} + x_{i2} &= 0 && (i = 1, 2, \ldots, n)\\
x_{i3} + x_{i4} &= 0 && (i = 1, 2, \ldots, n)\\
x_{i1} + x_{i4} + x_{i5} &= 0 && (i = 1, 2, \ldots, n)
\end{aligned}
$$

It immediately follows that for all i we have $x_{i2} = -x_{i1}$ and $x_{i4} = -x_{i3}$. Using these relations in the first n equations gives

$$
\begin{aligned}
x_{11} + x_{13} - x_{n1} - x_{n3} &= 0\\
x_{i1} + x_{i3} - x_{i-1,1} - x_{i-1,3} &= 0 && (i = 2, 3, \ldots, n).
\end{aligned}
$$

Hence $x_{i1} + x_{i3}$ does not depend on i; we call this quantity the **weight** of the solution. Since replacing each x_{ij} by $-x_{ij}$ does not change the solution set, we may assume that the weight is non-negative. Solutions for which $x_{13} = 0$ can be regarded as solutions to the problem instance obtained by discarding b_{13} from the $5n$ components b_{ij} and considering only the remaining $5n-1$

components. The weight of a solution is then equal to x_{11}, and hence the weight is either 0 or 1. We consider these two cases separately.

Weight 0: In this case, $x_{i1} + x_{i3} = 0$ for all i. Since we already know that $x_{i3} + x_{i4} = 0$, it follows that $x_{i1} = x_{i4}$. But now $x_{i1} + x_{i4} + x_{i5} = 0$ implies $x_{i5} = -x_{i1} - x_{i4} = -2x_{i1}$. Since $|x_{i5}| \le 1$, we obtain $x_{i5} = 0$ for all i, and combining this with the previous equations shows that $x_{ij} = 0$ for all i, j. But this is not a valid solution by the definition of WEAK PARTITION.

Weight 1: In this case, $x_{i1} + x_{i3} = 1$ for all i. Since $x_{i3} + x_{i4} = 0$, we get $x_{i4} = x_{i1} - 1$, and so

$$x_{i5} = -x_{i1} - x_{i4} = -x_{i1} - (x_{i1} - 1) = 1 - 2x_{i1}.$$

There are two subcases:

(a) $x_{i1} = 0$, which gives $x_{i2} = 0$, $x_{i3} = 1$, $x_{i4} = -1$, $x_{i5} = 1$.

(b) $x_{i1} = 1$, which gives $x_{i2} = -1$, $x_{i3} = 0$, $x_{i4} = 0$, $x_{i5} = -1$.

In subcase (a), we have $x_{i1} + x_{i4} = -1$; in subcase (b), we have $x_{i1} + x_{i4} = 1$. Using this in equation (13.3), we see that we have a solution of PARTITION for the original problem instance $[a_1, a_2, \ldots, a_n]$. Conversely, a solution to this instance of PARTITION corresponds by the above construction to a solution of WEAK PARTITION for $[b_{11}, b_{12}, b_{14}, b_{15}, b_{21}, \ldots, b_{n5}]$, recalling that b_{13} has been omitted from the problem instance.

The construction which converts the given instance of PARTITION to the corresponding instance of WEAK PARTITION is computable in polynomial time (Exercise 13.6), and this completes the proof. □

Theorem 13.4. (van Emde Boas [144], page 5) *The decision problem* BHLE *is* NP-*hard.*

Proof. This is a modification of the proof of Theorem 13.3. Let $[a_1, a_2, \ldots, a_n]$ be an instance of PARTITION, let K be a positive integer, and let the integers M and d be as before. For each a_i we define b_{ij} as follows; note that d is replaced by Kd, and M is replaced by KM for $1 \le j \le 4$ but not for $j = 5$. For $1 \le i \le n-1$ we set

$$
\begin{aligned}
b_{i1} &= a_i + KM\big((Kd)^{4i-4} + (Kd)^{4i-3} + 0 \qquad\quad + (Kd)^{4i-1} + 0 \qquad\ \big)\\
b_{i2} &= 0 \ + KM\big(0 \qquad\qquad + (Kd)^{4i-3} + 0 \qquad\quad + 0 \qquad\qquad + (Kd)^{4i} \big)\\
b_{i3} &= 0 \ + KM\big((Kd)^{4i-4} + 0 \qquad\qquad + (Kd)^{4i-2} + 0 \qquad\qquad + 0 \qquad\ \big)\\
b_{i4} &= a_i + KM\big(0 \qquad\qquad + 0 \qquad\qquad + (Kd)^{4i-2} + (Kd)^{4i-1} + (Kd)^{4i} \big)\\
b_{i5} &= 0 \ + \ M\big(0 \qquad\qquad + 0 \qquad\qquad + 0 \qquad\qquad + (Kd)^{4i-1} + 0 \qquad\ \big)
\end{aligned}
$$

For $i = n$ we set

$$
\begin{aligned}
b_{n1} &= a_n + KM\big((Kd)^{4n-4} + (Kd)^{4n-3} + 0 \qquad\qquad + (Kd)^{4n-1} + 0 \,\big)\\
b_{n2} &= 0 \ + KM\big(0 \qquad\qquad + (Kd)^{4n-3} + 0 \qquad\qquad + 0 \qquad\qquad + 1 \,\big)
\end{aligned}
$$

$$
\begin{aligned}
b_{n3} &= 0 \;\; + KM\big(\; (Kd)^{4n-4} + 0 && + (Kd)^{4n-2} + 0 && + 0\;\big) \\
b_{n4} &= a_n + KM\big(\; 0 && + 0 && + (Kd)^{4n-2} + (Kd)^{4n-1} + 1\;\big) \\
b_{n5} &= 0 \;\; + \;\; M\big(\; 0 && + 0 && + 0 && + (Kd)^{4n-1} + 0\;\big)
\end{aligned}
$$

Consider this instance of BHLE:

$$
\sum_{i=1}^{n}\sum_{j=1}^{5} x_{ij}b_{ij} = 0 \text{ where } x_{ij} \in \mathbb{Z}, \; |x_{ij}| \le K \text{ for all } i,j.
$$

Repeated application of Lemma 13.1 shows that this is equivalent to

$$
\sum_{i=1}^{n}(x_{i1} + x_{i4})a_i = 0,
$$

together with these $4n$ equations (note the change in the last n equations):

$$
\begin{aligned}
x_{11} + x_{13} + x_{n2} + x_{n4} &= 0 \\
x_{i1} + x_{i3} + x_{i-1,2} + x_{i-1,4} &= 0 && (i = 2,3,\ldots,n) \\
x_{i1} + x_{i2} &= 0 && (i = 1,2,\ldots,n) \\
x_{i3} + x_{i4} &= 0 && (i = 1,2,\ldots,n) \\
K(x_{i1} + x_{i4}) + x_{i5} &= 0 && (i = 1,2,\ldots,n)
\end{aligned}
$$

Since $|x_{i5}| \le K$ we have $|x_{i1}+x_{i4}| \le 1$. The rest of the argument is similar. $\square$

Theorem 13.5. (van Emde Boas [144], page 7) *The decision problem* SVPM *is NP-hard.*

Proof. By Theorem 13.4 we know that BHLE is NP-hard. It suffices to show that any instance of BHLE can be reduced in polynomial time to an instance of SVPM. Consider the instance of BHLE given by $\mathbf{a} = [a_1, a_2, \ldots, a_n]$ and $K \ge 1$. Define integers K' and K'' by

$$
K' = K + 1, \qquad K'' = K'\left(K\sum_{i=1}^{n}|a_i| + 1\right).
$$

We introduce $n+1$ linearly independent integer vectors $\mathbf{v}_1, \mathbf{v}_2, \ldots, \mathbf{v}_{n+1}$ in $\mathbb{R}^{n+1}$ given by the rows of the following $(n+1) \times (n+1)$ matrix (v_{ij}):

$$
\begin{bmatrix} \mathbf{v}_1 \\ \mathbf{v}_2 \\ \vdots \\ \mathbf{v}_n \\ \mathbf{v}_{n+1} \end{bmatrix}
=
\begin{bmatrix}
1 & 0 & \cdots & 0 & K'a_1 \\
0 & 1 & \cdots & 0 & K'a_2 \\
\vdots & \vdots & \ddots & \vdots & \vdots \\
0 & 0 & \cdots & 1 & K'a_n \\
0 & 0 & \cdots & 0 & K''
\end{bmatrix}
=
\begin{bmatrix} I_n & K'\mathbf{a}^t \\ O & K'' \end{bmatrix}
$$

Consider the instance of SVPM given by $\mathbf{v}_1, \mathbf{v}_2, \ldots, \mathbf{v}_{n+1}$ and K. Suppose that the vector

$$\mathbf{y} = x_1\mathbf{v}_1 + x_2\mathbf{v}_2 + \cdots + x_{n+1}\mathbf{v}_{n+1} = [y_1, y_2, \ldots, y_{n+1}] \in \mathbb{Z}^{n+1},$$

satisfies $|y_i| \leq K$ for all $i = 1, 2, \ldots, n+1$. We have

$$\mathbf{y} = \left[\, x_1, \ x_2, \ \ldots, \ x_n, \ K'\sum_{i=1}^{n} x_i a_i + K''x_{n+1}\,\right],$$

Hence

$$y_i = x_i \text{ for } i = 1, 2, \ldots, n, \qquad y_{n+1} = K'\sum_{i=1}^{n} x_i a_i + K''x_{n+1}.$$

It follows that $|x_i| \leq K$ for $i = 1, 2, \ldots, n$ and therefore

$$\left| K'\sum_{i=1}^{n} x_i a_i \right| \leq K'\sum_{i=1}^{n} |x_i|\,|a_i| \leq KK'\sum_{i=1}^{n} |a_i|.$$

On the other hand, if $x_{n+1} \neq 0$ then

$$\left| K''x_{n+1} \right| \geq K'' = KK'\sum_{i=1}^{n} |a_i| + K'.$$

This implies $|y_{n+1}| \geq K'$, contradicting $|y_{n+1}| \leq K$. Hence $x_{n+1} = 0$ and so

$$y_{n+1} = K'\sum_{i=1}^{n} x_i a_i,$$

but this implies that

$$\sum_{i=1}^{n} x_i a_i = 0.$$

Therefore the given instance of SVPM is solvable if and only if the original instance of BHLE is solvable, and this completes the proof. □

13.4   Projects

Project 13.1. Write a report and present a seminar talk based on the original papers of Cook [28] and Karp [74] which introduced the theory of NP-completeness. A useful reference is the book of Garey and Johnson [46].

Project 13.2. Write a report and present a seminar talk on the NP-completeness of the problem of finding a shortest nonzero lattice vector for the Euclidean norm,

$$|\mathbf{y}|_2 = \left(|y_1|^2 + |y_2|^2 + \cdots + |y_n|^2 \right)^{1/2} \quad \text{for} \quad \mathbf{y} = [y_1, y_2, \ldots, y_n] \in \mathbb{R}^n.$$

A very readable introduction to this topic is Kumar and Sivakumar [82] (especially §3). The fundamental results were first proved by Ajtai [5, 6]; see also Blömer and Seifert [17]. The original proof by Ajtai has been simplified by Micciancio [97, 98], whose Ph.D. thesis from MIT is available online [96]. The book by Micciancio and Goldwasser [100] develops this material from the point of view of its applications to cryptography.

Project 13.3. Write a report and give a seminar presentation on the complexity of lattice algorithms. There are many excellent survey papers on this topic; some examples are Micciancio [99], Khot [76], and Regev [120].

Project 13.4. A recent serious attempt to prove P $\neq$ NP was circulated in August 2010 by Vinay Deolalikar, a research scientist at HP Labs in Palo Alto, California. Write a report and present a seminar talk giving the basic ideas in Deolalikar's strategy, and discussing the objections that experts have made to his arguments. There is a lot of information about this available on the internet which can easily be found by googling "Vinay Deolalikar".

13.5 Exercises

Exercise 13.1. It is clear that PARTITION is at least as hard as POSITIVE PARTITION, since the latter is a special case of the former. Use this fact to show that the NP-completeness of POSITIVE PARTITION implies the NP-completeness of PARTITION.

Exercise 13.2. Suppose that $\mathbf{v}_1, \mathbf{v}_2, \ldots, \mathbf{v}_m$ are vectors in $\mathbb{Z}^n$ with $m > n$. Verify that integers $x_1, x_2, \ldots, x_m$, not all zero, satisfying

$$x_1 \mathbf{v}_1 + x_2 \mathbf{v}_2 + \cdots + x_m \mathbf{v}_m = \mathbf{0},$$

can be found in polynomial time in the size of the input.

Exercise 13.3. Suppose that $m < n$ and that $\mathbf{v}_1, \mathbf{v}_2, \ldots, \mathbf{v}_m$ are vectors in $\mathbb{Z}^n$, linearly independent in $\mathbb{R}^n$, and let K be a positive integer. Verify that vectors $\mathbf{v}_{m+1}, \ldots, \mathbf{v}_n$ in $\mathbb{Z}^n$ satisfying the following conditions can be found in polynomial time in the size of the input:

(i) The vectors $\mathbf{v}_1, \mathbf{v}_2, \ldots, \mathbf{v}_m, \mathbf{v}_{m+1}, \ldots, \mathbf{v}_n$ form a basis of $\mathbb{R}^n$.

(ii) For $m+1 \le i \le n$, the vector $\mathbf{v}_i$ is orthogonal to the other $n-1$ vectors $\mathbf{v}_1, \ldots, \mathbf{v}_{i-1}, \mathbf{v}_{i+1}, \ldots, \mathbf{v}_n$.

(iii) For $m+1 \le i \le n$, the Euclidean norm $|\mathbf{v}_i|$ is at least nK.

Exercise 13.4. Verify that the following pairs of integer vectors A and coefficient vectors X provide further examples of Lemma 13.1 for $K = 1$ and $M = 10$:

$A = [a_1, a_2, \ldots, a_{10}]:$ $X = [x_1, x_2, \ldots, x_{10}]:$

$[61, 48, -30, 0, -12, 80, -8, -90, 0, -62]$ $[0, -1, 0, 0, 1, 1, -1, 1, 0, -1]$
$[-78, -80, -60, 0, 12, 90, 11, 90, 70, -2]$ $[0, 1, 0, 0, 1, 0, 0, 0, 1, 1]$
$[18, 12, 69, 2, -10, -30, -61, 51, -30, -80]$ $[1, 0, 0, 1, 0, -1, 0, 0, -1, 1]$
$[22, 31, -70, 40, 70, 22, -10, 1, -58, -41]$ $[0, 0, 1, 1, -1, 0, 0, 1, -1, -1]$
$[59, 80, -12, -80, -71, 82, 40, 30, 19, 70]$ $[1, 0, -1, 0, -1, -1, 1, -1, 0, -1]$
$[90, 52, 22, 48, 12, 20, -50, 20, 21, -30]$ $[0, 0, 0, 0, 0, -1, -1, 0, 0, 1]$

Exercise 13.5. Write a computer program to generate further examples of integer vectors A and X satisfying the condition of Lemma 13.1:

$$M > K \sum_{i=1}^{n} |b_i|.$$

Exercise 13.6. Verify that the construction given in the proof of Theorem 13.3, which converts a given instance of PARTITION to a corresponding instance of WEAK PARTITION, is computable in polynomial time.

Exercise 13.7. Complete the proof of the NP-completeness of WEAK PARTITION by showing directly that this problem is in NP; that is, a "yes" instance can be verified in polynomial time on a deterministic Turing machine.

Exercise 13.8. Fill in the details in the proof of Theorem 13.4.

Exercise 13.9. Complete the proof of the NP-completeness of BHLE by showing that this problem is in NP: that is, a "yes" instance can be verified in polynomial time on a deterministic Turing machine.

Exercise 13.10. Complete the proof of the NP-completeness of SVPM by showing that this problem is in NP: that is, a "yes" instance can be verified in polynomial time on a deterministic Turing machine.

14

The Hermite Normal Form

CONTENTS

Our goal in this chapter is to present two different applications of lattice basis reduction to the problem of computing the Hermite normal form (HNF) of an arbitrary rectangular matrix with integer entries.

We begin by reviewing the algorithm, familiar from elementary linear algebra, which uses Gaussian elimination to compute the row canonical form (RCF) of a matrix with entries in a field. If we replace divisions in the field by computations of greatest common divisors in the integers, then we obtain an algorithm for the HNF of a matrix A with integer entries. This algorithm can be easily extended to compute also a transform matrix U for which $UA = H$, where H is the Hermite normal form of A. If instead we compute the HNF of the transpose A^t, then the bottom rows of the transform matrix U provide a lattice basis for the nullspace of A. Following an idea of Sims [131], we can then apply the LLL algorithm to this lattice basis to obtain a reduced basis for the nullspace of the integer matrix A, and then a transform matrix U with small entries.

We then consider two algorithms of Havas, Majewski and Matthews [56]. The first uses a modification of the LLL algorithm to compute short multiplier vectors for the greatest common divisor of a set of integers. The second applies the first to provide an alternative approach to computing a transform matrix with small entries for the Hermite normal form of an integer matrix.

14.1   The row canonical form over a field

We begin be recalling the familiar algorithm which uses Gaussian elimination to convert an arbitrary $m \times n$ matrix A with entries in a field $\mathbb{F}$ into a row-equivalent matrix R which is in row canonical form (RCF). This matrix canonical form is also called the reduced row-echelon form, or the Gauss-Jordan form.

Definition 14.1. Let A and B be $m \times n$ matrices over the field $\mathbb{F}$. We say that A and B are **row-equivalent** over $\mathbb{F}$ if there exists an invertible $m \times m$ matrix U with entries in $\mathbb{F}$ for which $B = UA$.

It is easy to verify that the relation of row-equivalence is reflexive, symmetric, and transitive; hence it is an equivalence relation on the set of all $m \times n$ matrices over $\mathbb{F}$. We now introduce representatives of the equivalence classes.

Definition 14.2. Let R be an $m \times n$ matrix over the field $\mathbb{F}$. We say that R is in **row canonical form** (reduced row-echelon form, Gauss-Jordan form) if the following conditions are satisfied:

> (1) For some r with $0 \leq r \leq m$ we have $R_{ij} = 0$ for $r < i \leq m$ and $1 \leq j \leq n$. (Any zero rows are at the bottom of the matrix.)

> (2) For some $j_1, j_2, \ldots, j_r$ with $1 \leq j_1 < j_2 < \cdots < j_r \leq n$ we have $R_{ij} = 0$ for $1 \leq j < j_i$ and $R_{ij_i} = 1$. (Each nonzero row has 1 as its leftmost nonzero entry, and these leading ones go from left to right.)

> (3) For all i and k with $1 \leq k < i \leq r$ we have $R_{kj_i} = 0$. (In any column which contains the leading one of some row, all the other entries are zero.)

The integer r is the **rank** of the matrix A.

The following result is well-known from elementary linear algebra.

Theorem 14.3. *If A is any $m \times n$ matrix over the field $\mathbb{F}$ then there is a unique $m \times n$ matrix R over $\mathbb{F}$ satisfying these two conditions:*

> *(1) R is row-equivalent to A.*

> *(2) R is in row canonical form.*

Definition 14.4. The matrix R of Theorem 14.3 is the **row canonical form** **(RCF)** of the matrix A, and is denoted RCF(A).

The row canonical form of a matrix with entries in a field can be efficiently computed using **Gaussian elimination**. This procedure uses three types of **elementary row operation**:

- *Input:* An $m \times n$ matrix A with entries in a field $\mathbb{F}$.

- *Output:* The $m \times n$ matrix $R = \text{RCF}(A)$ and an $m \times m$ matrix U for which $UA = R$.

    (1) Set $R \leftarrow A$ and set $U \leftarrow I_m$ *(identity matrix)*.

    (2) Set $i \leftarrow 1$ and set $j \leftarrow 1$.

    (3) While $i \leq m$ and $j \leq n$ do:

       If $R_{kj} = 0$ for all $k = i, \dots, m$ then

     (a) Set $j \leftarrow j+1$

        else

     (b) Set $k \leftarrow \min\{\, k \mid R_{kj} \neq 0,\, i \leq k \leq m \,\}$.

     (c) If $k \neq i$ then $R_i \leftrightarrow R_k$ and $U_i \leftrightarrow U_k$.

     (d) If $R_{ij} \neq 1$ then $\frac{1}{R_{ij}} R_i$ and $\frac{1}{R_{ij}} U_i$.

     (e) For $k = i+1$ to m do:

       If $R_{kj} \neq 0$ then $R_k - R_{kj} R_i$ and $U_k - R_{kj} U_i$.

     (f) For $k = 1$ to $i-1$ do:

       If $R_{kj} \neq 0$ then $R_k - R_{kj} R_i$ and $U_k - R_{kj} U_i$.

     (g) Set $i \leftarrow i+1$ and $j \leftarrow j+1$.

    (4) Return R and U.

FIGURE 14.1
Algorithm for the row canonical form

---

  (1) For $1 \leq i < j \leq m$, interchange rows i and j.

       Notation: $R_i \leftrightarrow R_j$.

  (2) For $1 \leq i \leq m$ and $a \in \mathbb{F} \setminus \{0\}$, multiply row i by a.

       Notation: aR_i.

  (3) For $1 \leq i \neq j \leq m$ and $a \in \mathbb{F} \setminus \{0\}$, add a times row i to row j.

       Notation: $R_j + aR_i$.

The algorithm for computing the row canonical form R of a matrix A over a field $\mathbb{F}$ is displayed in Figure 14.1. This algorithm also computes a transform matrix U for which $UA = R$. If X is a matrix then X_i denotes the i-th row of X.

We can use the row canonical form of A to find a basis of the nullspace of A. We regard A as the coefficient matrix of the homogeneous linear system $AX = O$ where $X = [x_1, x_2, \dots, x_n]^t$ is the column vector of variables. Let r be the rank of A and suppose that the leading ones in rows $1, 2, \dots, r$ of

RCF(A) occur in columns $j_1 < j_2 < \cdots < j_r$. Define $k_1 < k_2 < \cdots < k_{n-r}$ by

$$\{\, k_1, k_2, \ldots, k_{n-r} \,\} = \{\, 1, 2, \ldots, n \,\} \setminus \{\, j_1, j_2, \ldots, j_r \,\}.$$

For $\ell = 1, 2, \ldots, n-r$ we set the vector of free variables,

$$[\, x_{k_1}, x_{k_2}, \ldots, x_{k_{n-r}} \,],$$

equal to the ℓ-th standard basis vector in $\mathbb{F}^{n-r}$ ($x_{k_\ell} = 1$ and the other free variables are 0) and then use RCF(A) to solve for the leading variables,

$$[\, x_{j_1}, x_{j_2}, \ldots, x_{j_r} \,].$$

More precisely, for $i = 1, 2, \ldots, r$ we set

$$x_{j_i} = \begin{cases} -R_{ik_\ell} & \text{if } j_i < k_\ell \\ 0 & \text{otherwise} \end{cases}$$

where R_{ik_ℓ} denotes the (i, k_ℓ) entry of $R = \text{RCF}(A)$.

If $\mathbb{F} = \mathbb{Q}$ then we take each basis vector for the nullspace, multiply it by the least common multiple of the denominators of its components to obtain an integer vector, and divide the result by the greatest common divisor of its components. In this way we obtain a basis of integer vectors for the nullspace as a vector space over $\mathbb{Q}$.

Example 14.5. Consider this 3×6 matrix A with entries in the field $\mathbb{Q}$:

$$A = \begin{bmatrix} 0 & 9 & -8 & -3 & 8 & 4 \\ 0 & -9 & -2 & -2 & -7 & 3 \\ 5 & 3 & 1 & -1 & -5 & 4 \end{bmatrix}$$

Gaussian elimination produces the following matrices R and U:

$$R = \begin{bmatrix} 1 & 0 & 0 & -\dfrac{11}{30} & -\dfrac{73}{50} & \dfrac{157}{150} \\[2mm] 0 & 1 & 0 & \dfrac{1}{9} & \dfrac{4}{5} & -\dfrac{8}{45} \\[2mm] 0 & 0 & 1 & \dfrac{1}{2} & -\dfrac{1}{10} & -\dfrac{7}{10} \end{bmatrix} \qquad U = \begin{bmatrix} \dfrac{1}{150} & \dfrac{11}{150} & \dfrac{1}{5} \\[2mm] \dfrac{1}{45} & -\dfrac{4}{45} & 0 \\[2mm] -\dfrac{1}{10} & -\dfrac{1}{10} & 0 \end{bmatrix}$$

From the matrix R we obtain a basis for the nullspace of A, by setting the free variables equal to the standard basis vectors, and solving for the leading variables. If we consider instead the transpose of A,

$$A^t = \begin{bmatrix} 0 & 0 & 5 \\ 9 & -9 & 3 \\ -8 & -2 & 1 \\ -3 & -2 & -1 \\ 8 & -7 & -5 \\ 4 & 3 & 4 \end{bmatrix},$$

then Gaussian elimination produces the following matrices R and U:

$$R = \begin{bmatrix} 1 & 0 & 0 \\ 0 & 1 & 0 \\ 0 & 0 & 1 \\ 0 & 0 & 0 \\ 0 & 0 & 0 \\ 0 & 0 & 0 \end{bmatrix} \qquad U = \begin{bmatrix} \dfrac{1}{150} & \dfrac{1}{45} & -\dfrac{1}{10} & 0 & 0 & 0 \\[2mm] \dfrac{11}{150} & -\dfrac{4}{45} & -\dfrac{1}{10} & 0 & 0 & 0 \\[2mm] \dfrac{1}{5} & 0 & 0 & 0 & 0 & 0 \\[2mm] \dfrac{11}{30} & -\dfrac{1}{9} & -\dfrac{1}{2} & 1 & 0 & 0 \\[2mm] \dfrac{73}{50} & -\dfrac{4}{5} & \dfrac{1}{10} & 0 & 1 & 0 \\[2mm] -\dfrac{157}{150} & \dfrac{8}{45} & \dfrac{7}{10} & 0 & 0 & 1 \end{bmatrix}$$

The last three rows of U are the same basis of the nullspace of A.

Example 14.6. This example illustrates the limitations of using the row canonical form over $\mathbb{Q}$ to solve problems over $\mathbb{Z}$. Consider the matrix A with entries in $\mathbb{Z}$ and its RCF over $\mathbb{Q}$:

$$A = \begin{bmatrix} 1 & 0 & 1 & 2 \\ 0 & 2 & 3 & 5 \end{bmatrix}, \qquad R = \begin{bmatrix} 1 & 0 & 1 & 2 \\ 0 & 1 & \dfrac{3}{2} & \dfrac{5}{2} \end{bmatrix}.$$

We want to find integer vectors X_1 and X_2 which form a lattice basis of the integer nullspace of A. From R we obtain a basis for the nullspace of A as a vector space over $\mathbb{Q}$; we then clear denominators to obtain integer vectors:

$$\left[-1, -\frac{3}{2}, 1, 0 \right], \ \left[-2, -\frac{5}{2}, 0, 1 \right] \longrightarrow [-2, -3, 2, 0], \ [-4, -5, 0, 2].$$

However, the nullspace of A also contains the vector $[-1, -1, -1, 1]$, which is not an integer linear combination of the integer basis vectors. Using the RCF over $\mathbb{Q}$ does not provide a satisfactory solution to this problem.

14.2 The Hermite normal form over the integers

We now consider matrices with integer entries. We modify the definition of the RCF over a field to obtain a canonical form for matrices over $\mathbb{Z}$.

Definition 14.7. Let A and B be $m \times n$ matrices with entries in $\mathbb{Z}$. We say that A and B are **row-equivalent** over $\mathbb{Z}$ if there exists an $m \times m$ matrix U which is invertible over $\mathbb{Z}$ and which satisfies $B = UA$.

By invertible over $\mathbb{Z}$ we mean that U is invertible and that both U and U^{-1} have entries in $\mathbb{Z}$. This is equivalent to the condition that $\det(U) = \pm 1$.

Definition 14.8. Let H be an $m \times n$ matrix over $\mathbb{Z}$. We say that H is in **Hermite normal form** (or **HNF**) if the following conditions are satisfied:

(1) For some r with $0 \le r \le m$ we have $H_{ij} = 0$ for $r < i \le m$ and $1 \le j \le n$.

(2) For some $j_1, j_2, \ldots, j_r$ with $1 \le j_1 < j_2 < \cdots < j_r \le n$ we have $H_{ij} = 0$ for $1 \le j < j_i$ and $H_{ij_i} \ge 1$.

(3) For all i and k with $1 \le k < i \le r$ we have $0 \le H_{kj_i} < H_{ij_i}$.

Theorem 14.9. *If A is an $m \times n$ matrix over $\mathbb{Z}$ then there is a unique $m \times n$ matrix H over $\mathbb{Z}$ satisfying these two conditions:*

(1) H is row-equivalent to A over $\mathbb{Z}$.

(2) H is in Hermite normal form.

Proof. Adkins and Weintraub [3, §5.2]. $\square$

Definition 14.10. The matrix H of Theorem 14.9 is the **Hermite normal form** of the matrix A, and is denoted $\mathrm{HNF}(A)$.

This canonical form obtains its name from a paper of Hermite in 1851, which refers to the existence of such a canonical form in the following terms:

... "La définition précédente peut être simplifiée, en observant que toute forme Φ a une équivalente, dans laquelle le système:

$$
\begin{array}{ccccc}
m_1 & p_1 & \cdots & r_1 & s_1 \\
m_2 & p_2 & \cdots & r_2 & s_2 \\
\vdots & \vdots & & \vdots & \\
m_n & p_n & \cdots & r_n & s_n
\end{array}
$$

dont le déterminant a p' valeurs Δ, est remplacée par le suivant:

$$
\begin{array}{ccccc}
\delta & g & h & \cdots & l \\
0 & \delta_1 & h' & \cdots & l' \\
0 & 0 & \delta_2 & \cdots & l'' \\
\vdots & \vdots & \vdots & & \vdots \\
0 & 0 & 0 & \cdots & \delta_{n-1}
\end{array}
$$

Les nombres entiers, désignés par les lettres $g, h, \ldots, l$, sont positifs et vérifient tous les conditions

$$g < \delta_1, \quad h < \delta_2, \quad \ldots, l < \delta_{n-1},$$

et on a toujours

$$\delta \cdot \delta_1 \cdot \delta_2 \cdots \delta_{n-1} = \Delta.\text{"}$$

In order to compute the HNF, we modify the process of Gaussian elimination so that it preserves the property that the matrix entries are integers. We therefore use repeated division with remainder in $\mathbb{Z}$ instead of division in $\mathbb{Q}$. This amounts to using the Euclidean algorithm to compute the greatest common divisor of the matrix entries at and below the current position (i, j). To explain this in more detail, consider the following matrix:

$$
\begin{bmatrix}
& & a_{1j} & \\
H' & & \vdots & \cdots \\
& & a_{i-1,j} & \\
\hline
& & a_{ij} & \\
& & a_{i+1,j} & \\
O & & \vdots & \cdots \\
& & a_{m-1,j} & \\
& & a_{mj} &
\end{bmatrix}
$$

We assume that the upper left block H' has already been reduced to Hermite normal form, and that the lower left block O is the zero matrix. If the entry a_{ij} and the entries below it in column j are all 0, then we increment the column index j to $j+1$. Otherwise, we determine which of the elements $a_{ij}, \ldots, a_{mj}$ has the least positive absolute value; suppose this a_{kj}. We then we exchange rows i and k. If the new entry a_{ij} (the old entry a_{kj}) is negative, then we multiply row i by -1. At this point, the new entry a_{ij} is equal to the absolute value of the old entry a_{kj}. We then use division with remainder to replace each entry $a_{i+1,j}, \ldots, a_{mj}$ by its remainder after division by a_{ij}. We continue processing column j in this way until $a_{ij} > 0$ and $a_{i+1,j}, \ldots, a_{mj}$ are 0. We then replace the entries above a_{ij} by their remainders after division by a_{ij}. After we have finished processing column j, we increment both the row index i and the column index j.

This procedure is often called **integer Gaussian elimination**. The basic operations performed are the **integer elementary row operations**:

(1) For $1 \leq i < j \leq m$, interchange rows i and j.

(2) For $1 \leq i \leq m$, multiply row i by -1.

(3) For $1 \leq i \neq j \leq m$ and $a \in \mathbb{Z} \setminus \{0\}$, add a times row i to row j.

A formal statement of this algorithm is displayed in Figure 14.2.

Example 14.11. For the matrix A from Example 14.5, the algorithm of Figure 14.2 produces the following matrices H and U:

$$
H = \begin{bmatrix} 5 & 3 & 1 & -1 & -5 & 4 \\ 0 & 9 & 2 & 2 & 7 & -3 \\ 0 & 0 & 10 & 5 & -1 & -7 \end{bmatrix} \qquad U = \begin{bmatrix} 0 & 0 & 1 \\ 0 & -1 & 0 \\ -1 & -1 & 0 \end{bmatrix}
$$

- *Input:* An $m \times n$ matrix A with entries in $\mathbb{Z}$.

- *Output:* The $m \times n$ matrix $H = \text{HNF}(A)$ and an $m \times m$ integer matrix U for which $\det(U) = \pm 1$ and $UA = H$.

  (1) Set $H \leftarrow A$ and set $U \leftarrow I_m$.

  (2) Set $i \leftarrow 1$ and set $j \leftarrow 1$.

  (3) While $i \leq m$ and $j \leq n$ do:

     If $H_{kj} = 0$ for all $k = i, \ldots, m$ then

  (a) Set $j \leftarrow j+1$

        else

  (b) While not $\left(H_{ij} > 0 \text{ and } H_{kj} = 0 \text{ for } k = i+1, \ldots, m \right)$ do:

       (i) Set $s \leftarrow \min\{\, |H_{kj}| \mid H_{kj} \neq 0,\ i \leq k \leq m \,\}$.

       (ii) Set $k \leftarrow \min\{\, k \mid |H_{kj}| = s,\ i \leq k \leq m \,\}$.

       (iii) If $i \neq k$ then $H_i \leftrightarrow H_k$ and $U_i \leftrightarrow U_k$.

       (iv) If $H_{ij} < 0$ then $-H_i$ and $-U_i$.

       (v) For $k = i+1, \ldots, m$ do:

            Compute $q, r \in \mathbb{Z}$ such that $H_{kj} = qH_{ij} + r$ with $0 \leq r < H_{ij}$.

            If $q \neq 0$ then $H_k - qH_i$ and $U_k - qU_i$.

  (c) For $k = 1, 2, \ldots, i-1$ do:

       Compute $q, r \in \mathbb{Z}$ such that $H_{kj} = qH_{ij} + r$ with $0 \leq r < H_{ij}$.

       If $q \neq 0$ then $H_k - qH_i$ and $U_k - qU_i$.

  (d) Set $i \leftarrow i+1$ and $j \leftarrow j+1$.

  (4) Return H and U.

FIGURE 14.2

Algorithm for the Hermite normal form

If we consider the transpose A^t, then we obtain these matrices H and U:

$$H = \begin{bmatrix} 1 & 0 & 0 \\ 0 & 1 & 0 \\ 0 & 0 & 1 \\ 0 & 0 & 0 \\ 0 & 0 & 0 \\ 0 & 0 & 0 \end{bmatrix} \qquad U = \begin{bmatrix} 0 & 0 & -111 & 317 & -28 & 72 \\ 0 & 0 & 32 & -92 & 8 & -21 \\ 0 & 0 & -4 & 12 & -1 & 3 \\ 1 & 0 & 20 & -60 & 5 & -15 \\ 0 & 0 & -91 & 260 & -23 & 59 \\ 0 & 1 & 25 & -77 & 5 & -20 \end{bmatrix}$$

Since $UA^t = H$, and the last three rows of H are zero, we see that the last three rows of U are in the nullspace of A. We will see in the next section that these rows of U in fact form a lattice basis of the integer nullspace of A.

The Hermite normal form H is uniquely determined by the original matrix, but the transform matrix U is not unique: it depends on which sequence of row operations was used to obtain H. In many applications U is more important than H, and we often want to make the entries of U as small as possible. For example, the following matrix U' also satisfies the equation $U'A^t = H$, where A is the matrix from Example 14.11:

$$U' = \begin{bmatrix} 1 & 0 & 0 & -3 & 0 & -2 \\ 1 & -2 & 2 & 2 & 3 & 4 \\ -1 & 1 & 1 & -5 & -1 & -2 \\ 3 & 1 & -6 & 3 & -3 & -6 \\ 7 & -2 & -1 & -6 & 2 & -6 \\ 5 & -4 & 0 & 8 & 5 & 5 \end{bmatrix} \qquad (14.1)$$

The matrix U' was found using lattice basis reduction.

14.3 The HNF with lattice basis reduction

Definition 14.12. If A is an $m \times n$ matrix A with integer entries then its **nullspace lattice** is the set of all integer vectors in its nullspace:

$$N(A) = \{ X \in \mathbb{Z}^n \mid AX = O \}.$$

It is easy to verify that $N(A)$ is a lattice in $\mathbb{Z}^n$, since it is closed under taking integral linear combinations.

Lemma 14.13. *Let A be an $m \times n$ matrix over $\mathbb{Z}$, let H be the Hermite normal form of A^t, and let U be an $n \times n$ matrix over $\mathbb{Z}$ with $\det(U) = \pm 1$ and $H = UA^t$. If r is the rank of H, then the last $n-r$ rows of U form a lattice basis for $N(A)$.*

Proof. The last $n-r$ rows of H are zero, and so $VA^t = O$ where V is the $(n-r) \times n$ matrix consisting of the last $n-r$ rows of U. Hence $AV^t = O$, and so the rows of V are in $N(A)$. It remains to show that any vector $X \in N(A)$ is an integral linear combination of the last $n-r$ rows of U; for this we follow Cohen [26], Proposition 2.4.9. Suppose that $XA^t = O$ for some row vector $X \in \mathbb{Z}^n$, and set $Y = XU^{-1}$. We have

$$YH = (XU^{-1})(UA^t) = XA^t = O.$$

Solving the linear system $YH = O$ from left to right for the components of the row vector Y, using the fact that H is in Hermite normal form, we find that the first r components are zero, and the last $n-r$ components are arbitrary:

$$YH = [\,y_1, y_2, \ldots, y_r, y_{r+1}, \ldots, y_n\,] \begin{bmatrix} h_{11} & h_{1j_2} & \cdots & h_{1j_r} & \cdots & h_{1m} \\ 0 & h_{2j_2} & \cdots & h_{2j_r} & \cdots & h_{2m} \\ \vdots & \vdots & \ddots & \vdots & & \vdots \\ 0 & 0 & \cdots & h_{rj_r} & \cdots & h_{rm} \\ 0 & 0 & \cdots & 0 & \cdots & 0 \\ \vdots & \vdots & & \vdots & & \vdots \\ 0 & 0 & \cdots & 0 & \cdots & 0 \end{bmatrix}$$

It follows that the last $n-r$ standard basis vectors in $\mathbb{Z}^n$ form a lattice basis for the solutions of $YH = O$, which is equivalent to $H^t Y^t = O$. Thus these standard basis vectors form a lattice basis for for $N(H^t)$. Since $X = UY$, the vector X is an integral linear combination of the last $n-r$ rows of U. $\qquad\square$

It follows from Lemma 14.13 that we can apply lattice basis reduction to the last $n-r$ rows of the transform matrix U to obtain a reduced basis for the nullspace lattice $N(A)$. Using an idea from Sims [131], Chapter 8, we can then use this reduced lattice basis to size-reduce the first r rows of U.

Example 14.14. Consider the second matrix U from Example 14.11:

$$U = \begin{bmatrix} 0 & 0 & -111 & 317 & -28 & 72 \\ 0 & 0 & 32 & -92 & 8 & -21 \\ 0 & 0 & -4 & 12 & -1 & 3 \\ 1 & 0 & 20 & -60 & 5 & -15 \\ 0 & 0 & -91 & 260 & -23 & 59 \\ 0 & 1 & 25 & -77 & 5 & -20 \end{bmatrix}$$

The last three rows of U form the matrix V containing a basis for $N(A)$:

$$V = \begin{bmatrix} 1 & 0 & 20 & -60 & 5 & -15 \\ 0 & 0 & -91 & 260 & -23 & 59 \\ 0 & 1 & 25 & -77 & 5 & -20 \end{bmatrix}$$

Applying the LLL algorithm ($\alpha = 3/4$) to the rows of V gives the following matrix U_2 containing a reduced basis for $N(A)$:

$$U_2 = \begin{bmatrix} 3 & 1 & -6 & 3 & -3 & -6 \\ 7 & -2 & -1 & -6 & 2 & -6 \\ 5 & -4 & 0 & 8 & 5 & 5 \end{bmatrix}$$

We write U_1 for the matrix formed by the first three rows of U:

$$U_1 = \begin{bmatrix} 0 & 0 & -111 & 317 & -28 & 72 \\ 0 & 0 & 32 & -92 & 8 & -21 \\ 0 & 0 & -4 & 12 & -1 & 3 \end{bmatrix}$$

We stack U_2 on top of U_2 to obtain the matrix F:

$$F = \begin{bmatrix} U_2 \\ U_1 \end{bmatrix} = \begin{bmatrix} 3 & 1 & -6 & 3 & -3 & -6 \\ 7 & -2 & -1 & -6 & 2 & -6 \\ 5 & -4 & 0 & 8 & 5 & 5 \\ 0 & 0 & -111 & 317 & -28 & 72 \\ 0 & 0 & 32 & -92 & 8 & -21 \\ 0 & 0 & -4 & 12 & -1 & 3 \end{bmatrix}$$

We compute the Gram-Schmidt orthogonalization of the rows of F. We use procedure **reduce** from the LLL algorithm to size-reduce the bottom three rows of F using the top three rows of F in reverse order. Finally, we interchange the upper and lower 3×6 blocks; the result is the matrix U' from equation (14.1). We now verify that this matrix satisfies the equation $U'A^t = H$.

14.4   Systems of linear Diophantine equations

We return to the question raised in Example 14.6: Given an $m \times n$ matrix A with integer entries and nonzero nullspace, find a basis for the nullspace lattice $N(A)$ consisting of short vectors. In other words, find a set of short and linearly independent solutions of the system of linear Diophantine equations with coefficient matrix A, such that every solution is a linear combination of these basic solutions. We have already seen one example of this: for the matrix A of Example 14.5, a reduced basis for the nullspace lattice consists of the rows of the matrix U_2 of Example 14.14.

Example 14.15. Consider the 5×10 matrix A:

$$A = \begin{bmatrix} 5 & 3 & -8 & 1 & -8 & -7 & 4 & 8 & -8 & -8 \\ -9 & -9 & 0 & 7 & -6 & 9 & 0 & 3 & 9 & 5 \\ -5 & -3 & 0 & -4 & 0 & 5 & -1 & -7 & 3 & 3 \\ 7 & 9 & -4 & -6 & -9 & -7 & -7 & -2 & -6 & 8 \\ -4 & 6 & 1 & 5 & -1 & 7 & 6 & 8 & -1 & 1 \end{bmatrix}$$

Applying the algorithm of Figure 14.2 to A^t, we obtain the matrix U in Figure 14.3 which satisfies $UA^t = H$. In the Hermite normal form of A^t, the upper 5×5 block is I_5, and the lower 5×5 block is zero. So we take rows 6 to 10 of U and apply the LLL algorithm, obtaining the following reduced basis of the nullspace lattice $N(A)$:

$$U_2 = \begin{bmatrix} 5 & 1 & 1 & 1 & -2 & 7 & -5 & -1 & 0 & -5 \\ -2 & -5 & 1 & 0 & 2 & 1 & -2 & 2 & -9 & 3 \\ 0 & 3 & -8 & 4 & 4 & 3 & -6 & -1 & 1 & -2 \\ 7 & -4 & -4 & 6 & 2 & 3 & 6 & -5 & -5 & 6 \\ 15 & -8 & -10 & -14 & 8 & 11 & 2 & 13 & 4 & 7 \end{bmatrix}$$

We stack U_2 on top of the first five rows of the original matrix U, obtaining the matrix F in Figure 14.3. We compute the Gram-Schmidt orthogonalization of the rows of F, and call procedure **reduce** from the LLL algorithm to size-reduce the bottom five rows of F using the top five rows of F in reverse order. Finally, we interchange the upper and lower 5×10 blocks. The result of this computation is another transform matrix U' with much smaller entries:

$$U' = \begin{bmatrix} -2 & 6 & 0 & 10 & -3 & -1 & -2 & -9 & 1 & -5 \\ -2 & 0 & 5 & 0 & -3 & -2 & 1 & -1 & -1 & -1 \\ -2 & 2 & 1 & 0 & -1 & 0 & -3 & 0 & 1 & -3 \\ 4 & -3 & 2 & -1 & -1 & 2 & 3 & 0 & -2 & 2 \\ -2 & 2 & -2 & 8 & 0 & -4 & 4 & -7 & -1 & 2 \\ 5 & 1 & 1 & 1 & -2 & 7 & -5 & -1 & 0 & -5 \\ -2 & -5 & 1 & 0 & 2 & 1 & -2 & 2 & -9 & 3 \\ 0 & 3 & -8 & 4 & 4 & 3 & -6 & -1 & 1 & -2 \\ 7 & -4 & -4 & 6 & 2 & 3 & 6 & -5 & -5 & 6 \\ 15 & -8 & -10 & -14 & 8 & 11 & 2 & 13 & 4 & 7 \end{bmatrix}$$

Example 14.16. This example comes from a computational study of polynomial identities for nonassociative algebras. We consider a matrix E of size 120×250; each column contains 36 nonzero entries from the set $\{\pm 1, \pm 5, \pm 25\}$. (For the precise definition of E, see Bremner and Peresi [19].) We compute $\mathrm{RCF}(E)$ over $\mathbb{Q}$ and find that the rank is 109, so the nullspace has dimension 141. We use the RCF to find the canonical basis for the nullspace, and then clear denominators and cancel common factors to find integral basis vectors for the nullspace as a vector space over $\mathbb{Q}$. These integral basis vectors are very long; their squared Euclidean lengths have between 8 and 12 digits.

We now apply the algorithm of Figure 14.2 for the Hermite normal form, together with the LLL algorithm for lattice basis reduction, to find an integral basis of the nullspace with much smaller coefficients. The following table summarizes the results:

$$
U = \begin{bmatrix}
295789 & 0 & 190416 & -1035744 & 435 & 1341232 & -1856352 & 1128732 & 0 & -1108473 \\
-42822 & 0 & -27567 & 149947 & -63 & -194173 & 268748 & -163409 & 0 & 160476 \\
-2013 & 0 & -1296 & 7049 & -3 & -9128 & 12634 & -7682 & 0 & 7544 \\
732 & 0 & 471 & -2562 & 1 & 3318 & -4592 & 2792 & 0 & -2742 \\
61 & 0 & 39 & -212 & 0 & 275 & -380 & 231 & 0 & -227 \\
316 & 1 & 169 & -971 & 0 & 1307 & -1771 & 1061 & 0 & -1061 \\
58 & 0 & 34 & -191 & 0 & 254 & -349 & 209 & 0 & -209 \\
-153 & 0 & -95 & 521 & 0 & -680 & 937 & -568 & 0 & 560 \\
26 & 0 & 14 & -81 & 0 & 107 & -145 & 88 & 1 & -87 \\
263 & 0 & 136 & -779 & -2 & 1060 & -1427 & 851 & 0 & -857
\end{bmatrix}
$$

$$
F = \begin{bmatrix}
5 & 1 & 1 & 1 & -2 & 7 & -5 & -1 & 0 & -5 \\
-2 & -5 & 1 & 0 & 2 & 1 & -2 & 2 & -9 & 3 \\
0 & 3 & -8 & 4 & 4 & 3 & -6 & -1 & 1 & -2 \\
7 & -4 & -4 & 6 & 2 & 3 & 6 & -5 & -5 & 6 \\
15 & -8 & -10 & -14 & 8 & 11 & 2 & 13 & 4 & 7 \\
295789 & 0 & 190416 & -1035744 & 435 & 1341232 & -1856352 & 1128732 & 0 & -1108473 \\
-42822 & 0 & -27567 & 149947 & -63 & -194173 & 268748 & -163409 & 0 & 160476 \\
-2013 & 0 & -1296 & 7049 & -3 & -9128 & 12634 & -7682 & 0 & 7544 \\
732 & 0 & 471 & -2562 & 1 & 3318 & -4592 & 2792 & 0 & -2742 \\
61 & 0 & 39 & -212 & 0 & 275 & -380 & 231 & 0 & -227
\end{bmatrix}
$$

FIGURE 14.3
The matrices U and F from Example 14.15

α	x	xx	xxx	$xxxx$	shortest	longest	min:sec
3/4	4	5	22	110	8	2138	9:19
9/10	17	11	113	–	8	380	22:14
99/100	15	11	115	–	8	320	37:35
1	52	89	–	–	8	68	57:41

In this table,

- Column α gives the reduction parameter from the LLL algorithm.

- Columns x, xx, xxx, $xxxx$ give the number of reduced basis vectors whose squared Euclidean lengths have 1, 2, 3, 4 digits (respectively).

- Columns 'shortest' and 'longest' give the squared Euclidean lengths of the shortest and longest reduced basis vectors.

- Column 'min:sec' gives the total computation time in minutes and seconds, using Maple 11 on an IBM ThinkCentre M55 Tower 8811V2D with Intel Core 2 CPU 6700 at 2.66 GHz and 3 GB of memory.

Especially significant improvement occurred between $\alpha = 3/4$ and $\alpha = 9/10$, and again between $\alpha = 99/100$ and $\alpha = 1$. Increasing α from 3/4 to 1 did not improve the shortest vector, but it made the squared Euclidean length of the longest vector decrease to $\approx 3.18\%$ of its original value, at the cost of increasing the computation time by a factor of ≈ 6.18.

14.5   Using linear algebra to compute the GCD

Consider a column vector of m integers:

$$V = \begin{bmatrix} v_1 \\ v_2 \\ \vdots \\ v_m \end{bmatrix} \qquad (v_1, v_2, \ldots, v_m \in \mathbb{Z}).$$

To compute the greatest common divisor of $v_1, v_2, \ldots, v_m$, we can use the Euclidean algorithm repeatedly: first compute

$$\gcd(v_1, v_2),$$

then use this result to compute

$$\gcd(\gcd(v_1, v_2), v_3),$$

and finally, after $m-1$ applications of the Euclidean algorithm, we obtain

$$\gcd(\gcd(\cdots \gcd(\gcd(v_1, v_2), v_3), \ldots, v_{m-1}), v_m) = \gcd(v_1, v_2, \ldots, v_m).$$

We can express the Euclidean algorithm in terms of matrix multiplication. Each step in the Euclidean algorithm involves an integer division with remainder,

$$r_{i-1} = q_{i+1} r_i + r_{i+1},$$

or equivalently,

$$r_{i+1} = r_{i-1} - q_{i+1} r_i.$$

Considering consecutive remainders, we can express this as a matrix equation:

$$\begin{bmatrix} -q_{i+1} & 1 \\ 1 & 0 \end{bmatrix} \begin{bmatrix} r_i \\ r_{i-1} \end{bmatrix} = \begin{bmatrix} r_{i+1} \\ r_i \end{bmatrix}.$$

To compute $\gcd(v_1, v_2)$ we set $r_0 = v_2$, $r_1 = v_1$ (note the reverse order here) and calculate $r_2, r_3, \ldots, r_t$ where $r_t = \gcd(v_1, v_2)$ is the last nonzero remainder:

$$\begin{bmatrix} -q_{t+1} & 1 \\ 1 & 0 \end{bmatrix} \cdots \begin{bmatrix} -q_3 & 1 \\ 1 & 0 \end{bmatrix} \begin{bmatrix} -q_2 & 1 \\ 1 & 0 \end{bmatrix} \begin{bmatrix} v_1 \\ v_2 \end{bmatrix} = \begin{bmatrix} 0 \\ r_t \end{bmatrix}.$$

We can write this equation more concisely as

$$Q_1 \begin{bmatrix} v_1 \\ v_2 \end{bmatrix} = \begin{bmatrix} 0 \\ \gcd(v_1, v_2) \end{bmatrix},$$

where Q_1 is a 2×2 matrix of determinant $(-1)^t = \pm 1$.

Example 14.17. We use the Euclidean algorithm to compute $\gcd(546, 220)$:

$$546 = 2 \cdot 220 + 106, \quad 220 = 2 \cdot 106 + 8, \quad 106 = 13 \cdot 8 + 2, \quad 8 = 4 \cdot 2.$$

The matrix form of this calculation is

$$\begin{bmatrix} -4 & 1 \\ 1 & 0 \end{bmatrix} \begin{bmatrix} -13 & 1 \\ 1 & 0 \end{bmatrix} \begin{bmatrix} -2 & 1 \\ 1 & 0 \end{bmatrix} \begin{bmatrix} -2 & 1 \\ 1 & 0 \end{bmatrix} \begin{bmatrix} 220 \\ 546 \end{bmatrix} = \begin{bmatrix} 0 \\ 2 \end{bmatrix}.$$

Multiplying together the 2×2 matrices gives

$$\begin{bmatrix} 273 & -110 \\ -67 & 27 \end{bmatrix} \begin{bmatrix} 220 \\ 546 \end{bmatrix} = \begin{bmatrix} 0 \\ 2 \end{bmatrix}.$$

Using this matrix form of the Euclidean algorithm, we can express the computation of $\gcd(\gcd(v_1, v_2), v_3)$ as follows:

$$\begin{bmatrix} 1 & 0 \\ 0 & Q_2 \end{bmatrix} \begin{bmatrix} Q_1 & 0 \\ 0 & 1 \end{bmatrix} \begin{bmatrix} v_1 \\ v_2 \\ v_3 \end{bmatrix} = \begin{bmatrix} 1 & 0 \\ 0 & Q_2 \end{bmatrix} \begin{bmatrix} 0 \\ \gcd(v_1, v_2) \\ v_3 \end{bmatrix}$$

$$= \begin{bmatrix} 0 \\ 0 \\ \gcd(\gcd(v_1, v_2), v_3) \end{bmatrix}.$$

This generalizes to the computation of $d = \gcd(v_1, v_2, \ldots, v_m)$; we have

$$U \begin{bmatrix} v_1 \\ v_2 \\ \vdots \\ v_m \end{bmatrix} = \begin{bmatrix} 0 \\ \vdots \\ 0 \\ d \end{bmatrix},$$

where U is the $m \times m$ matrix of determinant ± 1 defined by

$$U = \begin{bmatrix} I_{m-2} & O \\ O & Q_{m-1} \end{bmatrix} \cdots \begin{bmatrix} I_{i-1} & O & O \\ O & Q_i & O \\ O & O & I_{m-i-1} \end{bmatrix} \cdots \begin{bmatrix} Q_1 & O \\ O & I_{m-2} \end{bmatrix}.$$

We can express this computation concisely as the equation

$$UV = \begin{bmatrix} 0 & \cdots & 0 & d \end{bmatrix}^t. \tag{14.2}$$

Example 14.18. We use this method to compute $\gcd(105, 70, 42, 30)$:

$$\begin{bmatrix} 1 & 0 & 0 & 0 \\ 0 & 1 & 0 & 0 \\ 0 & 0 & -30 & 7 \\ 0 & 0 & 13 & -3 \end{bmatrix} \begin{bmatrix} 1 & 0 & 0 & 0 \\ 0 & 6 & -5 & 0 \\ 0 & -1 & 1 & 0 \\ 0 & 0 & 0 & 1 \end{bmatrix} \begin{bmatrix} -2 & 3 & 0 & 0 \\ 1 & -1 & 0 & 0 \\ 0 & 0 & 1 & 0 \\ 0 & 0 & 0 & 1 \end{bmatrix} \begin{bmatrix} 105 \\ 70 \\ 42 \\ 30 \end{bmatrix} = \begin{bmatrix} 0 \\ 0 \\ 0 \\ 1 \end{bmatrix}$$

Multiplying together the three 4×4 matrices gives

$$\begin{bmatrix} -2 & 3 & 0 & 0 \\ 6 & -6 & -5 & 0 \\ 30 & -30 & -30 & 7 \\ -13 & 13 & 13 & -3 \end{bmatrix} \begin{bmatrix} 105 \\ 70 \\ 42 \\ 30 \end{bmatrix} = \begin{bmatrix} 0 \\ 0 \\ 0 \\ 1 \end{bmatrix}$$

From the last row of the matrix we obtain a multiplier vector; that is, a row vector whose scalar product with the original column vector gives the gcd:

$$\begin{bmatrix} -13 & 13 & 13 & -3 \end{bmatrix} \begin{bmatrix} 105 \\ 70 \\ 42 \\ 30 \end{bmatrix} = 1.$$

However, there exists a much shorter multiplier vector:

$$\begin{bmatrix} 1 & 1 & -2 & -3 \end{bmatrix} \begin{bmatrix} 105 \\ 70 \\ 42 \\ 30 \end{bmatrix} = 1.$$

Example 14.18 leads us to the application of lattice basis reduction to GCD computation. The last row of the transform matrix U gives a multiplier vector expressing $\gcd(v_1, v_2, \ldots, v_m)$ as an integral linear combination of $v_1, v_2, \ldots, v_m$. The next result explains the significance of the first $m-1$ rows of U.

Lemma 14.19. *Consider the $m \times 1$ integer column vectors $V, D \in \mathbb{Z}^m$:*

$$V = \begin{bmatrix} v_1 \\ v_2 \\ \vdots \\ v_m \end{bmatrix}, \qquad D = \begin{bmatrix} 0 \\ \vdots \\ 0 \\ d \end{bmatrix}, \qquad d = \gcd(v_1, v_2, \ldots, v_m).$$

If U is an $m \times m$ integer matrix with $\det(U) = \pm 1$ and $UV = D$, then the first $m-1$ rows of U form a basis for the lattice

$$\Lambda = \{ X \in \mathbb{Z}^m \mid X^t V = O \}.$$

Proof. This is essentially a special case of Lemma 14.13, since D is the 'upside-down' Hermite normal form of V. However, repeating the (simplified) proof may clarify the argument. We have $NV = O$ where N is the $(m-1) \times m$ matrix consisting of the first $m-1$ rows of U, and hence the (transposes of the) rows of N belong to Λ. It remains to show that if $X \in \Lambda$ then X^t is an integral linear combination of the rows of N. Suppose that $X^t V = O$ for some column vector $X \in \mathbb{Z}^m$. Setting $Y = (U^{-1})^t X$ so that $Y^t = X^t U^{-1}$, and noting that $V = U^{-1} D$, we obtain $Y^t D = X^t U^{-1} D = X^t V = O$. It follows that the first $m-1$ components of Y are arbitrary and the last component is 0. Hence the first $m-1$ standard basis vectors in $\mathbb{Z}^m$ form a basis for the lattice of solutions of $Y^t D = O$. Since $X^t = Y^t U$, the claim follows. $\qquad \square$

Lemma 14.19 shows that the general multiplier vector has the form

$$x_1 U_1 + \cdots + x_{m-1} U_{m-1} + U_m,$$

where $x_1, \ldots, x_{m-1} \in \mathbb{Z}$ and $U_1, \ldots, U_{m-1}, U_m$ are the rows of U. In order to obtain a multiplier vector of short Euclidean length, we first use the LLL algorithm to compute a reduced basis of the lattice spanned by the first $m-1$ rows of U, and then use this reduced basis to size-reduce the last row of U.

Example 14.20. Continuing from Example 14.18, we apply the LLL algorithm with $\alpha = \frac{3}{4}$ to the first three rows of the matrix U:

$$\begin{bmatrix} -2 & 3 & 0 & 0 \\ 6 & -6 & -5 & 0 \\ 30 & -30 & -30 & 7 \\ -13 & 13 & 13 & -3 \end{bmatrix} \xrightarrow{\ LLL\ } \begin{bmatrix} -2 & 3 & 0 & 0 \\ 2 & 0 & -5 & 0 \\ -2 & 0 & 0 & 7 \\ -13 & 13 & 13 & -3 \end{bmatrix}$$

We now size-reduce the last row using the first three rows, and obtain the much shorter multiplier vector $[1, 1, -2, -3]$.

Example 14.21. We use the Euclidean algorithm with least absolute remainders; that is, at each step we take

$$q_{i+1} = \left\lceil \frac{r_{i-1}}{r_i} \right\rfloor,$$

where as usual $\lceil x \rfloor = \lceil x - \frac{1}{2} \rceil$, the smallest integer greater than or equal to $x - \frac{1}{2}$. Consider Example 7.1 from Havas, Majewski and Matthews [56]:

$$V = \begin{bmatrix} v_1 \\ v_2 \\ v_3 \\ v_4 \end{bmatrix} = \begin{bmatrix} 116085838 \\ 181081878 \\ 314252913 \\ 10346840 \end{bmatrix}.$$

We obtain the following transform matrix U:

$$\begin{bmatrix} 1 & 0 & 0 & 0 \\ 0 & 1 & 0 & 0 \\ 0 & 0 & -10346840 & 1 \\ 0 & 0 & 1 & 0 \end{bmatrix} \begin{bmatrix} 1 & 0 & 0 & 0 \\ 0 & 314252913 & -2 & 0 \\ 0 & -157126456 & 1 & 0 \\ 0 & 0 & 0 & 1 \end{bmatrix} \times$$

$$\begin{bmatrix} -90540939 & 58042919 & 0 & 0 \\ -14327851 & 9185130 & 0 & 0 \\ 0 & 0 & 1 & 0 \\ 0 & 0 & 0 & 1 \end{bmatrix} =$$

$$\begin{bmatrix} -90540939 & 58042919 & 0 & 0 \\ -4502568913779963 & 2886453858783690 & -2 & 0 \\ -23293679995803545263040 & 14932838074590182275200 & -10346840 & 1 \\ 2251284449726056 & -1443226924799280 & 1 & 0 \end{bmatrix}$$

We have $UV = \begin{bmatrix} 0 & 0 & 0 & 1 \end{bmatrix}^t$, showing that $\gcd(v_1, v_2, v_3, v_4) = 1$. This result is unsatisfactory: the entries of U are extremely large compared to the components of V. In particular, the last row of U gives a multiplier vector expressing $\gcd(v_1, v_2, v_3, v_4)$ as a linear combination of v_1, v_2, v_3, v_4:

$$2251284449726056 \, v_1 - 1443226924799280 \, v_2 + v_3 = 1. \tag{14.3}$$

In contrast, the algorithm of Havas, Majewski and Matthews [56] produces this transform matrix U,

$$U = \begin{bmatrix} -103 & 146 & -58 & 362 \\ -603 & 13 & 220 & -144 \\ 15 & -1208 & 678 & 381 \\ -88 & 352 & -167 & -101 \end{bmatrix},$$

and the last row expresses $\gcd(v_1, v_2, v_3, v_4)$ as this linear combination,

$$-88 \, v_1 + 352 \, v_2 - 167 \, v_3 - 101 \, v_4 = 1. \tag{14.4}$$

14.6   The HMM algorithm for the GCD

In this section we will explain, following Havas, Majewski and Matthews [56], another way to use lattice basis reduction to compute a transform matrix U with very small entries satisfying equation (14.2). This approach is based on the following idea. We are given m integers $v_1, v_2, \ldots, v_m$ and we consider the lattice $L \subset \mathbb{Z}^{m+1}$ spanned by the rows of the following matrix which depends on the parameter γ, a positive integer:

$$
C_\gamma =
\begin{bmatrix}
1 & 0 & \cdots & 0 & \gamma v_1 \\
0 & 1 & \cdots & 0 & \gamma v_2 \\
\vdots & \vdots & \ddots & \vdots & \vdots \\
0 & 0 & \cdots & 1 & \gamma v_m
\end{bmatrix}
= \begin{bmatrix} I_m & \gamma V \end{bmatrix}, \qquad
V =
\begin{bmatrix}
v_1 \\
v_2 \\
\vdots \\
v_m
\end{bmatrix}.
$$

Lemma 14.22. *Let α $(\frac{1}{4} < \alpha < 1)$ be the reduction parameter of the LLL algorithm and let $\beta = 4/(4\alpha - 1)$. Let $\overline{C_\gamma}$ be the matrix obtained by applying the LLL algorithm with parameter α to the rows of C_γ. If*

$$
\gamma > \beta^{(m-2)/2} |V|,
$$

where $|V|$ is the Euclidean length of V, then the last column of $\overline{C_\gamma}$ has the form

$$
\begin{bmatrix}
0 \\
\vdots \\
0 \\
\pm \gamma d
\end{bmatrix}, \qquad
d = \gcd(v_1, v_2, \ldots, v_m).
$$

Proof. Exercise 14.3.   □

The first m entries of the last row of the matrix $\overline{C_\gamma}$ provide a multiplier vector of small Euclidean length. It is interesting to consider the limiting behavior of this approach as γ becomes arbitrarily large. (The reader will notice a similarity between this algorithm for the greatest common divisor and the MLLL algorithm for linearly dependent vectors; see Project 14.3.)

Example 14.23. We take $m = 6$ and consider the integer vector

$$
V =
\begin{bmatrix}
133 \\
43 \\
-277 \\
-46 \\
-617 \\
-833
\end{bmatrix}
$$

We use the Maple procedure `IntegerRelations[LLL]` to apply the LLL algorithm with $\alpha = \frac{3}{4}$ to C_γ for $\gamma = 1, 2, \ldots, 6$. We obtain the following matrices:

$$\overline{C_1} = \left[\begin{array}{cccccc|c} 1 & -2 & 0 & 1 & 0 & 0 & 1 \\ -1 & 1 & 0 & -2 & 0 & 0 & 2 \\ 2 & 1 & 0 & 2 & -1 & 1 & 1 \\ 0 & -1 & -3 & -1 & 0 & 1 & 1 \\ 2 & 1 & 0 & -2 & 2 & -1 & 0 \\ 0 & 0 & 1 & -2 & -3 & 2 & 0 \end{array}\right]$$

$$\overline{C_2} = \left[\begin{array}{cccccc|c} 1 & -2 & 0 & 1 & 0 & 0 & 2 \\ 1 & 3 & 0 & 1 & -1 & 1 & 0 \\ -1 & 1 & -3 & -2 & 0 & 1 & 0 \\ -1 & 2 & 3 & -1 & 0 & -1 & 2 \\ 2 & 1 & 0 & -2 & 2 & -1 & 0 \\ 0 & 0 & 1 & -2 & -3 & 2 & 0 \end{array}\right]$$

$$\overline{C_3} = \left[\begin{array}{cccccc|c} 1 & -2 & 0 & 1 & 0 & 0 & 3 \\ 1 & 3 & 0 & 1 & -1 & 1 & 0 \\ -1 & 1 & -3 & -2 & 0 & 1 & 0 \\ 2 & 1 & 0 & -2 & 2 & -1 & 0 \\ 0 & 0 & 1 & -2 & -3 & 2 & 0 \\ -1 & 2 & 3 & -1 & 0 & -1 & 3 \end{array}\right]$$

$$\overline{C_4} = \left[\begin{array}{cccccc|c} 1 & 3 & 0 & 1 & -1 & 1 & 0 \\ 1 & -2 & 0 & 1 & 0 & 0 & 4 \\ -1 & 1 & -3 & -2 & 0 & 1 & 0 \\ 2 & 1 & 0 & -2 & 2 & -1 & 0 \\ 0 & 0 & 1 & -2 & -3 & 2 & 0 \\ 3 & -1 & -3 & 3 & -1 & 2 & 0 \end{array}\right]$$

$$\overline{C_5} = \left[\begin{array}{cccccc|c} 1 & 3 & 0 & 1 & -1 & 1 & 0 \\ -1 & 1 & -3 & -2 & 0 & 1 & 0 \\ 2 & 1 & 0 & -2 & 2 & -1 & 0 \\ 0 & 0 & 1 & -2 & -3 & 2 & 0 \\ 1 & -2 & 0 & 1 & 0 & 0 & 5 \\ 3 & -1 & -3 & 3 & -1 & 2 & 0 \end{array}\right]$$

$$\overline{C_6} = \left[\begin{array}{cccccc|c} 1 & 3 & 0 & 1 & -1 & 1 & 0 \\ -1 & 1 & -3 & -2 & 0 & 1 & 0 \\ 2 & 1 & 0 & -2 & 2 & -1 & 0 \\ 0 & 0 & 1 & -2 & -3 & 2 & 0 \\ 3 & -1 & -3 & 3 & -1 & 2 & 0 \\ 1 & -2 & 0 & 1 & 0 & 0 & 6 \end{array}\right]$$

For $\gamma \geq 7$, the matrix $\overline{C_\gamma}$ is the same as $\overline{C_6}$, except that the entry in the bottom right corner is γ.

As $\gamma \to \infty$, the matrices $\overline{C_\gamma}$ eventually converge (except for the entry in the bottom right corner), and in fact the reduce and exchange steps performed by the LLL algorithm also settle down to the same sequence of operations. Our next task is to identify this limiting sequence of operations, and perform them not on the matrix C_γ but instead on the matrix $\begin{bmatrix} I_m & V \end{bmatrix}$. In this way, we will achieve the same result but without the complications introduced by multiplying the input vector V by an arbitrarily large integer γ.

Recall that we are given an input vector

$$V = \begin{bmatrix} v_1 & v_2 & \cdots & v_m \end{bmatrix}^t.$$

We define the $(m-1)$-dimensional lattice $\Lambda \subset \mathbb{Z}^m$ as follows:

$$\Lambda = \{\, X \in \mathbb{Z}^m \mid V^t X = 0 \,\}.$$

For each positive integer γ, we define the m-dimensional lattice $L_\gamma \subset \mathbb{Z}^{m+1}$ to be the lattice spanned by the (transposes of the) rows of the following matrix:

$$C_\gamma = \begin{bmatrix} I_m & \gamma V \end{bmatrix}.$$

Thus L_γ consists of all (column) vectors of the form

$$\begin{bmatrix} X \\ a \end{bmatrix} = \begin{bmatrix} x_1 \\ \vdots \\ x_m \\ \gamma \sum_{i=1}^m x_i v_i \end{bmatrix} \qquad (x_1, \ldots, x_m \in \mathbb{Z}).$$

It is clear that

$$X \in \Lambda \quad \Longleftrightarrow \quad \begin{bmatrix} X \\ 0 \end{bmatrix} \in L_\gamma.$$

Furthermore, if

$$\begin{bmatrix} X \\ a \end{bmatrix} \in L_\gamma, \qquad X \notin \Lambda,$$

then $a \neq 0$ and so the Euclidean length of $\begin{bmatrix} X & a \end{bmatrix}^t$ is at least γ.

From our analysis of the LLL algorithm we know that if $\mathbf{b}_1, \mathbf{b}_2, \ldots, \mathbf{b}_{m-1}$ are the first $m-1$ vectors in an LLL-reduced basis (with parameter α) for the m-dimensional lattice L, and $X_1, X_2, \ldots, X_{m-1}$ are any $m-1$ linearly independent vectors in L, then for $j = 1, 2, \ldots, m-1$ we have

$$|\mathbf{b}_j| \leq \beta^{(m-2)/2} \max\big(|X_1|, |X_2|, \ldots, |X_{m-1}|\big).$$

But it is clear that these $m-1$ vectors belong to L:

$$X_1 = \begin{bmatrix} -v_2 & v_1 & 0 & \cdots & 0 & 0 \end{bmatrix},$$
$$X_2 = \begin{bmatrix} -v_3 & 0 & v_1 & \cdots & 0 & 0 \end{bmatrix},$$
$$\vdots$$

$$X_{m-1} = [\ -v_m \quad 0 \quad 0 \quad \cdots \quad v_1 \quad 0 \ ].$$

For $j = 1, 2, \ldots, m-1$ we have $|X_j| \leq |V|$ and so

$$|\mathbf{b}_j| \leq \beta^{(m-2)/2} |V|.$$

From the last two paragraphs it follows that if

$$\gamma > \beta^{(m-2)/2} |V|,$$

then the first $m-1$ vectors in an LLL-reduced basis for L_γ must belong to Λ. The last row of the reduced matrix $\overline{C_\gamma}$ then has the form

$$[\ b_{m1} \quad \cdots \quad b_{mm} \quad \gamma g \ ] \qquad (g \in \mathbb{Z}).$$

But we know that for some unimodular matrix U we have

$$\overline{C_\gamma} = U C_\gamma = U [\ I_m \quad \gamma V \ ] = [\ U \quad \gamma U V \ ] = [\ U \quad \gamma D \ ],$$

where

$$D = [\ 0 \quad \cdots \quad 0 \quad d \ ]^t, \qquad d = \gcd(v_1, v_2, \ldots, v_m).$$

Hence $g = \pm d$.

We now verify that the sequence of operations performed by the LLL algorithm on the rows of C_γ converges to a limiting sequence as γ tends to infinity. The following argument closely follows Section 3 of Havas, Majewski and Matthews [56]. We apply the LLL algorithm to the rows of the matrix

$$C_\gamma = [\ I_m \quad \gamma V \ ]$$

At an intermediate stage of the computation, we have the matrix

$$C = [\ B \quad \gamma A \ ],$$

where B is an $m \times m$ integer matrix and A is an $m \times 1$ integer vector:

$$A = [\ a_1 \quad a_2 \quad \cdots \quad a_m \ ]^t.$$

The rows of $C = (c_{ij})$ will be denoted $\mathbf{c}_1, \mathbf{c}_2, \ldots, \mathbf{c}_m$ so that

$$\mathbf{c}_i = [\ c_{i1} \quad \cdots \quad c_{im} \quad c_{i,m+1} \ ].$$

At the end of the computation, we have the matrix described above,

$$\overline{C_\gamma} = [\ U \quad \gamma D \ ].$$

We proceed by induction on k, the index variable in the LLL algorithm. We assume that $2 \leq k \leq m$ and that the first k rows of C are LLL-reduced. As we have already seen, this implies that the first $k-2$ components of A are zero:

$$a_1 = \cdots = a_{k-2} = 0.$$

Recall the Gram-Schmidt orthogonalization,

$$\mathbf{c}_i^* = \mathbf{c}_i - \sum_{j=1}^{i-1} \mu_{ij} \mathbf{c}_j^*, \qquad \mu_{ij} = \frac{\mathbf{c}_i \cdot \mathbf{c}_j^*}{\mathbf{c}_j^* \cdot \mathbf{c}_j^*}.$$

This implies that

$$c_{i,m+1}^* = 0 \quad (i = 1, \ldots, k-2), \qquad c_{k-1,m+1}^* = \gamma a_{k-1}.$$

Using this and $c_{k,m+1} = \gamma a_k$ we see that for $j = 1, \ldots, k-2$ we have

$$\mu_{kj} = \frac{\mathbf{c}_k \cdot \mathbf{c}_j^*}{\mathbf{c}_j^* \cdot \mathbf{c}_j^*} = \frac{\sum_{\ell=1}^{m} c_{k\ell} c_{j\ell}^* + \gamma a_k c_{j,m+1}^*}{\sum_{\ell=1}^{m} (c_{j\ell}^*)^2 + (c_{j,m+1}^*)^2} = \frac{\sum_{\ell=1}^{m} c_{k\ell} c_{j\ell}^*}{\sum_{\ell=1}^{m} (c_{j\ell}^*)^2}.$$

Thus in the range $1 \le j \le k-2$, the Gram-Schmidt coefficients μ_{kj} for C are the same as those for the submatrix B (that is, C with the last column deleted); they do not depend on γ. For $j = k-1$ the equation $c_{k-1,m+1}^* = \gamma a_{k-1}$ gives

$$\mu_{k,k-1} = \frac{\mathbf{c}_k \cdot \mathbf{c}_{k-1}^*}{\mathbf{c}_{k-1}^* \cdot \mathbf{c}_{k-1}^*} = \frac{\sum_{\ell=1}^{m} c_{k\ell} c_{k-1,\ell}^* + \gamma^2 a_{k-1} a_k}{\sum_{\ell=1}^{m} (c_{k-1,\ell}^*)^2 + \gamma^2 a_{k-1}^2}.$$

From this we see that as $\gamma \to \infty$ we have

$$\mu_{k,k-1} \to \frac{a_k}{a_{k-1}} \qquad \text{(assuming } a_{k-1} \neq 0\text{)}.$$

Therefore, as $\gamma \to \infty$ we have

$$\lceil \mu_{k,k-1} \rfloor \in \{ t, t+1 \}, \qquad t = \left\lceil \frac{a_k}{a_{k-1}} \right\rfloor \qquad \text{(assuming } a_{k-1} \neq 0\text{)}.$$

Hence the reduce step of the LLL algorithm performs either $\mathbf{c}_k \leftarrow \mathbf{c}_k - t\mathbf{c}_{k-1}$ or $\mathbf{c}_k \leftarrow \mathbf{c}_k - (t+1)\mathbf{c}_{k-1}$. If $a_{k-1} = 0$ then the reduce step is the same as that for the submatrix B; it does not depend on γ.

We next consider the exchange step of the LLL algorithm. This takes place if the exchange condition is satisfied:

$$|\mathbf{c}_k^*|^2 < (\alpha - \mu_{k,k-1}^2) |\mathbf{c}_{k-1}^*|^2.$$

There are three cases:

(1) If $a_{k-1} = a_k = 0$ then the exchange condition for C is the same as that for the submatrix B; it does not depend on γ.

(2) If $a_{k-1} = 0$ but $a_k \neq 0$ then $c_{k,m+1}^* = \gamma a_k$, and hence the exchange condition fails for γ sufficiently large; no exchange is performed.

(3)   If $a_{k-1} \neq 0$ then we consider the equation

$$\mathbf{c}_k^* = \mathbf{c}_k - \sum_{j=1}^{k-2} \mu_{kj} \mathbf{c}_j^* - \mu_{k,k-1} \mathbf{c}_{k-1}^*.$$

The orthogonal vectors $\mathbf{c}_j^*$ do not change during the reduce step, and so

$$c_{k,m+1} = \gamma a_k, \quad c_{j,m+1}^* = 0 \ (j = 1, \ldots, k-2), \quad c_{k-1,m+1}^* = \gamma a_{k-1}.$$

Combining this with the previous estimate for $\mu_{k,k-1}$ gives

$$c_{k,m+1}^* = \gamma a_k - \mu_{k,k-1} \gamma a_{k-1} \approx \gamma a_k - \left(\frac{a_k}{a_{k-1}} \right) \gamma a_{k-1} = 0.$$

Therefore the exchange condition will be satisfied for large γ, and an exchange will be performed.

This completes the proof that the LLL algorithm applied to the matrix C_γ converges to a limiting sequence of operations as γ becomes arbitrarily large.

Maple code for the Havas-Majewski-Matthews (HMM) algorithm for the GCD is given in Figures 14.4 and 14.5. The structure of the algorithm is very similar to that of the original LLL algorithm, except that the initial computation of the Gram-Schmidt orthogonalization is not required, since the original lattice basis is given by the rows of a matrix of the very special form $\begin{bmatrix} I_m & V \end{bmatrix}$. Furthermore, following an idea of de Weger [37], the algorithm exclusively uses integer arithmetic: the Gram-Schmidt coefficients are stored in terms of the integers d_i and ν_{ij} defined by

$$|\mathbf{b}_i^*|^2 = d_i / d_{i-1} \ (d_0 = 1), \qquad \nu_{ij} = d_j \mu_{ij}.$$

Similarly, the reduction parameter is stored as the rational number

$$\alpha = a_1 / a_2 \qquad (a_1, a_2 \in \mathbb{Z}).$$

Example 14.24. We conclude this section with a detailed example of the execution of the Havas-Majewski-Matthews algorithm for the GCD. We compute the greatest common divisor of the five integers 561, 909, 258, 549, 756. At the start of the computation, the variables are initialized as follows:

$$\left[\begin{array}{c|c} U \mid VV \mid \nu \\ \hline d \end{array} \right] = \left[\begin{array}{ccccc|c|ccccc} 1 & 0 & 0 & 0 & 0 & 561 & 0 & 0 & 0 & 0 & 0 \\ 0 & 1 & 0 & 0 & 0 & 909 & 0 & 0 & 0 & 0 & 0 \\ 0 & 0 & 1 & 0 & 0 & 258 & 0 & 0 & 0 & 0 & 0 \\ 0 & 0 & 0 & 1 & 0 & 549 & 0 & 0 & 0 & 0 & 0 \\ 0 & 0 & 0 & 0 & 1 & 756 & 0 & 0 & 0 & 0 & 0 \\ \hline & & & & & & 1 & 1 & 1 & 1 & 1 \end{array} \right]$$

```
with( LinearAlgebra ):
a1 := 3: a2 := 4: # reduction parameter: alpha = a1/a2
near := proc( x ) ceil( x - 1/2 ) end:

reducegcd := proc( k, i )
  global d, nu, U, VV:
  local j, q:
  if VV[i] <> 0 then
    q := near( VV[k]/VV[i] )
  else
    if 2*abs(nu[k,i]) > d[i] then
      q := near( nu[k,i]/d[i] )
    else
      q := 0
    fi
  fi:
  if q <> 0 then
    VV[k] := VV[k] - q*VV[i]:
    U := RowOperation( U, [k,i], -q ):
    nu[k,i] := nu[k,i] - q*d[i]:
    for j to i-1 do nu[k,j] := nu[k,j] - q*nu[i,j] od
  fi
end:

swapgcd := proc( k )
  global d, m, nu, U, VV:
  local i, j, t:
  t := VV[k]: VV[k] := VV[k-1]: VV[k-1] := t:
  U := RowOperation( U, [k,k-1] ):
  for j to k-2 do
    t := nu[k,j]: nu[k,j] := nu[k-1,j]: nu[k-1,j] := t
  od:
  for i from k+1 to m do
    t := nu[i,k-1]*d[k] - nu[i,k]*nu[k,k-1]:
    nu[i,k-1] :=
      ( nu[i,k-1]*nu[k,k-1] + nu[i,k]*d[k-2] ) / d[k-1]:
    nu[i,k] := t/d[k-1]
  od:
  d[k-1] := ( d[k-2]*d[k] + nu[k,k-1]^2 ) / d[k-1]
end:
```

FIGURE 14.4
HMM algorithm for GCD: Maple code, part 1

```
HMMGCD := proc( V )
  global a1, a2, d, m, nu, U, VV:
  local i, k:
  VV := copy( V ):
  m := Dimension( VV ):
  U := IdentityMatrix( m ):
  nu := Matrix( m, m ):
  for i from 0 to m do d[i] := 1 od:
  k := 2:
  while k <= m do
    reducegcd( k, k-1 ):
    if ( VV[k-1] <> 0 ) or
       ( VV[k-1] = 0 and VV[k] = 0 and
         a2*(d[k-2]*d[k]+nu[k,k-1]^2) < a1*d[k-1]^2 )
    then
      swapgcd( k ):
      if k > 2 then k := k - 1 fi
    else
      for i from k-2 to 1 by -1 do reducegcd( k, i ) od:
      k := k + 1
    fi
  od:
  if VV[m] < 0 then
    VV[m] := -VV[m]: U := RowOperation( U, m, -1 )
  fi
end:
```

FIGURE 14.5
HMM algorithm for GCD: Maple code, part 2

The upper left block is the (current state of the) transform matrix U, the next column is the copy VV of the input vector V, the upper right block is the integral Gram-Schmidt coefficient matrix ν, and the bottom row is the denominator vector d. We will display this data after each call to swapgcd.

At the start of the computation, the index variable has the value $k = 2$. After calls to reducegcd$(2, 1)$ and swapgcd(2), we have

$$\left[\begin{array}{c|c} U \mid VV \mid \nu \\ \hline d \end{array}\right] = \left[\begin{array}{ccccc|c|ccccc} -2 & 1 & 0 & 0 & 0 & -213 & 0 & 0 & 0 & 0 & 0 \\ 1 & 0 & 0 & 0 & 0 & 561 & -2 & 0 & 0 & 0 & 0 \\ 0 & 0 & 1 & 0 & 0 & 258 & 0 & 0 & 0 & 0 & 0 \\ 0 & 0 & 0 & 1 & 0 & 549 & 0 & 0 & 0 & 0 & 0 \\ 0 & 0 & 0 & 0 & 1 & 756 & 0 & 0 & 0 & 0 & 0 \\ \hline & & & & & & 5 & 1 & 1 & 1 & 1 \end{array}\right]$$

The same procedures are called five more times, giving these results:

$$\left[\begin{array}{c|c} U \mid VV \mid \nu \\ \hline d \end{array}\right] = \left[\begin{array}{ccccc|c|ccccc} -5 & 3 & 0 & 0 & 0 & -78 & 0 & 0 & 0 & 0 & 0 \\ -2 & 1 & 0 & 0 & 0 & -213 & 13 & 0 & 0 & 0 & 0 \\ 0 & 0 & 1 & 0 & 0 & 258 & 0 & 0 & 0 & 0 & 0 \\ 0 & 0 & 0 & 1 & 0 & 549 & 0 & 0 & 0 & 0 & 0 \\ 0 & 0 & 0 & 0 & 1 & 756 & 0 & 0 & 0 & 0 & 0 \\ \hline & & & & & & 34 & 1 & 1 & 1 & 1 \end{array}\right]$$

$$\left[\begin{array}{c|c} U \mid VV \mid \nu \\ \hline d \end{array}\right] = \left[\begin{array}{ccccc|c|ccccc} 13 & -8 & 0 & 0 & 0 & 21 & 0 & 0 & 0 & 0 & 0 \\ -5 & 3 & 0 & 0 & 0 & -78 & -89 & 0 & 0 & 0 & 0 \\ 0 & 0 & 1 & 0 & 0 & 258 & 0 & 0 & 0 & 0 & 0 \\ 0 & 0 & 0 & 1 & 0 & 549 & 0 & 0 & 0 & 0 & 0 \\ 0 & 0 & 0 & 0 & 1 & 756 & 0 & 0 & 0 & 0 & 0 \\ \hline & & & & & & 233 & 1 & 1 & 1 & 1 \end{array}\right]$$

$$\left[\begin{array}{c|c} U \mid VV \mid \nu \\ \hline d \end{array}\right] = \left[\begin{array}{ccccc|c|ccccc} 47 & -29 & 0 & 0 & 0 & 6 & 0 & 0 & 0 & 0 & 0 \\ 13 & -8 & 0 & 0 & 0 & 21 & 843 & 0 & 0 & 0 & 0 \\ 0 & 0 & 1 & 0 & 0 & 258 & 0 & 0 & 0 & 0 & 0 \\ 0 & 0 & 0 & 1 & 0 & 549 & 0 & 0 & 0 & 0 & 0 \\ 0 & 0 & 0 & 0 & 1 & 756 & 0 & 0 & 0 & 0 & 0 \\ \hline & & & & & & 3050 & 1 & 1 & 1 & 1 \end{array}\right]$$

$$\left[\begin{array}{c|c} U \mid VV \mid \nu \\ \hline d \end{array}\right] = \left[\begin{array}{ccccc|c|ccccc} -128 & 79 & 0 & 0 & 0 & 3 & 0 & 0 & 0 & 0 & 0 \\ 47 & -29 & 0 & 0 & 0 & 6 & -8307 & 0 & 0 & 0 & 0 \\ 0 & 0 & 1 & 0 & 0 & 258 & 0 & 0 & 0 & 0 & 0 \\ 0 & 0 & 0 & 1 & 0 & 549 & 0 & 0 & 0 & 0 & 0 \\ 0 & 0 & 0 & 0 & 1 & 756 & 0 & 0 & 0 & 0 & 0 \\ \hline & & & & & & 22625 & 1 & 1 & 1 & 1 \end{array}\right]$$

$$\left[\begin{array}{c|c} U \mid VV \mid \nu \\ \hline d \end{array}\right] = \left[\begin{array}{ccccc|c|ccccc} 303 & -187 & 0 & 0 & 0 & 0 & 0 & 0 & 0 & 0 & 0 \\ -128 & 79 & 0 & 0 & 0 & 3 & -53557 & 0 & 0 & 0 & 0 \\ 0 & 0 & 1 & 0 & 0 & 258 & 0 & 0 & 0 & 0 & 0 \\ 0 & 0 & 0 & 1 & 0 & 549 & 0 & 0 & 0 & 0 & 0 \\ 0 & 0 & 0 & 0 & 1 & 756 & 0 & 0 & 0 & 0 & 0 \\ \hline & & & & & & 126778 & 1 & 1 & 1 & 1 \end{array}\right]$$

At this point the algorithm has computed the GCD of the first two input values.

After one more call to `reducegcd(2, 1)`, we increase the index to $k = 3$, call `reducegcd(3, 2)` and `swap(3)`, and obtain

$$\left[\begin{array}{ccccc|c|cccc} 303 & -187 & 0 & 0 & 0 & 0 & 0 & 0 & 0 & 0 \\ 11008 & -6794 & 1 & 0 & 0 & 0 & 4605902 & 0 & 0 & 0 \\ -128 & 79 & 0 & 0 & 0 & 3 & -53557 & -86 & 0 & 0 \\ 0 & 0 & 0 & 1 & 0 & 549 & 0 & 0 & 0 & 0 \\ 0 & 0 & 0 & 0 & 1 & 756 & 0 & 0 & 0 & 0 \\ \hline & & & & & & 126778 & 134174 & 1 & 1 & 1 \end{array}\right]$$

The algorithm has now computed the GCD of the first three input values.

The index now decreases to $k = 2$; the algorithm calls `reducegcd(2, 1)` and `swapgcd(2)` twice, giving

$$\left[\begin{array}{ccccc|c|cccc} 100 & -62 & 1 & 0 & 0 & 0 & 0 & 0 & 0 & 0 \\ 303 & -187 & 0 & 0 & 0 & 0 & 41894 & 0 & 0 & 0 \\ -128 & 79 & 0 & 0 & 0 & 3 & -17698 & -56653 & 0 & 0 \\ 0 & 0 & 0 & 1 & 0 & 549 & 0 & 0 & 0 & 0 \\ 0 & 0 & 0 & 0 & 1 & 756 & 0 & 0 & 0 & 0 \\ \hline & & & & & & 13845 & 134174 & 1 & 1 & 1 \end{array}\right]$$

$$\left[\begin{array}{ccccc|c|cccc} 3 & -1 & -3 & 0 & 0 & 0 & 0 & 0 & 0 & 0 \\ 100 & -62 & 1 & 0 & 0 & 0 & 359 & 0 & 0 & 0 \\ -128 & 79 & 0 & 0 & 0 & 3 & -463 & -170045 & 0 & 0 \\ 0 & 0 & 0 & 1 & 0 & 549 & 0 & 0 & 0 & 0 \\ 0 & 0 & 0 & 0 & 1 & 756 & 0 & 0 & 0 & 0 \\ \hline & & & & & & 19 & 134174 & 1 & 1 & 1 \end{array}\right]$$

The algorithm calls `reducegcd(2, 1)` again, increases the index to $k = 3$, calls `reducegcd(3, 2)` and `reducegcd(3, 1)`, increases the index to $k = 4$, and then calls `reducegcd(4, 3)` and `swapgcd(4)`:

$$\left[\begin{array}{ccccc|c|cccc} 3 & -1 & -3 & 0 & 0 & 0 & 0 & 0 & 0 & 0 \\ 43 & -43 & 58 & 0 & 0 & 0 & -2 & 0 & 0 & 0 \\ 2379 & -2196 & 2562 & 1 & 0 & 0 & 1647 & 6564393 & 0 & 0 \\ -13 & 12 & -14 & 0 & 0 & 3 & -9 & -35871 & -183 & 0 \\ 0 & 0 & 0 & 0 & 1 & 756 & 0 & 0 & 0 & 0 \\ \hline & & & & & & 19 & 134174 & 167663 & 1 & 1 \end{array}\right]$$

The algorithm has now computed the GCD of the first four input values.

The index now decreases to $k = 3$, and after calls to `reducegcd(3, 2)` and `swapgcd(3)` we have

3	−1	−3	0	0	0	0	0	0	0	0
272	−89	−280	1	0	0	1745	0	0	0	0
43	−43	58	0	0	0	−2	−10133	0	0	0
−13	12	−14	0	0	3	−9	2709	−44838	0	0
0	0	0	0	1	756	0	0	0	0	0
						19	789	167663	1	1

The index now decreases to $k = 2$. The algorithm calls `reducegcd(2, 1)`, increases the index to $k = 3$, calls `reducegcd(3, i)` ($i = 2, 1$), increases the index to $k = 4$, calls `reducegcd(4, i)` ($i = 3, 2, 1$), increases the index to $k = 5$, and then calls `reducegcd(5, 4)` and `swapgcd(5)`. The current state is now

3	−1	−3	0	0	0	0	0	0	0	0
−4	3	−4	1	0	0	−3	0	0	0	0
−3	−6	0	13	0	0	−3	124	0	0	0
252	−756	504	756	1	0	0	−86184	11299176	0	0
−1	3	−2	−3	0	3	0	342	−44838	−252	0
						19	789	167663	231167	1

At this point the algorithm has computed the GCD of all five input values, but the transformation matrix U requires further reduction.

The index decreases to $k = 4$, and `reducegcd(4, 3)` and `swapgcd(4)` produce

3	−1	−3	0	0	0	0	0	0	0	0
−4	3	−4	1	0	0	−3	0	0	0	0
453	−354	504	−115	1	0	201	−94492	0	0	0
−3	−6	0	13	0	0	−3	124	65755	0	0
−1	3	−2	−3	0	3	0	342	−17586	−61722	0
						19	789	26876	231167	1

The algorithm decreases the index to $k = 3$, calls `reducegcd(3, i)` ($i = 2, 1$), increases the index to $k = 4$, and calls `reducegcd(4, 3)` and `swapgcd(4)`:

3	−1	−3	0	0	0	0	0	0	0	0
−4	3	−4	1	0	0	−3	0	0	0	0
3	−2	0	3	−2	0	11	−252	0	0	0
−3	−2	0	5	1	0	−7	188	12003	0	0
−1	3	−2	−3	0	3	0	342	−9666	−123696	0
						19	789	12147	231167	1

The index decreases to $k = 3$, and reducegcd$(3, 2)$ and swapgcd(3) produce:

$$\left[\begin{array}{ccccc|c|cccccc}
3 & -1 & -3 & 0 & 0 & 0 & 0 & 0 & 0 & 0 & 0 \\
3 & -2 & 0 & 3 & -2 & 0 & 11 & 0 & 0 & 0 & 0 \\
-4 & 3 & -4 & 1 & 0 & 0 & -3 & -252 & 0 & 0 & 0 \\
-3 & -2 & 0 & 5 & 1 & 0 & -7 & 229 & 6728 & 0 & 0 \\
-1 & 3 & -2 & -3 & 0 & 3 & 0 & -342 & 2178 & -123696 & 0 \\
 & & & & & & 19 & 373 & 12147 & 231167 & 1
\end{array}\right]$$

The index now decreases to $k = 2$. The algorithm calls reducegcd$(2, 1)$, increases the index to $k = 3$, calls reducegcd$(3, i)$ $(i = 2, 1)$, increases the index to $k = 4$, calls reducegcd$(4, i)$ $(i = 3, 2, 1)$, increases the index to $k = 5$, and then calls reducegcd$(5, i)$ $(i = 4, 3, 2, 1)$. At this point the algorithm terminates:

$$\left[\begin{array}{ccccc|c|cccccc}
3 & -1 & -3 & 0 & 0 & 0 & 0 & 0 & 0 & 0 & 0 \\
0 & -1 & 3 & 3 & -2 & 0 & -8 & 0 & 0 & 0 & 0 \\
-1 & 1 & -4 & 4 & -2 & 0 & 8 & 121 & 0 & 0 & 0 \\
1 & -4 & 1 & 1 & 3 & 0 & 4 & 108 & -5419 & 0 & 0 \\
0 & -2 & 2 & 1 & 1 & 3 & -4 & 139 & -3241 & 107471 & 0 \\
 & & & & & & 19 & 373 & 12147 & 231167 & 1
\end{array}\right]$$

We have $\gcd(v_1, v_2, v_3, v_4, v_5) = 3$ and the transformation matrix is

$$U = \left[\begin{array}{ccccc}
3 & -1 & -3 & 0 & 0 \\
0 & -1 & 3 & 3 & -2 \\
-1 & 1 & -4 & 4 & -2 \\
1 & -4 & 1 & 1 & 3 \\
0 & -2 & 2 & 1 & 1
\end{array}\right]$$

The last row gives the short multiplier vector $-2v_2 + 2v_3 + v_4 + v_5 = 3$.

14.7 The HMM algorithm for the HNF

We can regard a column vector of dimension m as a matrix of size $m \times 1$:

$$V = \left[\begin{array}{c}
v_1 \\
v_2 \\
\vdots \\
v_m
\end{array}\right].$$

If the matrix entries are integers, then its Hermite normal form is

$$H = \begin{bmatrix} d \\ 0 \\ \vdots \\ 0 \end{bmatrix}, \qquad D = \gcd(v_1, v_2, \ldots, v_m).$$

If we turn this vector upside down, then we have the vector computed by the GCD algorithms described in the previous sections. Therefore, using elementary row operations to compute the greatest common divisor of the components of an $m \times 1$ column vector is equivalent to computing the Hermite normal form of an $m \times 1$ matrix.

In this section, we consider an integer matrix of size $m \times n$, and follow Havas, Majewski and Matthews [56] to extend the algorithm for the GCD to an algorithm for the HNF. In the previous section, we considered the limiting behavior of the LLL algorithm applied to a matrix of this form:

$$C_\gamma = \begin{bmatrix} I_m & \gamma V \end{bmatrix}.$$

We replace the single column V by n columns $V_1, V_2, \ldots, V_n$, and consider the limiting behavior of the LLL algorithm applied to a matrix of this form:

$$C_\gamma = \begin{bmatrix} I_m & \gamma^n V_1 & \gamma^{n-1} V_2 & \cdots & \gamma V_n \end{bmatrix}.$$

This approach leads to the Maple code for the HMM algorithm for the HNF given in Figures 14.6–14.8.

Example 14.25. We use the Maple program of Figures 14.6–14.8 to compute the Hermite normal form of this 4×4 matrix:

$$M = \begin{bmatrix} 9 & 6 & 2 & 4 \\ -8 & 2 & 0 & -7 \\ -6 & -4 & 0 & -7 \\ 6 & 9 & -1 & -8 \end{bmatrix}$$

We display the current state of the computation as follows:

$$\left[\begin{array}{c|c|c} U & MM & \nu \\ \hline & d & \end{array} \right] = \left[\begin{array}{cccc|cccc|cccc} 1 & 0 & 0 & 0 & 9 & 6 & 2 & 4 & 0 & 0 & 0 & 0 \\ 0 & 1 & 0 & 0 & -8 & 2 & 0 & -7 & 0 & 0 & 0 & 0 \\ 0 & 0 & 1 & 0 & -6 & -4 & 0 & -7 & 0 & 0 & 0 & 0 \\ 0 & 0 & 0 & 1 & 6 & 9 & -1 & -8 & 0 & 0 & 0 & 0 \\ \hline & & & & & & & & 1 & 1 & 1 & 1 \end{array} \right]$$

In this representation of the initial values of the variables, U is the transformation matrix, MM is a copy of the input matrix, ν is the matrix of integral Gram-Schmidt coefficients, and d is the vector of denominators.

```
with( LinearAlgebra ):
a1 := 3: a2 := 4: # alpha = a1/a2 (reduction parameter)
near := proc( x ) ceil( x - 1/2 ) end:

plusminus := proc( ij )
  global m, nu: local i, j:
  for i from 2 to m do for j to i-1 do
    if i = ij or j = ij then nu[i,j] := -nu[i,j] fi
  od od
end:


reducehnf := proc( k, i )
  global c1, c2, d, m, MM, n, nu, U: local j, jj, q:
  jj := 0:
  for j from n to 1 by -1 do if MM[i,j]<>0 then jj := j fi od:
  if jj = 0 then c1 := n + 1 else
    c1 := jj:
    if MM[i,c1] < 0 then
      plusminus(i):
      MM := RowOperation( MM, i, -1 ):
      U  := RowOperation( U, i, -1 )
    fi
  fi:
  jj := 0:
  for j from n to 1 by -1 do if MM[k,j]<>0 then jj := j fi od:
  if jj = 0 then c2 := n + 1 else c2 := jj fi:
  if c1 <= n then q := floor( MM[k,c1]/MM[i,c1] ) else
    if 2*abs(nu[k,i]) > d[i] then
      q := near( nu[k,i]/d[i] )
    else
      q := 0
    fi
  fi:
  if q <> 0 then
    MM := RowOperation( MM, [k,i], -q ):
    U  := RowOperation( U, [k,i], -q ):
    nu[k,i] := nu[k,i] - q*d[i]:
    for j to i-1 do nu[k,j] := nu[k,j] - q*nu[i,j] od
  fi
end:
```

FIGURE 14.6
HMM algorithm for HNF: Maple code, part 1

```
swaphnf := proc( k )
  global d, m, MM, n, nu, U: local i, j, t:
  MM := RowOperation( MM, [k,k-1] ):
  U := RowOperation( U, [k,k-1] ):
  for j to k-2 do
    t := nu[k,j]: nu[k,j] := nu[k-1,j]: nu[k-1,j] := t
  od:
  for i from k+1 to m do
    t := nu[i,k-1]*d[k] - nu[i,k]*nu[k,k-1]:
    nu[i,k-1] :=
      (nu[i,k-1]*nu[k,k-1]+nu[i,k]*d[k-2]) / d[k-1]:
    nu[i,k] := t / d[k-1]
  od:
  d[k-1] := ( d[k-2]*d[k] + nu[k,k-1]^2 ) / d[k-1]
end:
```

FIGURE 14.7
HMM algorithm for HNF: Maple code, part 2

The algorithm calls reducehnf$(2, 1)$ and swaphnf(2) twice, giving

$$
\left[
\begin{array}{cccc|cccc|cccc}
1 & 1 & 0 & 0 & 1 & 8 & 2 & -3 & 0 & 0 & 0 & 0 \\
1 & 0 & 0 & 0 & 9 & 6 & 2 & 4 & 1 & 0 & 0 & 0 \\
0 & 0 & 1 & 0 & -6 & -4 & 0 & -7 & 0 & 0 & 0 & 0 \\
0 & 0 & 0 & 1 & 6 & 9 & -1 & -8 & 0 & 0 & 0 & 0 \\
\hline
 & & & & & & & & 2 & 1 & 1 & 1
\end{array}
\right]
$$

$$
\left[
\begin{array}{cccc|cccc|cccc}
-8 & -9 & 0 & 0 & 0 & -66 & -16 & 31 & 0 & 0 & 0 & 0 \\
1 & 1 & 0 & 0 & 1 & 8 & 2 & -3 & -17 & 0 & 0 & 0 \\
0 & 0 & 1 & 0 & -6 & -4 & 0 & -7 & 0 & 0 & 0 & 0 \\
0 & 0 & 0 & 1 & 6 & 9 & -1 & -8 & 0 & 0 & 0 & 0 \\
\hline
 & & & & & & & & 145 & 1 & 1 & 1
\end{array}
\right]
$$

After another call to reducehnf$(2, 1)$, the index increases to $k = 3$, and the algorithm calls reducehnf$(3, 2)$ and swaphnf(3):

$$
\left[
\begin{array}{cccc|cccc|cccc}
8 & 9 & 0 & 0 & 0 & 66 & 16 & -31 & 0 & 0 & 0 & 0 \\
6 & 6 & 1 & 0 & 0 & 44 & 12 & -25 & 102 & 0 & 0 & 0 \\
1 & 1 & 0 & 0 & 1 & 8 & 2 & -3 & 17 & 6 & 0 & 0 \\
0 & 0 & 0 & 1 & 6 & 9 & -1 & -8 & 0 & 0 & 0 & 0 \\
\hline
 & & & & & & & & 145 & 181 & 1 & 1
\end{array}
\right]
$$

The index is now $k = 2$; reducehnf$(2, 1)$ and swaphnf(2) are called three

```
HMMHNF := proc( M )
  global a1, a2, c1, c2, d, m, MM, n, nu, U:
  local i, j, jj, k, nz:
  MM := copy( M ):
  m := RowDimension( MM ):
  n := ColumnDimension( MM ):
  U := Matrix( m, m, (i,j) -> if i=j then 1 else 0 fi ):
  nu := Matrix( m, m ):
  for i from 0 to m do d[i] := 1 od:
  jj := 0:
  for j from n to 1 by -1 do
    if not Equal( Column(MM,j), Vector(m) ) then jj := j fi
  od:
  if jj <> 0 then
    nz := 0:
    for i to m do
      if MM[i,jj] <> 0 then nz := nz+1 fi
    od:
    if nz = 1 and i = m and MM[m,jj] < 0 then
      MM := RowOperation( MM, m, -1 ):
      U[m,m] := -1
    fi
  fi:
  k := 2:
  while k <= m do
    reducehnf( k, k-1 ):
    if ( c1 <= min(c2,n) ) or
       ( c1 = n+1 and c2 = n+1 and
         a2*(d[k-2]*d[k]+nu[k,k-1]^2) < a1*d[k-1]^2 )
    then
      swaphnf( k ):
      if k > 2 then k := k - 1 fi
    else
      for i from k-2 to 1 by -1 do reducehnf(k,i) od:
      k := k+1
    fi
  od
end:
```

FIGURE 14.8
HMM algorithm for HNF: Maple code, part 3

times:

$$
\left[
\begin{array}{cccc|cccc|cccc}
6 & 6 & 1 & 0 & 0 & 44 & 12 & -25 & 0 & 0 & 0 & 0 \\
8 & 9 & 0 & 0 & 0 & 66 & 16 & -31 & 102 & 0 & 0 & 0 \\
1 & 1 & 0 & 0 & 1 & 8 & 2 & -3 & 12 & 17 & 0 & 0 \\
0 & 0 & 0 & 1 & 6 & 9 & -1 & -8 & 0 & 0 & 0 & 0 \\
\hline
 & & & & & & & & 73 & 181 & 1 & 1
\end{array}
\right]
$$

$$
\left[
\begin{array}{cccc|cccc|cccc}
2 & 3 & -1 & 0 & 0 & 22 & 4 & -6 & 0 & 0 & 0 & 0 \\
6 & 6 & 1 & 0 & 0 & 44 & 12 & -25 & 29 & 0 & 0 & 0 \\
1 & 1 & 0 & 0 & 1 & 8 & 2 & -3 & 5 & 23 & 0 & 0 \\
0 & 0 & 0 & 1 & 6 & 9 & -1 & -8 & 0 & 0 & 0 & 0 \\
\hline
 & & & & & & & & 14 & 181 & 1 & 1
\end{array}
\right]
$$

$$
\left[
\begin{array}{cccc|cccc|cccc}
2 & 0 & 3 & 0 & 0 & 0 & 4 & -13 & 0 & 0 & 0 & 0 \\
2 & 3 & -1 & 0 & 0 & 22 & 4 & -6 & 1 & 0 & 0 & 0 \\
1 & 1 & 0 & 0 & 1 & 8 & 2 & -3 & 2 & 63 & 0 & 0 \\
0 & 0 & 0 & 1 & 6 & 9 & -1 & -8 & 0 & 0 & 0 & 0 \\
\hline
 & & & & & & & & 13 & 181 & 1 & 1
\end{array}
\right]
$$

The algorithm calls `reducehnf(2,1)`, increments to $k = 3$, calls `reducehnf(3,i)` ($i = 2,1$), increments to $k = 4$, calls `reducehnf(4,3)` and then `swaphnf(4)`:

$$
\left[
\begin{array}{cccc|cccc|cccc}
2 & 0 & 3 & 0 & 0 & 0 & 4 & -13 & 0 & 0 & 0 & 0 \\
0 & 3 & -4 & 0 & 0 & 22 & 0 & 7 & -12 & 0 & 0 & 0 \\
-6 & -6 & 0 & 1 & 0 & -39 & -13 & 10 & -12 & -378 & 0 & 0 \\
1 & 1 & 0 & 0 & 1 & 8 & 2 & -3 & 2 & 63 & -6 & 0 \\
\hline
 & & & & & & & & 13 & 181 & 217 & 1
\end{array}
\right]
$$

The index is now $k = 3$; the algorithm calls `reducehnf(3,2)` and then `swaphnf(3)`:

$$
\left[
\begin{array}{cccc|cccc|cccc}
2 & 0 & 3 & 0 & 0 & 0 & 4 & -13 & 0 & 0 & 0 & 0 \\
-6 & 0 & -8 & 1 & 0 & 5 & -13 & 24 & -36 & 0 & 0 & 0 \\
0 & 3 & -4 & 0 & 0 & 22 & 0 & 7 & -12 & -16 & 0 & 0 \\
1 & 1 & 0 & 0 & 1 & 8 & 2 & -3 & 2 & -6 & 75 & 0 \\
\hline
 & & & & & & & & 13 & 17 & 217 & 1
\end{array}
\right]
$$

The index is now $k = 2$; the algorithm calls `reducehnf(2,1)`, increments to $k = 3$, calls `reducehnf(3,2)` and then `swaphnf(3)`:

$$
\left[
\begin{array}{cccc|cccc|cccc}
2 & 0 & 3 & 0 & 0 & 0 & 4 & -13 & 0 & 0 & 0 & 0 \\
-8 & 3 & -20 & -4 & 0 & 2 & -12 & 119 & -76 & 0 & 0 & 0 \\
2 & 0 & 4 & 1 & 0 & 5 & 3 & -28 & 16 & -84 & 0 & 0 \\
1 & 1 & 0 & 0 & 1 & 8 & 2 & -3 & 2 & 87 & 294 & 0 \\
\hline
 & & & & & & & & 13 & 581 & 217 & 1
\end{array}
\right]
$$

The same operations are repeated twice more, giving

$$
\left[
\begin{array}{rrrr|rrrr|rrrr}
2 & 0 & 3 & 0 & 0 & 0 & 4 & -13 & 0 & 0 & 0 & 0 \\
6 & -6 & 26 & 9 & 0 & 1 & 3 & -188 & 90 & 0 & 0 & 0 \\
-2 & 3 & -11 & -4 & 0 & 2 & 0 & 80 & -37 & -1246 & 0 & 0 \\
1 & 1 & 0 & 0 & 1 & 8 & 2 & -3 & 2 & -180 & 663 & 0 \\
\hline
 & & & & & & & & 13 & 2677 & 217 & 1
\end{array}
\right]
$$

$$
\left[
\begin{array}{rrrr|rrrr|rrrr}
2 & 0 & 3 & 0 & 0 & 0 & 4 & -13 & 0 & 0 & 0 & 0 \\
-14 & 15 & -63 & -22 & 0 & 0 & -6 & 456 & -217 & 0 & 0 & 0 \\
6 & -6 & 26 & 9 & 0 & 1 & 3 & -188 & 90 & -6600 & 0 & 0 \\
1 & 1 & 0 & 0 & 1 & 8 & 2 & -3 & 2 & 447 & 1620 & 0 \\
\hline
 & & & & & & & & 13 & 16273 & 217 & 1
\end{array}
\right]
$$

The index is now $k = 2$; `reducehnf(2, 1)` and `swaphnf(2)` are called twice:

$$
\left[
\begin{array}{rrrr|rrrr|rrrr}
-10 & 15 & -57 & -22 & 0 & 0 & 2 & 430 & 0 & 0 & 0 & 0 \\
2 & 0 & 3 & 0 & 0 & 0 & 4 & -13 & -191 & 0 & 0 & 0 \\
6 & -6 & 26 & 9 & 0 & 1 & 3 & -188 & -1830 & 15690 & 0 & 0 \\
1 & 1 & 0 & 0 & 1 & 8 & 2 & -3 & 5 & 9071 & 1620 & 0 \\
\hline
 & & & & & & & & 4058 & 16273 & 217 & 1
\end{array}
\right]
$$

$$
\left[
\begin{array}{rrrr|rrrr|rrrr}
22 & -30 & 117 & 44 & 0 & 0 & 0 & -873 & 0 & 0 & 0 & 0 \\
-10 & 15 & -57 & -22 & 0 & 0 & 2 & 430 & -8307 & 0 & 0 & 0 \\
6 & -6 & 26 & 9 & 0 & 1 & 3 & -188 & 3750 & 24780 & 0 & 0 \\
1 & 1 & 0 & 0 & 1 & 8 & 2 & -3 & -8 & 18589 & 1620 & 0 \\
\hline
 & & & & & & & & 17009 & 16273 & 217 & 1
\end{array}
\right]
$$

After `reducehnf(2, 1)`, `reducehnf(3, i)` $(i = 2, 1)$, and `reducehnf(4, i)` $(i = 3, 2, 1)$, we obtain the final result:

$$
\left[
\begin{array}{rrrr|rrrr|rrrr}
-22 & 30 & -117 & -44 & 0 & 0 & 0 & 873 & 0 & 0 & 0 & 0 \\
-10 & 15 & -57 & -22 & 0 & 0 & 2 & 430 & 8307 & 0 & 0 & 0 \\
-6 & 9 & -34 & -13 & 0 & 1 & 1 & 255 & 4952 & 8507 & 0 & 0 \\
-3 & 4 & -16 & -6 & 1 & 0 & 0 & 120 & 2322 & -648 & -116 & 0 \\
\hline
 & & & & & & & & 17009 & 16273 & 217 & 1
\end{array}
\right]
$$

We now turn the matrices U and H upside down, to obtain

$$
\left[
\begin{array}{rrrr}
-3 & 4 & -16 & -6 \\
-6 & 9 & -34 & -13 \\
-10 & 15 & -57 & -22 \\
-22 & 30 & -117 & -44
\end{array}
\right]
\left[
\begin{array}{rrrr}
9 & 6 & 2 & 4 \\
-8 & 2 & 0 & -7 \\
-6 & -4 & 0 & -7 \\
6 & 9 & -1 & -8
\end{array}
\right]
=
\left[
\begin{array}{rrrr}
1 & 0 & 0 & 120 \\
0 & 1 & 1 & 255 \\
0 & 0 & 2 & 430 \\
0 & 0 & 0 & 873
\end{array}
\right]
$$

The computation is complete.

14.8   Projects

Project 14.1. Consider the product of 2×2 matrices of the form occurring in the computation of the GCD of two integers:

$$\begin{bmatrix} b & 1 \\ 1 & 0 \end{bmatrix} \begin{bmatrix} a & 1 \\ 1 & 0 \end{bmatrix} = \begin{bmatrix} ab+1 & b \\ a & 1 \end{bmatrix}$$

$$\begin{bmatrix} c & 1 \\ 1 & 0 \end{bmatrix} \begin{bmatrix} b & 1 \\ 1 & 0 \end{bmatrix} \begin{bmatrix} a & 1 \\ 1 & 0 \end{bmatrix} = \begin{bmatrix} abc+a+c & bc+1 \\ ab+1 & b \end{bmatrix}$$

$$\begin{bmatrix} d & 1 \\ 1 & 0 \end{bmatrix} \begin{bmatrix} c & 1 \\ 1 & 0 \end{bmatrix} \begin{bmatrix} b & 1 \\ 1 & 0 \end{bmatrix} \begin{bmatrix} a & 1 \\ 1 & 0 \end{bmatrix} = \begin{bmatrix} abcd+ab+ad+cd+1 & bcd+b+d \\ abc+a+c & bc+1 \end{bmatrix}$$

The entries of the matrices on the right sides of these equations are called 'continuants': these polynomials are important in the theory of continued fractions. See Graham et al. [50], §6.7; Knuth [78], §4.5.3; von zur Gathen and Gerhard [147], Exercise 3.20. In particular, the entry in the upper left corner is obtained from the product of the variables by deleting adjacent pairs of letters in all possible ways and adding the results. Write a report and present a seminar talk on the theory of continuant polynomials.

Project 14.2. In the algorithm of Figure 14.2, the entries of H and U tend to increase exponentially during the calculation. For the worst-case complexity of this problem, see Fang and Havas [41]. An active area of research in computer algebra is the search for HNF algorithms which control the space required for storage of the intermediate results. A famous paper on this topic is Kannan and Bachem [73] (see also Sims [131]). For recent work see Storjohann [135], Micciancio and Warinschi [101], Pernet and Stein [116]. Write a report and present a seminar talk on algorithms for computing the Hermite normal form.

Project 14.3. Compare the MLLL algorithm for linearly dependent vectors with the algorithm for the greatest common divisor outlined in the statement of Lemma 14.22. Both are based on the idea of embedding linearly dependent vectors into a space of higher dimension as linearly independent vectors.

Project 14.4. Write a report on the complexity of the Havas-Majewski-Matthews algorithm for the HNF, based on the paper of van der Kallen [143].

14.9   Exercises

Exercise 14.1. Consider the following matrix A:

$$
A = \begin{bmatrix}
-7 & -6 & -4 & -3 & -5 & 3 & -2 & 2 & -2 & 7 \\
2 & 7 & 5 & 8 & 1 & 4 & 1 & 6 & -4 & 9 \\
-5 & -2 & 5 & 2 & -9 & 4 & -3 & 6 & 8 & -2 \\
0 & -6 & 3 & 2 & -2 & -6 & -8 & -1 & 1 & -2 \\
5 & 7 & 9 & -9 & 6 & 7 & -8 & -2 & 7 & -7
\end{bmatrix}
$$

(a) Find a transform matrix U for which $UA^t = H$ is the Hermite normal form of the transpose of A.

(b) Apply lattice basis reduction to the result of part (a) to find a reduced basis of the nullspace lattice of A.

(c) Use the result of part (b) to find a transform matrix U' with much smaller entries for which $U'A^t = H$.

Exercise 14.2. Apply the LLL algorithm to the first three rows of the matrix U in Example 14.21, and then use the reduced basis to size-reduce the last row of U. Compare the resulting multiplier vector with equations (14.3) and (14.4).

Exercise 14.3. Prove Lemma 14.22.

Exercise 14.4. For each of the following integer column vectors, compute the greatest common divisor in two ways: first, using the method of Example 14.21; second, using the Havas-Majewski-Matthews algorithm of Figures 14.4 and 14.5:

(a) $V = \begin{bmatrix} 2 \\ 3 \\ 5 \end{bmatrix}$ (b) $V = \begin{bmatrix} 46 \\ 52 \\ 64 \end{bmatrix}$ (c) $V = \begin{bmatrix} 243 \\ 704 \\ 337 \end{bmatrix}$ (d) $V = \begin{bmatrix} 3650 \\ 5985 \\ 6781 \end{bmatrix}$

(e) $V = \begin{bmatrix} 2 \\ 3 \\ 5 \\ 7 \end{bmatrix}$ (f) $V = \begin{bmatrix} 70 \\ 67 \\ 35 \\ 53 \end{bmatrix}$ (g) $V = \begin{bmatrix} 671 \\ 272 \\ 391 \\ 826 \end{bmatrix}$ (h) $V = \begin{bmatrix} 9009 \\ 7458 \\ 5049 \\ 6156 \end{bmatrix}$

(i) $V = \begin{bmatrix} 2 \\ 3 \\ 5 \\ 7 \\ 11 \end{bmatrix}$ (j) $V = \begin{bmatrix} 96 \\ 68 \\ 81 \\ 25 \\ 21 \end{bmatrix}$ (k) $V = \begin{bmatrix} 112 \\ 771 \\ 312 \\ 605 \\ 836 \end{bmatrix}$ (l) $V = \begin{bmatrix} 7047 \\ 7478 \\ 2418 \\ 3929 \\ 7909 \end{bmatrix}$

Exercise 14.5. For each of the following integer matrices, compute the HNF using the Havas-Majewski-Matthews algorithm of Figures 14.6–14.8:

(a) $V = \begin{bmatrix} 5 & 9 & 1 \\ 9 & 2 & 7 \\ 4 & 4 & 1 \end{bmatrix}$ (b) $V = \begin{bmatrix} 36 & 31 & 67 \\ 41 & 16 & 35 \\ 92 & 70 & 53 \end{bmatrix}$

(c) $V = \begin{bmatrix} 836 & 891 & 909 \\ 726 & 295 & 258 \\ 698 & 561 & 549 \end{bmatrix}$

(d) $V = \begin{bmatrix} 3 & 6 & 8 & 4 \\ 2 & 4 & 6 & 6 \\ 2 & 3 & 5 & 6 \\ 8 & 9 & 7 & 2 \end{bmatrix}$

(e) $V = \begin{bmatrix} 14 & 52 & 90 & 47 \\ 27 & 76 & 42 & 16 \\ 18 & 52 & 99 & 49 \\ 76 & 85 & 36 & 40 \end{bmatrix}$

(f) $V = \begin{bmatrix} 691 & 877 & 512 & 326 \\ 479 & 725 & 602 & 113 \\ 461 & 901 & 952 & 291 \\ 294 & 652 & 608 & 606 \end{bmatrix}$

(g) $V = \begin{bmatrix} 8 & 8 & 7 & 7 & 6 \\ 9 & 1 & 1 & 8 & 8 \\ 7 & 2 & 6 & 8 & 9 \\ 3 & 3 & 6 & 7 & 1 \\ 9 & 3 & 4 & 3 & 7 \end{bmatrix}$

(h) $V = \begin{bmatrix} 35 & 40 & 56 & 17 & 14 \\ 65 & 98 & 80 & 51 & 85 \\ 87 & 31 & 19 & 50 & 90 \\ 87 & 44 & 60 & 21 & 81 \\ 26 & 26 & 86 & 62 & 81 \end{bmatrix}$

(i) $V = \begin{bmatrix} 415 & 655 & 857 & 638 & 635 \\ 837 & 100 & 308 & 256 & 994 \\ 306 & 842 & 570 & 372 & 385 \\ 479 & 898 & 774 & 382 & 207 \\ 670 & 146 & 103 & 133 & 161 \end{bmatrix}$

15

Polynomial Factorization

CONTENTS

Our goal in this chapter is to present the first important application of the LLL algorithm, which explains the title of the original paper by Lenstra, Lenstra and Lovász [88]: a polynomial-time algorithm for factoring polynomials in one variable with coefficients in the field of rational numbers.

The first few sections of this chapter provide an brief introduction to the general problem of polynomial factorization. We begin with the case of coefficients in a finite field; the most important original references are Knuth [78] (§4.6), Berlekamp [14, 15], and Cantor and Zassenhaus [21]. We then use Hensel lifting to extend a factorization over the field with p elements to a factorization over the ring of congruence classes modulo p^n; the classic paper is by Hensel [59], and the modern algorithmic treatment is by Zassenhaus [149]. We conclude by showing how the LLL algorithm can be used to provide a polynomial-time algorithm for factoring (squarefree) primitive polynomials with integer coefficients. Our presentation of these topics mostly ignores complexity questions in favor of providing the clearest possible description of the basic algorithms. We also assume familiarity with the basic concepts of abstract algebra, especially the theory of polynomial rings.

Most of the topics in this chapter are discussed in standard textbooks on computer algebra; we mention in particular von zur Gathen and Gerhard [147], especially Chapters 14, 15 and 16. For background material on polynomial rings and finite fields, there are many suitable textbooks; a classic reference is the book by Herstein [61].

15.1   The Euclidean algorithm for polynomials

Let $\mathbb{F}$ be an arbitrary field, and let $\mathbb{F}[x]$ be the ring of polynomials in one variable x with coefficients in $\mathbb{F}$.

Definition 15.1. The **polynomial ring** $\mathbb{F}[x]$ consists of the zero polynomial together with the elements of the form

$$f = a_n x^n + a_{n-1} x^{n-1} + \cdots + a_1 x + a_0 \qquad (n \geq 0), \qquad (15.1)$$

where $a_n, a_{n-1}, \ldots, a_1, a_0 \in \mathbb{F}$ and $a_n \neq 0$. Addition and multiplication of polynomials are defined as usual. The integer n is the **degree** of the polynomial f, and we write $\deg(f) = n$. The element $a_n \in \mathbb{F}$ is the **leading coefficient** of f, and we write $\ell(f) = a_n$. If $\ell(f) = 1$ then f is **monic**. We say f is **irreducible** if any factorization $f = gh$ has either $g \in \mathbb{F}$ or $h \in \mathbb{F}$.

Theorem 15.2. *The polynomial algebra $\mathbb{F}[x]$ is a unique factorization domain (UFD): every monic polynomial $f \in \mathbb{F}[x]$ is the product of monic irreducible factors, and this factorization is unique up to the order of the factors.*

The Euclidean algorithm for the GCD of two integers a, b generalizes directly to the case of two polynomials $f, g \in \mathbb{F}[x]$. The main difference is that the absolute value $|a|$ of the integer a is replaced by the degree $\deg(f)$ of the nonzero polynomial f. The polynomial version of the Euclidean algorithm is presented in Figure 15.1.

Example 15.3. Let $\mathbb{F}_3 = \{0, 1, 2\}$ be the field with 3 elements, and consider these two polynomials in $\mathbb{F}_3[x]$:

$$r_0 = x^{12} + x^8 + x^6 + x^4 + x^2 + 1, \qquad r_1 = x^3 + 2x.$$

Dividing r_0 by r_1 gives

$$r_0 = q_2 r_1 + r_2, \qquad q_2 = x^9 + x^7 + 2x^5 + x, \qquad r_2 = 2x^2 + 1.$$

We divide r_2 by its leading coefficient, replacing r_2 by a monic polynomial:

$$r_2 = x^2 + 2.$$

Dividing r_1 by r_2 gives

$$r_1 = q_3 r_2 + r_3, \qquad q_3 = x, \qquad r_3 = 0.$$

From the last nonzero remainder we obtain $\gcd(r_0, r_1) = x^2 + 2$.

Definition 15.4. A polynomial $f \in \mathbb{F}[x]$ is called **squarefree** if f is not divisible by the square of any irreducible polynomial; in other words, every irreducible factor of f occurs with multiplicity 1.

- Input: Nonzero polynomials $f, g \in \mathbb{F}[x]$.

- Output: The monic GCD of f and g.

  (1) If $\deg(f) \geq \deg g$ then

   Set $r_0 \leftarrow \dfrac{1}{\ell(f)} f$ and $r_1 \leftarrow \dfrac{1}{\ell(g)} g$

   else

   Set $r_0 \leftarrow \dfrac{1}{\ell(g)} g$ and $r_1 \leftarrow \dfrac{1}{\ell(f)} f$.

  (2) Set $i \leftarrow 1$.

  (3) While $r_i \neq 0$ do:

   (a) Compute the unique polynomials q_{i+1} and r_{i+1} such that

   $$r_{i-1} = q_{i+1} r_i + r_{i+1},$$

   such that

   $$r_{i+1} = 0 \quad \text{or} \quad \deg(r_{i+1}) < \deg(r_i).$$

   (b) If $r_{i+1} \neq 0$ then set $r_{i+1} \leftarrow \dfrac{1}{\ell(r_{i+1})} r_{i+1}$.

   (c) Set $i \leftarrow i + 1$.

  (4) Return r_{i-1}.

FIGURE 15.1
The Euclidean algorithm **Euclid**(f, g) for $f, g \in \mathbb{F}[x]$

Definition 15.5. We can write equation (15.1) more compactly as follows:

$$f = \sum_{i=0}^{n} a_i x^i.$$

The **(formal) derivative** of f is the polynomial f' defined by

$$f' = \sum_{i=1}^{n} i a_i x^{i-1}.$$

Lemma 15.6. *If $f \in \mathbb{F}[x]$ and $\gcd(f, f') = 1$ then f is squarefree.*

Proof. Suppose that f is not square free; then some irreducible factor g occurs with multiplicity ≥ 2. Hence $f = g^2 h$ for some polynomial h. Since the formal derivative satisfies the product rule, we have

$$f' = 2gg'h + g^2 h' = g\big(2g'h + gh'\big).$$

Hence g is a factor of both f and f', and so $\gcd(f, f') \neq 1$.   □

The Euclidean algorithm of Figure 15.1 can be extended to compute not only the monic GCD h of the polynomials f and g but also polynomials s and t satisfying $sf + tg = h$; see Figure 15.2. This algorithm does not normalize the polynomials; that is, none of the polynomials are assumed to be monic. We initialize s_i and t_i for $i = 0, 1$ by

$$s_0 = 1, \quad t_0 = 0, \qquad s_1 = 0, \quad t_1 = 1.$$

We then verify by induction on i that $s_i f + t_i g = r_i$ for all i:

$$s_0 f + t_0 g = f = r_0,$$
$$s_1 f + t_1 g = g = r_1,$$
$$s_{i+1}f + t_{i+1}g = (s_{i-1} - q_i s_i)f + (t_{i-1} - q_i t_i)g$$
$$= (s_{i-1}f + t_{i-1}g) - q_i(s_i f + t_i g)$$
$$= r_{i-1} - q_i r_i$$
$$= r_{i+1}.$$

It can also be shown that $\deg(s) < \deg(g)$ and $\deg(t) < \deg(f)$.

15.2   Structure theory of finite fields

To understand polynomial factorization over the field $\mathbb{Q}$ of rational numbers, we first need to understand polynomial factorization over the finite field $\mathbb{F}_p$ where p is a prime number. In this section we recall the basic theory of finite fields; a detailed exposition may be found in most introductory textbooks on abstract algebra, and a specialized reference is Lidl and Niederreiter [90].

- Input: Polynomials $f, g \in \mathbb{F}[x]$, where $\mathbb{F}$ is a field.

- Output: A gcd h of f and g, and polynomials $s, t \in \mathbb{F}[x]$ for which $sf + tg = h$.

  (1) Set $r_0 \leftarrow f$, $r_1 \leftarrow g$.

  (2) Set $s_0 \leftarrow 1$, $t_0 \leftarrow 0$; set $s_1 \leftarrow 0$, $t_1 \leftarrow 1$; set $i \leftarrow 1$.

  (3) While $r_i \neq 0$ do:

   (a) Compute q_i and r_{i+1} in $\mathbb{F}[x]$ such that

$$r_{i-1} = q_i r_i + r_{i+1}, \qquad \deg(r_{i+1}) < \deg(r_i).$$

   (b) Set $s_{i+1} \leftarrow s_{i-1} - q_i s_i$.

   (c) Set $t_{i+1} \leftarrow t_{i-1} - q_i t_i$.

   (d) Set $i \leftarrow i + 1$.

  (4) Return $h = r_{i-1}$, $s = s_{i-1}$, $t = t_{i-1}$.

FIGURE 15.2

The extended Euclidean algorithm **XEuclid**(f, g)

Theorem 15.7. *A finite field with q elements exists if and only if $q = p^n$ for some prime number p and some positive integer n. For each such q there is, up to isomorphism, a unique field with q elements, denoted $\mathbb{F}_q$. Furthermore, $\mathbb{F}_{p^m}$ is isomorphic to a subfield of $\mathbb{F}_{p^n}$ if and only if m is a divisor of n.*

Finite fields may be constructed as follows. For each prime number p and positive integer n there exists at least one irreducible polynomial $f \in \mathbb{F}_p$ with $\deg(f) = n$. Let $\langle f \rangle$ be the principal ideal generated by f; that is, the set of all multiples of f by elements of $\mathbb{F}_p[x]$. We recall that the quotient ring $\mathbb{F}_p[x]/\langle f \rangle$ can be identified with the set of cosets $g + \langle f \rangle$ where either $g = 0$ or $\deg(g) < n$. Addition and multiplication in $\mathbb{F}_p[x]/\langle f \rangle$ are defined by

$$\bigl(g + \langle f \rangle\bigr) + \bigl(h + \langle f \rangle\bigr) = (g + h) + \langle f \rangle,$$
$$\bigl(g + \langle f \rangle\bigr)\bigl(h + \langle f \rangle\bigr) = k + \langle f \rangle,$$

where k is the remainder obtained after dividing gh by f. The quotient ring $\mathbb{F}_p[x]/\langle f \rangle$ is a field with p^n elements.

Lemma 15.8. *Let $q = p^n$ as above. In the polynomial ring $\mathbb{F}_q[x]$ we have*

$$x^q - x = \prod_{a \in \mathbb{F}_q} (x - a).$$

The last lemma is the special case $i = 1$ of the next theorem.

Theorem 15.9. *Let* $q = p^n$ *as above, and let* $i \geq 1$. *In* $\mathbb{F}_q[x]$ *the polynomial* $x^{q^i} - x$ *is the product of all monic irreducibles whose degrees are divisors of* i:

$$x^{q^i} - x = \prod_{d \,|\, i} \;\; \prod_{\substack{\deg(f) = d \\ f \text{ monic irreducible}}} f. \tag{15.2}$$

Proof. We write $g = x^{q^i} - x$ for the left side of equation (15.2). We first observe that g is squarefree by Lemma 15.6: clearly $g' = -1$ and so $\gcd(g, g') = 1$.

Suppose that $f \in \mathbb{F}_q[x]$ is a monic irreducible with $\deg(f) = d$. We will show that if f is a divisor of g then d is a divisor of i. Since g is squarefree, this will imply that g is a divisor of the right side of equation (15.2). If f is a divisor of g then Lemma 15.8 applied to q^i (in place of q) shows that

$$f = \prod_{a \in S} (x - a) \quad \text{for some subset } S \subseteq \mathbb{F}_{q^i}.$$

We choose some $a \in S$ and write $\mathbb{F}_q(a)$ for the smallest subfield of $\mathbb{F}_{q^i}$ which contains $\mathbb{F}_q$ and a. Since f is irreducible, this subfield is isomorphic to $\mathbb{F}_q[x]/\langle f \rangle$; since $\deg(f) = d$, we see that $\mathbb{F}_q(a)$ has q^d elements. Since $\mathbb{F}_{q^i}$ has q^i elements, and is a field extension of $\mathbb{F}_q(a)$, it follows that $q^i = (q^d)^j$ for some $j \geq 1$. Hence d is a divisor of i, as required.

We next show that if d is a divisor of i then f is a divisor of $x^{q^i} - x$; this implies that the right side of equation (15.2) is a divisor of g. Since $d \,|\, i$ we know that $\mathbb{F}_{q^i}$ has a subfield isomorphic to $\mathbb{F}_{q^d}$. Since f is irreducible of degree d, we know that $\mathbb{F}_q[x]/\langle f \rangle$ is a field with q^d elements, which we may identify with $\mathbb{F}_{q^d}$. We have

$$q^i - 1 = (q^d - 1)(q^{i-d} + q^{i-2d} + \cdots + q^d + 1). \tag{15.3}$$

If r, s, t are positive integers with $r = st$, then

$$x^r - 1 = (x^s - 1)\big(x^{s(t-1)} + x^{s(t-2)} + \cdots + x^s + 1\big).$$

Combining this with equation (15.3) we see that $x^{q^d - 1} - 1$ is a divisor of $x^{q^i - 1} - 1$, and hence that $x^{q^d} - x$ is a divisor of $x^{q^i} - x$. Let $a \in \mathbb{F}_{q^d} \subseteq \mathbb{F}_{q^i}$ be a root of f, so that $x - a$ is a divisor of f in $\mathbb{F}_{q^d}[x]$. Lemma 15.8 implies that $a^{q^d} = a$ and hence $x - a$ is a divisor of $x^{q^d} - x$ in $\mathbb{F}_{q^d}[x]$. Let $h \in \mathbb{F}_{q^i}[x]$ be the monic GCD of f and g; we have just shown that $x - a$ is a divisor of h. Since $x - a$ is a divisor of f, and f is irreducible, it follows that $f | h$. But $h | f$, so $f = h$. Hence f is a divisor of g, as required. $\qquad \square$

Example 15.10. For $p = 3$ we have $\mathbb{F}_3 = \{0, 1, 2\}$ and

$$x^3 - x = x(x + 1)(x + 2),$$
$$x^9 - x = x(x + 1)(x + 2)(x^2 + 1)(x^2 + x + 2)(x^2 + 2x + 2),$$

$$x^{27} - x = x(x+1)(x+2)(x^2+1)(x^3+2x+1)(x^3+2x+2) \cdot$$
$$(x^3+x^2+2)(x^3+x^2+x+2)(x^3+x^2+2x+1) \cdot$$
$$(x^3+2x^2+1)(x^3+2x^2+x+1)(x^3+2x^2+2x+2).$$

Remark 15.11. Let $I(d,q)$ denote the number of distinct monic irreducible polynomials of degree d over the field with q elements. Taking the degree of both sides of Theorem 15.9 we obtain the equation

$$q^i = \sum_{d \mid i} d\, I(d,q).$$

We can apply Möbius inversion to this equation; this gives

$$d\, I(d,q) = \sum_{i \mid d} \mu\left(\frac{d}{i}\right) q^i,$$

where μ is the Möbius function. Dividing by d gives a formula for $I(d,q)$.

15.3 Distinct-degree decomposition of a polynomial

Theorem 15.9 and the Euclidean algorithm of Figure 15.1 are the crucial ingredients in the computation of the distinct-degree decomposition of a squarefree polynomial over a finite field. We consider a squarefree monic nonconstant polynomial $f \in \mathbb{F}[x]$ where $\mathbb{F}$ is any field, not necessarily finite. Since $\mathbb{F}[x]$ is a unique factorization domain, we know that

$$f = \prod_{i=1}^{k} g_i,$$

where the factors $g_i \in \mathbb{F}[x]$ are distinct, monic and irreducible. Let δ be the maximum of the degrees of the irreducible factors:

$$\delta = \max\{\, \deg(g_1),\, \deg(g_2),\, \ldots,\, \deg(g_k)\,\}.$$

For each degree $d = 1, 2, \ldots, \delta$ we write ℓ_d for the number of distinct irreducible factors of degree d, and we collect these factors according to degree:

$$f = \prod_{d=1}^{\delta} h_d, \qquad h_d = \prod_{j=1}^{\ell_d} h_{dj},$$

where $\deg(h_{dj}) = d$. Thus h_d is the product of the irreducible factors of f of degree d; if $\ell_d = 0$ then the product is empty and $h_d = 1$.

Definition 15.12. The **distinct-degree decomposition** of f is the sequence
$$\text{ddd}(f) = [\, h_1, \, h_2, \, \ldots, \, h_\delta \,].$$

Now suppose that $\mathbb{F}$ is the finite field $\mathbb{F}_q$. Theorem 15.9 implies that
$$h_1 = \gcd(\, f, \, x^q - x \,).$$

We remove the irreducible factors of degree 1 from f by setting
$$f_1 = \frac{f}{h_1}.$$

(Here we use the assumption that f is squarefree; otherwise, f_1 could still contain irreducible factors of degree 1, namely those with multiplicity ≥ 2.) Applying Theorem 15.9 again we obtain
$$h_2 = \gcd(\, f_1, \, x^{q^2} - x \,).$$

We then define
$$f_2 = \frac{f_1}{h_2},$$

and compute
$$h_3 = \gcd(\, f_2, \, x^{q^3} - x \,).$$

We continue until we have removed all the irreducible factors from f.

A formal statement of this algorithm is given in Figure 15.3. During each iteration of step (2) we consider irreducible factors of degree d; we set $g_d = x^{q^d}$ so that $g_d - x$ is the left side of equation (15.2). We use the Euclidean algorithm to compute the product h_d of the irreducible factors of f of degree d; we then set f_d equal to the product of the remaining factors of degree $> d$.

Example 15.13. Choosing two irreducible polynomials in each degree ≤ 3 from Example 15.10 we consider the following polynomial in $\mathbb{F}_3[x]$:
$$f = (x + 1)(x + 2)(x^2 + x + 2)(x^2 + 2x + 2)(x^3 + 2x + 1)(x^3 + 2x + 2).$$

For this polynomial we have
$$\begin{aligned}
h_1 &= (x + 1)(x + 2) = x^2 + 2, \\
h_2 &= (x^2 + x + 2)(x^2 + 2x + 2) = x^4 + 1, \\
h_3 &= (x^3 + 2x + 2)(x^3 + 2x + 1) = x^6 + x^4 + x^2 + 2.
\end{aligned}$$

Consider the complete expansion f_0 of the factored polynomial f:
$$f_0 = x^{12} + x^8 + x^6 + x^4 + x^2 + 1.$$

- Input: A squarefree monic nonconstant polynomial $f \in \mathbb{F}_q[x]$.

- Output: The distinct-degree decomposition $\mathrm{ddd}(f)$.

  (1) Set $f_0 \leftarrow f$, set $g_0 \leftarrow x$, set $d \leftarrow 0$.

  (2) While $f_d \neq 1$ do:

   (a) Set $d \leftarrow d + 1$,

   (b) Set $g_d \leftarrow g_{d-1}^q$.

   (c) Set $h_d \leftarrow \mathbf{Euclid}(f_{d-1}, g_d - x)$.

   (d) Set $f_d \leftarrow f_{d-1}/h_d$.

  (3) Return $h_1, h_2, \ldots, h_d$.

FIGURE 15.3
Algorithm $\mathbf{DDD}(f)$: distinct-degree decomposition in $\mathbb{F}_q[x]$

We show how to recover h_1, h_2, h_3 from f_0 using Algorithm $\mathbf{DDD}(f)$. To find h_1 we follow Example 15.3 and obtain

$$h_1 = \gcd(f_0, x^3 - x)$$
$$= \gcd(x^{12} + x^8 + x^6 + x^4 + x^2 + 1, x^3 + 2x) = x^2 + 2.$$

We now divide f_0 by h_1 to find f_1:

$$f_1 = \frac{f_0}{h_1} = x^{10} + x^8 + 2x^6 + x^2 + 2.$$

To find h_2 we apply Algorithm $\mathbf{Euclid}(f, g)$ again:

$$h_2 = \gcd(f_1, x^9 - x)$$
$$= \gcd(x^{10} + x^8 + 2x^6 + x^2 + 2, x^9 + 2x) = x^4 + 1.$$

We now divide f_1 by h_2 to find f_2:

$$f_2 = \frac{f_1}{h_2} = x^6 + x^4 + x^2 + 2.$$

To find h_3 we apply Algorithm $\mathbf{Euclid}(f, g)$ one more time:

$$h_3 = \gcd(f_2, x^{27} - x)$$
$$= \gcd(x^6 + x^4 + x^2 + 2, x^{27} + 2x) = x^6 + x^4 + x^2 + 2.$$

Since $h_3 = f_2$, we have $f_3 = 1$, and so the algorithm terminates and returns the discrete-degree decomposition:

$$\mathrm{ddd}(f_0) = [\, x^2 + 2,\ x^4 + 1,\ x^6 + x^4 + x^2 + 2 \,].$$

15.4   Equal-degree decomposition of a polynomial

Our next algorithm splits each of the polynomials h_d from the distinct-degree decomposition into its irreducible factors. The result is the equal-degree decomposition of a squarefree polynomial all of whose irreducible factors have the same degree. We consider a squarefree monic nonconstant polynomial $h_d \in \mathbb{F}[x]$, and assume that every irreducible factor of h_d has degree d:

$$h_d = \prod_{j=1}^{\ell_d} h_{dj}, \qquad \deg(h_{dj}) = d,$$

where $\ell_d \geq 1$ is the number of distinct monic irreducible factors.

Definition 15.14. The **equal-degree decomposition** of h_d is the sequence

$$\text{edd}(h_d) = \begin{bmatrix} h_{d1}(x), \ \ldots, \ h_{d\ell_d} \end{bmatrix}.$$

To compute the equal-degree decomposition we use a probabilistic algorithm based on the following lemma; note that here we assume $p \neq 2$.

Lemma 15.15. *Let $q = p^n$ ($p \neq 2$) and let $\mathbb{F}_q$ be the field with q elements. Let $\mathbb{F}_q^{\times}$ be the multiplicative group of nonzero elements, and let $S = \{\, a^2 \mid a \in \mathbb{F}_q^{\times} \,\}$ be the set of squares in $\mathbb{F}_q^{\times}$. We have:*

(a) S is a subgroup of $\mathbb{F}_q^{\times}$ of order $(q-1)/2$.

(b) $S = \{\, a \in \mathbb{F}_q^{\times} \mid a^{(q-1)/2} = 1 \,\}$.

(c) For every $a \in \mathbb{F}_q^{\times}$ we have $a^{(q-1)/2} = \pm 1$.

Proof. Since $\mathbb{F}_q^{\times}$ is an Abelian group, the squaring map $\mathbb{F}_q^{\times} \to \mathbb{F}_q^{\times}$ ($a \mapsto a^2$) is a homomorphism; since the image of this map is S, it follows that S is a subgroup of $\mathbb{F}_q^{\times}$. The kernel of this map is the set of all $a \in \mathbb{F}_q^{\times}$ for which $a^2 = 1$; that is, the set of roots of the polynomial $x^2 - 1$ in $\mathbb{F}_q[x]$. A polynomial of degree d over a field has at most d roots, and so the kernel of the squaring map has 1 or 2 elements. We need to show that the kernel has 2 elements.

Let $e = (q-1)/2$ and consider the e-power map $\mathbb{F}_q^{\times} \to \mathbb{F}_q^{\times}$ ($a \mapsto a^e$). As before, this is a homomorphism, and the same reasoning shows that its kernel has $\leq e$ elements. Since $\mathbb{F}_q^{\times}$ has order $q-1$, for every $a \in \mathbb{F}_q^{\times}$ we have $(a^2)^e = a^{q-1} = 1$; in particular, S is a subset of the kernel, and so $|S| \leq e$.

The First Isomorphism Theorem implies that any homomorphism satisfies

$$|\text{kernel}| \, |\text{image}| = |\text{domain}|.$$

Applying this to the squaring map, we obtain

$$q - 1 = |\text{domain}| = |\text{kernel}| \, |\text{image}| = |\text{kernel}| \, |S| \leq 2|S| \leq q - 1.$$

It follows that the two inequalities must be equalities. In particular, the kernel has 2 elements, which proves (a). Furthermore, S has e elements, and so it equals the kernel of the e-power map, which proves (b). Finally, we see that the image of the e-power map has two elements; it clearly contains ± 1, and these elements are distinct since $p \neq 2$, and which proves (c). $\qquad\square$

Let $q = p^n$ $(p \neq 2)$. Suppose that $d \geq 1$ and that $h \in \mathbb{F}_q[x]$ has $\ell \geq 2$ distinct monic irreducible factors each of degree d:

$$h = \prod_{j=1}^{\ell} h_j, \qquad \deg(h_j) = d, \qquad \deg(h) = d\ell.$$

We consider the quotient ring $\mathbb{F}_q[x]/\langle h \rangle$. Since $h_1, h_2, \ldots, h_\ell$ are relatively prime, the Chinese Remainder Theorem gives the isomorphism

$$\mathbb{F}_q[x]/\langle h \rangle \approx \mathbb{F}_q[x]/\langle h_1 \rangle \times \mathbb{F}_q[x]/\langle h_2 \rangle \times \cdots \times \mathbb{F}_q[x]/\langle h_\ell \rangle. \tag{15.4}$$

Since each h_i is irreducible, each factor $\mathbb{F}_q[x]/\langle h_j \rangle$ is isomorphic to the field $\mathbb{F}_{q^d}$ with q^d elements. We have $|\mathbb{F}_q[x]/\langle h \rangle| = q^{d\ell}$, and we identify the elements with their coset representatives from the set of polynomials of degree $< d\ell$:

$$g = \sum_{k=0}^{d\ell-1} a_k x^k + \langle h \rangle \in \mathbb{F}_q[x]/\langle h \rangle.$$

An explicit form of the isomorphism (15.4) is given by the following equations:

$$\epsilon : \mathbb{F}_q[x]/\langle h \rangle \to \underbrace{\mathbb{F}_{q^d} \times \mathbb{F}_{q^d} \times \cdots \times \mathbb{F}_{q^d}}_{\ell \text{ factors}},$$

$$\epsilon(g) = \big(\epsilon_1(g), \ldots, \epsilon_2(g), \ldots, \epsilon_\ell(g) \big),$$

$$\epsilon_j(g) = g + \langle h_j \rangle \in \mathbb{F}_q[x]/\langle h_j \rangle \cong \mathbb{F}_{q^d} \quad (j = 1, 2, \ldots, \ell).$$

Clearly h_j is a divisor of g if and only if $\epsilon_j(g) = 0$. A proper factor of h corresponds to a polynomial g for which $\epsilon_j(g) = 0$ for some j and $\epsilon_j(g) \neq 0$ for some other j. We do not actually compute ϵ (this would require knowing $h_1, h_2, \ldots, h_\ell$ already); we merely use ϵ to prove the following lemma.

Lemma 15.16. *Let h be as above, and let g be a uniformly distributed pseudorandom nonconstant polynomial with $\deg(g) < \deg(h)$. If $\gcd(g, h) \neq 1$ then $\gcd(g, h)$ is a proper factor of h. If $\gcd(g, h) = 1$ then*

$$\widetilde{g} = \gcd(\, \overline{g^e} - 1,\, h\,), \qquad e = (q^d - 1)/2,$$

where the bar denotes the remainder modulo h, is a proper factor of h with probability $\geq \frac{1}{2}$.

Proof. As before we represent g as (the coset of) a polynomial of degree $< d\ell$. The statement for the case $\gcd(g, h) \neq 1$ is clear. If $\gcd(g, h) = 1$ then Lemma 15.15 with q^d (in place of q) and the isomorphism (15.4) imply that

$$\epsilon(\overline{g^e}) = (\pm 1, \pm 1, \ldots, \pm 1) \quad (\ell \text{ components}),$$

where each component $\epsilon_j(\overline{g^e})$ is a uniformly distributed element of $\{\pm 1\}$. Since ϵ is a ring isomorphism, we have

$$\epsilon(\overline{g^e} - 1) = (\pm 1 - 1, \pm 1 - 1, \ldots, \pm 1 - 1),$$

and so $\widetilde{g} = \gcd(\overline{g^e} - 1, h)$ is a proper factor of h except when all the components of $\epsilon(\overline{g^e})$ are equal. (If all components are 1, then $\overline{g^e} = 1$ and $\widetilde{g} = h$; if all components are -1, then $\overline{g^e} = -1$ and $\widetilde{g} = 1$.) These two exceptional cases occur with probability

$$2\left(\tfrac{1}{2}\right)^\ell = 2^{1-\ell} \leq \tfrac{1}{2} \text{ since } \ell \geq 2.$$

Thus we obtain a proper factor of h with probability $\geq \tfrac{1}{2}$. $\square$

A formal statement of this probabilistic algorithm is given in Figure 15.4.

Example 15.17. Suppose $q = 5$. In $\mathbb{F}_5[x]$ we consider the product of $\ell = 6$ from the 10 distinct irreducible polynomials of degree $d = 2$:

$$h = (x^2 + 2)(x^2 + 3)(x^2 + x + 2)(x^2 + x + 1)(x^2 + 2x + 3)(x^2 + 2x + 4)$$
$$= x^{12} + x^{11} + 3x^{10} + 4x^8 + 2x^7 + 4x^5 + 2x^4 + x^3 + 2x^2 + 4x + 4.$$

We consider three separate calls to **TrialSplit**(h) to illustrate the three possible cases. The first call generates the pseudorandom polynomial

$$g_1 = 2x^{11} + 2x^9 + x^7 + 3x^6 + 3x^5 + x^3 + 2x + 3.$$

We have

$$g_2 = \gcd(g_1, h) = x^2 + 2x + 4.$$

The algorithm returns the proper factor g_2. The second call generates the pseudorandom polynomial

$$g_1 = 3x^{11} + 4x^{10} + 4x^8 + 2x^7 + 3x^5 + 3x^4 + 4x^3 + x.$$

We have

$$g_2 = \gcd(g_1, h) = 1.$$

Since $(q^d - 1)/2 = 12$, the algorithm computes

$$g_3 = \overline{g_1^{12}} = 3x^{10} + 4x^8 + x^7 + x^5 + 2x^4 + x^3 + 2x^2 + x + 2.$$

From this we obtain

$$g_4 = \gcd(g_3 - 1, h) = x^6 + x^4 + 4x^3 + 2x^2 + x + 2.$$

- Input: A squarefree monic nonconstant polynomial $h \in \mathbb{F}_q[x]$ of degree $d\ell$ which is the product of ℓ monic irreducible factors of degree d.

- Output: If $\ell \geq 2$ then with probability $\geq \frac{1}{2}$ the algorithm returns a proper factor g of h (in case of failure the algorithm returns $g = 0$).

  (1) If $\deg(h) = 1$ then

      (a) Set $g \leftarrow 0$.

      else

      (b) Set $g_1 \leftarrow 0$.
      (c) While $g_1 \in \mathbb{F}_q$ do:

          Generate pseudorandom $g_1 \in \mathbb{F}_q[x]$ with $\deg(g_1) < \deg(h)$.
      (d) Set $g_2 \leftarrow \gcd(g_1, h)$.
      (e) If $g_2 \neq 1$ then

          Set $g \leftarrow g_2$.

          else

          Set $e \leftarrow (q^d - 1)/2$.
          Set $g_3 \leftarrow \overline{g_1^e}$ where the bar denotes remainder modulo h.
          Set $g_4 \leftarrow \gcd(g_3 - 1, h)$.
          If $0 < \deg(g_4) < \deg(h)$ then set $g \leftarrow g_4$ else set $g \leftarrow 0$.
  (2) Return g.

FIGURE 15.4
Algorithm **TrialSplit**(h) for $h \in \mathbb{F}_q[x]$ where $q = p^n$ $(p \neq 2)$

- Input: A squarefree monic nonconstant polynomial $h \in \mathbb{F}_q[x]$ of degree $d\ell$ which is the product of ℓ monic irreducible factors of degree d, and an integral termination parameter $s \geq 1$.

- Output: With probability $\geq 1 - 2^{-s}$, a proper factor g of h.

  (1) Set $g \leftarrow 0$, set $k \leftarrow 0$.
  (2) While $g = 0$ and $k < s$ do

      (a) Set $g \leftarrow$ **TrialSplit**(h).
      (b) Set $k \leftarrow k + 1$.
  (3) Return g.

FIGURE 15.5
Algorithm **Split**(h, s) for $h \in \mathbb{F}_q[x]$ where $q = p^n$ $(p \neq 2)$

- Input: A squarefree monic nonconstant polynomial $h \in \mathbb{F}_q[x]$ of degree $d\ell$ which is the product of $\ell \geq 2$ monic irreducible factors each of degree d, and an integral termination parameter $s \geq 1$.

- Output: The equal-degree decomposition of h.

- Remark: Before this procedure is called, the global variable `factorlist` is initialized to the empty list.

  (1) Set $g_1 \leftarrow \textbf{Split}(h, s)$.

  (2) If $g_1 = 0$ then

   (a) Append h to `factorlist`.

   else

   (b) Recursively call $\textbf{EDD}(g_1, s)$.

   (c) Recursively call $\textbf{EDD}(h/g_1, s)$.

FIGURE 15.6
Algorithm $\textbf{EDD}(h, s)$ for $h \in \mathbb{F}_q[x]$ where $q = p^n$ $(p \neq 2)$

The algorithm returns the proper factor g_4. The third call generates the pseudorandom polynomial

$$g_1 = 3x^{10} + 2x^9 + x^8 + x^7 + 2x^6 + x^5 + 2x^4 + 2x^3 + 2x^2 + 4x + 3.$$

We have

$$g_2 = \gcd(g_1, h) = 1.$$

The algorithm computes

$$g_3 = \overline{g_1^{12}} = 1.$$

From this we obtain

$$g_4 = \gcd(g_3 - 1, h) = h.$$

The algorithm returns $g = 0$ indicating failure.

If we make s calls to **TrialSplit**, the probability of failure is $\leq 2^{-s}$, and the probability of finding a proper factor is $\geq 1 - 2^{-s}$, which converges to 1 very rapidly as s increases. To increase the probability that we find a proper factor of h, we use algorithm $\textbf{Split}(h, s)$ displayed in Figure 15.5. The input to $\textbf{Split}(h, s)$ includes a termination parameter $s \geq 1$; this is the upper bound on the number of trials **Split** will perform before concluding that h is irreducible. Algorithm **Split** does not provide the complete equal-degree decomposition of h, it merely splits h into the product of two proper factors. To find the complete equal-degree decomposition, we use the algorithm $\textbf{EDD}(h)$ given in Figure 15.6 which recursively calls **Split** until no further splitting is possible.

- Input: A monic nonconstant polynomial $f \in \mathbb{F}_q[x]$, and an integral termination parameter s to control the calls to **Split**(h, s).

- Output: The complete factorization of f into monic irreducible factors.

    (1) Set `complete` $\leftarrow [\ ]$.

    (2) Set $f_0 \leftarrow f$, set $g_0 \leftarrow x$, set $d \leftarrow 0$.

    (3) While $f_d \neq 1$ do:

        (a) Set $d \leftarrow d+1$.

        (b) Set $g_d \leftarrow g_{d-1}^q$.

        (c) Set $h_d \leftarrow$ **Euclid**$(\ f_{d-1},\ g_d - x\ )$.

        (d) If $h_d \neq 1$ then

            - Set `factorlist` $\leftarrow [\ ]$ *(empty list)*.
            - Call **EDD**(h_d, s).
            - For k in `factorlist` do
                Set $m \leftarrow 0$.
                While $k \mid f$ do: Set $m \leftarrow m+1$, set $f \leftarrow f/k$.
                Append $[k, m]$ to `complete`.

    (4) Return `complete`.

FIGURE 15.7
Algorithm **Factor**(f, s) for $f \in \mathbb{F}_q[x]$ where $q = p^n$ $(p \neq 2)$

We now give an algorithm which finds the complete factorization of a polynomial f (not necessarily squarefree) in one variable x with coefficients in the finite field $\mathbb{F}_q$ where $q = p^n$ $(p \neq 2)$. The last step is to compute the multiplicity of each irreducible factor. We start as in algorithm **DDD**, but after step 2(c) which determines the factor h_d, we check whether $h_d \neq 1$; if so, we call algorithm **EDD** to find the irreducible factors of h_d, and then use repeated division to eliminate the correct power of each of these irreducible factors from f. The algorithm **Factor**(f) is given in Figure 15.7.

15.5 Hensel lifting of polynomial factorizations

We now begin our study of polynomials with integer coefficients. Suppose that $f \in \mathbb{Z}[x]$ and that p is a prime number which does not divide the leading coefficient $\ell(f)$. Let $\overline{f}$ denote the polynomial in $\mathbb{F}_p[x]$ obtained by reducing the coefficients of f modulo p. Suppose further that we have polynomials

$g_1, h_1 \in \mathbb{Z}[x]$ such that $\overline{f} = \overline{g_1}\overline{h_1}$ in $\mathbb{F}_p[x]$ and this factorization is proper in the sense that neither $\overline{g_1}$ nor $\overline{h_1}$ is a constant. (We also need to assume that $\overline{g_1}$ and $\overline{h_1}$ are relatively prime in $\mathbb{F}_p[x]$, but we ignore this for the moment.) We want to lift this factorization from the modulus p to the modulus p^2: that is, we want to find polynomials $g_2, h_2 \in \mathbb{Z}[x]$ such that $\deg(g_1) = \deg(g_2)$, $\deg(h_1) = \deg(h_2)$, and $f \equiv g_2 h_2 \pmod{p^2}$. In other words, we want to lift the factorization from the polynomial ring $(\mathbb{Z}/p\mathbb{Z})[x]$ to the polynomial ring $(\mathbb{Z}/p^2\mathbb{Z})[x]$. It is important to realize that $\mathbb{Z}/p^2\mathbb{Z}$ is not a field, so in particular it is not isomorphic to $\mathbb{F}_{p^2}$. In fact, $\mathbb{Z}/p^2\mathbb{Z}$ is not even an integral domain, since it has zero divisors: $p^2 \equiv 0 \pmod{p^2}$ but $p \not\equiv 0 \pmod{p^2}$.

Example 15.18. Consider the polynomial

$$f = x^4 - 10x^3 + 35x^2 - 50x + 24 \in \mathbb{Z}[x].$$

Modulo $p = 5$ we have

$$\overline{f} \equiv x^4 + 4,$$

and in $\mathbb{F}_5[x]$ we have the factorization $\overline{f} \equiv \overline{g_1}\overline{h_1} \pmod 5$ where

$$g_1 = x^2 + 2x + 2 \in \mathbb{Z}[x], \qquad h_1 = x^2 + 3x + 2 \in \mathbb{Z}[x].$$

Modulo $p^2 = 25$ we have

$$\overline{f} \equiv x^4 + 15x^3 + 10x^2 + 24,$$
$$\overline{g_1}\overline{h_1} \equiv x^4 + 5x^3 + 10x^2 + 10x + 4,$$

and so $\overline{f} \not\equiv \overline{g_1}\overline{h_1} \pmod{25}$.

We need to understand how to obtain $g_2, h_2 \in \mathbb{Z}[x]$ satisfying $\overline{f} \equiv \overline{g_2}\overline{h_2} \pmod{p^2}$ from $g_1, h_1 \in \mathbb{Z}[x]$ satisfying $\overline{f} \equiv \overline{g_1}\overline{h_1} \pmod{p}$. We can then repeat this process indefinitely, to obtain a lifting of the factorization $\overline{f} \equiv \overline{g_1}\overline{h_1}$ to the modulus p^n for any $n \geq 1$. This process is called **Hensel lifting**. We denote the modulus by m; we may assume that $m = p^n$. Given a factorization

$$\overline{f} \equiv \overline{g_1}\overline{h_1} \pmod{m},$$

we will show how to find polynomials $g_2, h_2 \in \mathbb{Z}[x]$ satisfying

$$\overline{f} \equiv \overline{g_2}\overline{h_2} \pmod{m^2};$$

furthermore $\deg(g_1) = \deg(g_2)$ and $\deg(h_1) = \deg(h_2)$.

We assume that we have polynomials $f, g_1, h_1, s_1, t_1 \in \mathbb{Z}[x]$, and an integer $m \geq 2$ which does not divide $\ell(f)$, such that

$$\overline{f} \equiv \overline{g_1}\overline{h_1} \pmod{m}, \qquad \overline{s_1}\,\overline{g_1} + \overline{t_1}\,\overline{h_1} \equiv 1 \pmod{m}.$$

The second condition comes from the extended Euclidean algorithm for polynomials over a field. If $m = p$ is prime, as at the start of the lifting, then this condition is equivalent to g_1 and h_1 being relatively prime in $\mathbb{F}_p[x]$. Since the Euclidean algorithm is not available for general m, we must use this condition in the general case. We further assume that h_1 is monic, and that

$$\deg(f) = \deg(g_1) + \deg(h_1), \qquad \deg(s_1) < \deg(h_1), \qquad \deg(t_1) < \deg(g_1).$$

The basic idea is as follows. We have

$$f = g_1 h_1 + mv,$$

for some polynomial v, and hence

$$e = f - g_1 h_1 = mv.$$

We also have

$$s_1 g_1 + t_1 h_1 = 1 + mw,$$

for some polynomial w, and hence

$$s_1 g_1 + t_1 h_1 - 1 = mw.$$

We now calculate

$$\begin{aligned}
(g_1 &+ t_1 e)(h_1 + s_1 e) \\
&= g_1 h_1 + (s_1 g_1 + t_1 h_1)e + s_1 t_1 e^2 \\
&= g_1 h_1 + (s_1 g_1 + t_1 h_1)(f - g_1 h_1) + s_1 t_1 e^2 \\
&= g_1 h_1 + (s_1 g_1 + t_1 h_1)f - (s_1 g_1 + t_1 h_1)g_1 h_1 + s_1 t_1 e^2 \\
&= f + (s_1 g_1 + t_1 h_1 - 1)f - (s_1 g_1 + t_1 h_1 - 1)g_1 h_1 + s_1 t_1 e^2 \\
&= f + (s_1 g_1 + t_1 h_1 - 1)(f - g_1 h_1) + s_1 t_1 e^2 \\
&= f + (mw)(mv) + s_1 t_1 (mv)^2.
\end{aligned}$$

From this we see that $(g_1 + t_1 e)(h_1 + s_1 e)$ is congruent to f modulo m^2. This suggests that we define $g_2 = g_1 + t_1 e$ and $h_2 = h_1 + s_1 e$. The problem with this definition is that the degrees of g_2 and h_2 are too large. We avoid this problem as follows; in this discussion, all polynomials are in $\mathbb{Z}[x]$:

Step 1: We compute

$$e \equiv f - g_1 h_1 \pmod{m^2}.$$

Since h_1 is monic, we can do division with remainder of $s_1 e$ by h_1 in $\mathbb{Z}[x]$:

$$s_1 e \equiv q h_1 + r \pmod{m^2}, \qquad \deg(r) < \deg(h_1).$$

By definition, $e \equiv 0 \pmod{m}$ and so $s_1 e \equiv 0 \pmod{m}$; hence $q \equiv 0 \pmod{m}$ and $r \equiv 0 \pmod{m}$. We now define

$$g_2 \equiv g_1 + t_1 e + q g_1 \pmod{m^2}, \qquad h_2 \equiv h_1 + r \pmod{m^2}.$$

It is clear that $g_2 \equiv g_1 \pmod{m}$ and $h_2 \equiv h_1 \pmod{m}$. Since $\deg(r) < \deg(h_1)$ we see that $\deg(h_2) = \deg(h_1)$ and that h_2 is monic. We calculate modulo m^2:

$$
\begin{aligned}
&f - g_2 h_2 \\
&\equiv f - (g_1 + t_1 e + q g_1)(h_1 + r) \\
&\equiv f - (g_1 + t_1 e + q g_1)(h_1 + s_1 e - q h_1) \\
&\equiv f - (g_1 h_1 + s_1 g_1 e - g_1 h_1 q + t_1 h_1 e + s_1 t_1 e^2 - t_1 h_1 q e \\
&\qquad + g_1 h_1 q + s_1 g_1 q e - g_1 h_1 q^2) \\
&\equiv f - g_1 h_1 - s_1 g_1 e - t_1 h_1 e - s_1 t_1 e^2 + t_1 h_1 q e - s_1 g_1 q e + g_1 h_1 q^2 \\
&\equiv f - g_1 h_1 - (s_1 g_1 + t_1 h_1) e - s_1 t_1 e^2 - (s_1 g_1 - t_1 h_1) q e + g_1 h_1 q^2 \\
&\equiv e - (s_1 g_1 + t_1 h_1) e - s_1 t_1 e^2 - (s_1 g_1 - t_1 h_1) q e + g_1 h_1 q^2 \\
&\equiv -(s_1 g_1 + t_1 h_1 - 1) e - s_1 t_1 e^2 - (s_1 g_1 - t_1 h_1) q e + g_1 h_1 q^2 \\
&\equiv 0 \pmod{m^2}.
\end{aligned}
$$

The conditions

$$
\ell(f) \not\equiv 0 \pmod{m}, \qquad h_2 \text{ is monic}, \qquad f \equiv g_2 h_2 \pmod{m^2},
$$

together imply that

$$
\deg(g_2) = \deg(f) - \deg(h_2) = \deg(f) - \deg(h_1) = \deg(g_1).
$$

Step 2: We compute

$$
e^* = s_1 g_2 + t_1 h_2 - 1 \pmod{m^2}.
$$

Since h_2 is monic, we do division with remainder of $s_1 e^*$ by h_2 in $\mathbb{Z}[x]$:

$$
s_1 e^* \equiv q^* h_2 + r^* \pmod{m^2}, \qquad \deg(r^*) < \deg(h_2).
$$

By definition, $e^* \equiv 0 \pmod{m}$ and so $s_1 e^* \equiv 0 \pmod{m}$; hence $q^* \equiv 0 \pmod{m}$ and $r^* \equiv 0 \pmod{m}$. We define

$$
s_2 \equiv s_1 - r^* \pmod{m^2}, \qquad t_2 \equiv t_1 - t_1 e^* - q^* g_2 \pmod{m^2}.
$$

It can be shown that

$$
\begin{aligned}
s_2 &\equiv s_1 \pmod{m}, & \deg(s_2) &< \deg(h_2), \\
t_2 &\equiv t_1 \pmod{m}, & \deg(t_2) &< \deg(g_2).
\end{aligned}
$$

A formal statement of this algorithm is given in Figure 15.8.

Example 15.19. We continue from Example 15.18. We have

$$
f = x^4 - 10x^3 + 35x^2 - 50x + 24, \quad g_1 = x^2 + 2x + 2, \quad h_1 = x^2 + 3x + 2,
$$

- Input:

  - The modulus $m \in \mathbb{Z}$ $(m \geq 2)$.
  - Polynomials $f, g_1, h_1 \in \mathbb{Z}[x]$ such that h_1 is monic and

  $$f \equiv g_1 h_1 \pmod{m}, \qquad \deg(f) = \deg(g_1) + \deg(h_1).$$

  - Polynomials $s_1, t_1 \in \mathbb{Z}[x]$ such that

  $$s_1 g_1 + t_1 h_1 \equiv 1 \pmod{m}, \quad \deg(s_1) < \deg(h_1), \quad \deg(t_1) < \deg(g_1).$$

- Output:

  - Polynomials $g_2, h_2 \in \mathbb{Z}[x]$ such that h_2 is monic and

  $$f \equiv g_2 h_2 \pmod{m^2}, \quad g_2 \equiv g_1 \pmod{m}, \quad \deg(g_2) = \deg(g_1),$$
  $$h_2 \equiv h_1 \pmod{m}, \quad \deg(h_2) = \deg(h_1).$$

  - Polynomials $s_2, t_2 \in \mathbb{Z}[x]$ such that

  $$s_2 g_2 + t_2 h_2 \equiv 1 \pmod{m^2}, \quad s_2 \equiv s_1 \pmod{m}, \quad \deg(s_2) < \deg(h_2),$$
  $$t_2 \equiv t_1 \pmod{m}, \quad \deg(t_2) < \deg(g_2).$$

- Algorithm:

  (1) Set $e \leftarrow f - g_1 h_1 \pmod{m^2}$.
  (2) Compute $q, r \in \mathbb{Z}[x]$ such that

  $$s_1 e \equiv q h_1 + r \pmod{m^2}, \qquad \deg(r) < \deg(h_1).$$

  (3) Set $g_2 \leftarrow g_1 + t_1 e + q g_1 \pmod{m^2}$.
  (4) Set $h_2 \leftarrow h_1 + r \pmod{m^2}$.
  (5) Set $e^* \leftarrow s_1 g_2 + t_1 h_2 - 1 \pmod{m^2}$.
  (6) Compute $q^*, r^* \in \mathbb{Z}[x]$ such that

  $$s_1 e^* = q^* h_2 + r^* \pmod{m^2}, \qquad \deg(r^*) < \deg(h_2).$$

  (7) Set $s_2 \leftarrow s_1 - r^* \pmod{m^2}$.
  (8) Set $t_2 \leftarrow t_1 - t_1 e^* - q^* g_2 \pmod{m^2}$.
  (9) Return g_2, h_2, s_2, t_2.

FIGURE 15.8

The Hensel lifting algorithm **Hensel**$(m, f, g_1, h_1, s_1, t_1)$

such that $f \equiv g_1 h_1 \pmod 5$. The extended Euclidean algorithm gives

$$s_1 = 3x + 4, \qquad t_1 = 2x + 4,$$

such that

$$s_1 g_1 + t_1 h_1 = 5x^3 + 20x^2 + 30x + 16 \equiv 1 \pmod 5.$$

We compute

$$e = f - g_1 h_1 = 10x^3 + 15x + 20,$$

such that $e \equiv 0 \pmod 5$ but $e \not\equiv 0 \pmod{25}$. We have

$$s_1 e = 5x^4 + 15x^3 + 20x^2 + 20x + 5,$$

and division with remainder gives

$$s_1 e = q h_1 + r, \qquad q = 5x^2 + 10, \qquad r = 15x + 10.$$

From this we obtain

$$g_2 = x^2 + 22x + 2, \qquad h_2 = x^2 + 18x + 12,$$

and we verify that

$$f \equiv g_2 h_2 \pmod{25}.$$

This is the lifting $\pmod{25}$ of the original factorization $f \equiv g_1 h_1 \pmod 5$. We now compute

$$e^* = 5x^3 + 10x^2 + 15x + 5, \qquad s_1 e^* = 15x^4 + 10x^2 + 20,$$

and division with remainder gives

$$s_1 e^* = q^* h_1 + r^*, \qquad q^* = 15x^2 + 5x + 15, \qquad r^* = 20x + 15.$$

From this we obtain

$$s_2 = 8x + 14, \qquad t_2 = 17x + 4,$$

and we verify that

$$s_2 g_2 + t_2 h_2 \equiv 1 \pmod{25}.$$

Example 15.20. We continue for the next 3 steps beyond Example 15.19. Modulo $25^2 = 625$ we obtain

$$g_3 = x^2 + 622x + 2, \qquad h_3 = x^2 + 618x + 12,$$

together with

$$s_3 = 208x + 314, \qquad t_3 = 417x + 104.$$

Modulo $625^2 = 390625$ we obtain

$$g_4 = x^2 + 390622x + 2, \qquad h_4 = x^2 + 390618x + 12,$$

together with

$$s_4 = 130208x + 195314, \qquad t_4 = 260417x + 65104.$$

Modulo $390625^2 = 152587890625$ we obtain

$$g_5 = x^2 + 152587890622x + 2, \qquad h_5 = x^2 + 152587890618x + 12,$$

together with

$$s_5 = 50862630208x + 76293945314, \quad t_5 = 101725260417x + 25431315104.$$

Theorem 15.21. Hensel's Lemma. *Let p be a prime number. Let $f, g_1, h_1, s_1, t_1 \in \mathbb{Z}[x]$ be such that h_1 is monic and*

$$f \equiv g_1 h_1 \;(\text{mod } p), \qquad \deg(f) = \deg(g_1) + \deg(h_1),$$
$$s_1 g_1 + t_1 h_1 \equiv 1 \;(\text{mod } p), \quad \deg(s_1) < \deg(h_1), \quad \deg(t_1) < \deg(g_1).$$

Then for any $n \geq 1$ there exist $g_n, h_n, s_n, t_n \in \mathbb{Z}[x]$ such that h_n is monic and

$$f \equiv g_n h_n \;\left(\text{mod } p^{2^n}\right), \qquad g_n \equiv g_1 \;(\text{mod } p), \quad \deg(g_n) = \deg(g_1),$$
$$h_n \equiv h_1 \;(\text{mod } p), \quad \deg(h_n) = \deg(h_1),$$
$$s_n g_n + t_n h_n \equiv 1 \;\left(\text{mod } p^{2^n}\right), \quad s_n \equiv s_1 \;(\text{mod } p), \quad \deg(s_n) < \deg(h_n),$$
$$t_n \equiv t_1 \;(\text{mod } p), \quad \deg(t_n) < \deg(g_n).$$

Proof. Perform n iterations of algorithm **Hensel** in Figure 15.8. □

The next result establishes the uniqueness of the Hensel lifting.

Theorem 15.22. *Suppose that $m \geq 2$, $n \geq 1$ and $g, h, s, t, G, H \in \mathbb{Z}[x]$ satisfy*

(a) $sg + th \equiv 1 \;(\text{mod } m)$

(b) $\ell(g)$ and $\ell(h)$ are not zero-divisors modulo m

(c) $\ell(g) = \ell(G)$, $\deg(g) = \deg(G)$, $g \equiv G \;(\text{mod } m)$

(d) $\ell(h) = \ell(H)$, $\deg(h) = \deg(H)$, $h \equiv H \;(\text{mod } m)$

(e) $gh \equiv GH \;(\text{mod } m^n)$

Then $g \equiv G \;(\text{mod } m^n)$ and $h \equiv H \;(\text{mod } m^n)$.

Proof. By contradiction. Suppose that $g \not\equiv G \;(\text{mod } m^n)$ or $h \not\equiv H \;(\text{mod } m^n)$. Let k be the largest integer with $1 \leq k < n$ such that

$$g \equiv G \;(\text{mod } m^k), \qquad h \equiv H \;(\text{mod } m^k).$$

By assumption, this holds for $k = 1$. Thus for some $v, w \in \mathbb{Z}[x]$ we have

$$G - g = m^k v, \qquad H - h = m^k w,$$

and either $v \not\equiv 0 \pmod{m}$ or $w \not\equiv 0 \pmod{m}$. We may assume without loss of generality that $v \not\equiv 0 \pmod{m}$. Calculating modulo m^n we obtain

$$0 \equiv GH - gh \equiv G(H - h) + h(G - g) \equiv Gm^k w + hm^k v$$
$$\equiv (Gw + hv)m^k \pmod{m^n}.$$

Hence m^{n-k} is a divisor of $Gw + hv$. Calculating modulo m we obtain

$$0 \equiv t(Gw + hv) \equiv tgw + thv \equiv tgw + (1 - sg)v$$
$$\equiv (tw - sv)g + v \pmod{m}.$$

Hence $\overline{g}$ is a divisor of $\overline{v}$, where the bar denotes reduction modulo m. Since $G - g = m^k v$, the assumptions $\ell(g) = \ell(G)$ and $\deg(g) = \deg(G)$ imply $\deg(\overline{v}) < \deg(\overline{g})$. But then $\overline{v}$ must be the zero polynomial, contradicting the assumption that $v \not\equiv 0 \pmod{m}$.   □

The Hensel lifting algorithm of Figure 15.8 applies to a factorization of f into a product of two factors g and h. We can extend this to any number of factors by recursively calling the algorithm for two factors, as follows.

Suppose that $p \in \mathbb{Z}$ is a prime number, and that $f \in \mathbb{Z}[x]$ is a nonconstant polynomial with $\ell(f) \not\equiv 0 \pmod{p}$. Let $\overline{f} \in \mathbb{F}_p[x]$ be the polynomial obtained by reducing the coefficients of f modulo p. Suppose that we have a factorization of $\overline{f}$ modulo p; that is, we have monic nonconstant polynomials $f_1, f_2, \dots, f_r \in \mathbb{Z}[x]$ for which

$$\overline{f} = \overline{\ell(f)} \, \overline{f_1} \cdots \overline{f_r} \text{ in } \mathbb{F}_p[x].$$

We further assume that for all i, j with $1 \le i < j \le r$ we have polynomials $u_{ij}, v_{ij} \in \mathbb{Z}[x]$ for which

$$\overline{u_{ij}} \overline{f_i} + \overline{v_{ij}} \overline{f_j} = 1 \text{ in } \mathbb{F}_p[x].$$

Given $n \ge 2$, we want to lift this factorization of f to the modulus p^n.

We first split the factorization as evenly as possible by defining

$$k = \left\lfloor \frac{r}{2} \right\rfloor, \qquad g_0 = \ell(f) f_1 \cdots f_k, \qquad h_0 = f_{k+1} \cdots f_r.$$

Using the extended Euclidean algorithm we compute $s_0, t_0 \in \mathbb{Z}[x]$ for which

$$\overline{s_0} \, \overline{g_0} + \overline{t_0} \, \overline{h_0} = 1 \text{ in } \mathbb{F}_p[x].$$

We repeatedly apply Hensel lifting d times for $d = \lceil \log n \rceil$, so that p^n is a divisor of p^{2^d}. We obtain polynomials $g_d, h_d, s_d, t_d \in \mathbb{Z}[x]$ for which

$$f \equiv g_d h_d \pmod{p^{2^d}}, \qquad s_d g_d + t_d h_d \equiv 1 \pmod{p^{2^d}}.$$

We now make two recursive calls to this algorithm: first, to split and lift the factorization of g_d; and second, to split and lift the factorization of h_d. These recursive calls terminate when it is no longer possible to split the factorizations; we have then obtained polynomials $F_1, \ldots, F_r$ in $\mathbb{Z}[x]$ for which

$$F_i \equiv f_i \pmod{p}, \qquad f \equiv F_1 \cdots F_r \pmod{p^{2^d}},$$

together with polynomials $U_{ij}, V_{ij} \in \mathbb{Z}[x]$ for which

$$U_{ij} \equiv u_{ij} \pmod{p}, \quad V_{ij} \equiv v_{ij} \pmod{p}, \quad \overline{U_{ij}}\,\overline{F_i} + \overline{V_{ij}}\,\overline{F_j} \equiv 1 \pmod{p^{2^d}}.$$

This gives a lifting of the original factorization into r factors modulo p to a factorization into r corresponding factors modulo p^{2^d} (hence modulo p^n).

15.6 Polynomials with integer coefficients

Consider an arbitrary polynomial with integer coefficients, $f \in \mathbb{Z}[x]$.

Definition 15.23. The **content** of f is the GCD of its coefficients:

$$f = a_n x^n + \cdots + a_1 x + a_0, \qquad \text{cont}(f) = \gcd(a_n, \ldots, a_1, a_0).$$

We say that f is **primitive** if $\text{cont}(f) = 1$; that is, no prime $p \in \mathbb{Z}$ divides every coefficient of f. If we divide f by its content, then we obtain a primitive polynomial, called the **primitive part** of f:

$$\text{prim}(f) = \frac{1}{\text{cont}(f)}\, f.$$

Lemma 15.24. *Every polynomial $f \in \mathbb{Z}[x]$ can be written in the form*

$$f = \text{cont}(f)\,\text{prim}(f).$$

Factoring a polynomial $f \in \mathbb{Z}[x]$ therefore involves two distinct steps: first, we need to factor the integer $\text{cont}(f)$; second, we need to factor the primitive polynomial $\text{prim}(f)$. The first step corresponds to the irreducible elements of $\mathbb{Z}[x]$ of degree 0: these are just the primes in $\mathbb{Z}$. The second step corresponds to the irreducible elements of $\mathbb{Z}[x]$ with degree ≥ 1. We will be exclusively concerned with this second step.

Factoring polynomials in $\mathbb{Q}[x]$ is equivalent to factoring primitive polynomials in $\mathbb{Z}[x]$. To prove this, we need to the following famous result.

Lemma 15.25. Gauss's Lemma. *If $f, g \in \mathbb{Z}[x]$ are primitive polynomials, then so is their product fg.*

Suppose that $f \in \mathbb{Z}[x]$ is primitive. Consider the factorization of f regarded as an element of $\mathbb{Q}[x]$:

$$f = f_1 f_2 \cdots f_r,$$

where $f_i \in \mathbb{Q}[x]$ is irreducible for $i = 1, 2, \ldots, r$. We multiply f_i by the least common multiple d_i of the denominators of its coefficients, and get $d_i f_i \in \mathbb{Z}[x]$. We divide $d_i f_i$ by its content c_i, and get the primitive polynomial

$$\widetilde{f_i} = \frac{d_i}{c_i} f_i \in \mathbb{Z}[x],$$

which is irreducible in $\mathbb{Z}[x]$. We then have

$$f = a \widetilde{f_1} \widetilde{f_2} \cdots \widetilde{f_r},$$

for some $a \in \mathbb{Q}$. But f is primitive, and by Gauss's Lemma $\widetilde{f_1} \widetilde{f_2} \cdots \widetilde{f_r}$ is also primitive; hence $a = 1$. This gives the factorization in $\mathbb{Z}[x]$:

$$f = \widetilde{f_1} \widetilde{f_2} \cdots \widetilde{f_r}.$$

Conversely, suppose that $f \in \mathbb{Q}[x]$. Let d be the least common multiple of the denominators of its coefficients; then $df \in \mathbb{Z}[x]$. Let $c = \operatorname{cont}(df) \in \mathbb{Z}$; then $(d/c)f \in \mathbb{Z}[x]$ is primitive, and the irreducible factors of $(d/c)f$ in $\mathbb{Z}[x]$ are also irreducible regarded as elements of $\mathbb{Q}[x]$.

Consider an arbitrary polynomial $f \in \mathbb{Z}[x]$. Since $\mathbb{Z}[x]$ is a unique factorization domain, we have the factorization of f into distinct irreducible factors $f_1, f_2, \ldots, f_r \in \mathbb{Z}[x]$ with multiplicities $e_1, e_2, \ldots, e_r \geq 1$:

$$f = f_1^{e_1} f_2^{e_2} \cdots f_r^{r_r}.$$

Computing the formal derivative of f, we obtain

$$f' = \sum_{i=1}^{r} f_1^{e_1} \cdots f_{i-1}^{e_{i-1}} \left(e_i f_i^{e_i - 1} f_i' \right) f_{i+1}^{e_{i+1}} \cdots f_r^{e_r} = \sum_{i=1}^{r} e_i \frac{f}{f_i} f_i'.$$

Consider $\gcd(f, f')$. The only possible irreducible factors are $f_1, f_2, \ldots, f_r$. For all i, the polynomial $f_i^{e_i - 1}$ is a divisor of f/f_i, and hence of f'. The following conditions are equivalent, since $f_1, f_2, \ldots, f_r$ are distinct and irreducible:

$$f_i^{e_i} \text{ divides } f' \iff f_i^{e_i - 1} \text{ divides } e_i \frac{f}{f_i} f_i' \iff f_i \text{ divides } e_i f_i'.$$

But since $\deg(f_i') < \deg(f_i)$ and $e_i f_i' \neq 0$, we see that f_i does not divide $e_i f_i'$. (In characteristic p, this argument breaks down: we may have $e_i = 0$, and this happens if and only if $p \mid e_i$.) It follows that

$$\gcd(f, f') = \prod_{i=1}^{r} f_i^{e_i - 1}.$$

Definition 15.26. The **squarefree part** of $f \in \mathbb{Z}[x]$ is

$$\mathrm{squarefree}(f) = \frac{f}{\gcd(f, f')} = \prod_{i=1}^{r} f_i.$$

If we can find the irreducible factors $f_1, f_2, \ldots, f_r$ of $\mathrm{squarefree}(f)$, then we can recover the factorization of f simply by repeated division of f by each f_i in order to determine each multiplicity e_i. Therefore, in order to factor a polynomial $f \in \mathbb{Z}[x]$, if we ignore the problem of factoring the content of f, it suffices to factor the squarefree part of the primitive part of f. So we now assume that $f \in \mathbb{Z}[x]$ is a squarefree primitive polynomial.

Given a squarefree primitive polynomial $f \in \mathbb{Z}[x]$, we first need to find a prime p for which $\ell(f) \not\equiv 0 \pmod{p}$ and for which the reduced polynomial $\bar{f} \in \mathbb{F}_p[x]$ is also squarefree. To do this we need to understand the discriminant of f, and for this we need to define the resultant of two polynomials $f, g \in \mathbb{Z}[x]$:

$$f = \sum_{i=0}^{n} a_i x^i, \qquad g = \sum_{j=0}^{m} b_j x^j.$$

From the coefficients of f and g we construct the **Sylvester matrix** $S(f, g)$ given in Figure 15.9; it has size $(m+n) \times (m+n)$ and represents the linear map

$$\lambda_{f,g} \colon P_m \times P_n \to P_{m+n}, \qquad \lambda_{f,g}(s, t) = sf + tg,$$

where P_m is the m-dimensional vector space over $\mathbb{Q}$ spanned by $1, x, \ldots, x^{m-1}$.

Lemma 15.27. *We have* $\gcd(f, g) = 1$ *if and only if* $\det S(f, g) \neq 0$.

Proof. Suppose that $h = \gcd(f, g) \neq 1$. For $s = g/h$ and $t = -f/h$ we have

$$\deg(s) < \deg(g), \qquad \deg(t) < \deg(f), \qquad sf + tg = 0.$$

Hence $\lambda_{f,g}$ is not injective and so $\det S(f, g) = 0$. Conversely, suppose that there exists $(s, t) \in P_m \times P_n$ with $(s, t) \neq (0, 0)$ and $\lambda_{f,g}(s, t) = 0$; that is, $sf + tg = 0$. If $s \neq 0$ then $tg = -sf$, and $\gcd(f, g) = 1$ would imply $g|s$, contradicting $\deg(s) < \deg(g)$; similarly if $t \neq 0$. Hence $\gcd(f, g) = 1$. $\square$

Definition 15.28. The integer $\det S(f, g)$ is the **resultant** of f and g, and is denoted $\mathrm{res}(f, g)$. The **discriminant** of a polynomial $f \in \mathbb{Z}[x]$ is the resultant of f and its formal derivative f'; thus $\mathrm{disc}(f) = \mathrm{res}(f, f')$.

Example 15.29. For the quadratic polynomial $f = ax^2 + bx + c$ we have

$$\mathrm{disc}(f) = -a(b^2 - 4ac).$$

Lemma 15.30. *Suppose* $f, g \in \mathbb{Z}[x]$ *with* $n = \deg(f)$ *and* $m = \deg(g)$. *Then*

$$|\mathrm{res}(f, g)| \leq (n+1)^{m/2}(m+1)^{n/2} |f|_\infty^m |g|_\infty^n,$$

where $|f|_\infty$ *is the maximum of the absolute values of the coefficients.*

$$S(f,g) = \begin{bmatrix} a_n & & & & b_m & & \\ a_{n-1} & \ddots & & & b_{m-1} & \ddots & \\ \vdots & \ddots & a_n & & \vdots & \ddots & b_m \\ a_0 & & a_{n-1} & & b_0 & & b_{m-1} \\ & \ddots & \vdots & & & \ddots & \vdots \\ & & a_0 & & & & b_0 \end{bmatrix}$$

$$\underbrace{\qquad\qquad}_{m \text{ columns}} \quad \underbrace{\qquad\qquad}_{n \text{ columns}}$$

FIGURE 15.9
The Sylvester matrix $S(f,g)$ of polynomials $f, g \in \mathbb{Z}[x]$

Proof. Apply Hadamard's inequality to the Sylvester matrix. $\square$

If $g = f'$ in Lemma 15.30, then $m = n-1$ and $|f'|_\infty \le n|f|_\infty$, and hence

$$\begin{aligned} |\text{res}(f, f')| &\le (n+1)^{(n-1)/2} n^{n/2} |f|_\infty^{n-1} |f'|_\infty^n \\ &\le (n+1)^{(n-1)/2} n^{n/2} |f|_\infty^{n-1} (n|f|_\infty)^n \\ &= (n+1)^{(n-1)/2} n^{n/2} n^n |f|_\infty^{2n-1} \\ &\le (n+1)^{n/2} (n+1)^{n/2} (n+1)^n |f|_\infty^{2n-1} \\ &= (n+1)^{2n} |f|_\infty^{2n-1}. \end{aligned}$$

Therefore

$$|\text{disc}(f)| \le (n+1)^{2n} |f|_\infty^{2n-1}. \tag{15.5}$$

Lemma 15.31. *Let $\mathbb{F}_q$ ($q = p^n$) be a finite field, and let $\overline{f} \in \mathbb{F}_q[x]$ be a nonconstant polynomial. The following conditions are equivalent:*

$$\overline{f} \text{ is squarefree} \iff \gcd(\overline{f}, \overline{f}') = 1 \iff \text{disc}(\overline{f}) \ne 0.$$

Proof. von zur Gathen and Gerhard [147], page 387. $\square$

Lemma 15.32. *Let $f \in \mathbb{Z}[x]$ ($f \ne 0$) be squarefree, and let $p \in \mathbb{Z}$ be a prime for which $\ell(f) \not\equiv 0 \pmod{p}$. Then the reduced polynomial $\overline{f} \in \mathbb{F}_p[x]$ is squarefree if and only if p does not divide the discriminant of f.*

Proof. von zur Gathen and Gerhard [147], page 423. $\square$

Our next task is to find a prime p which does not divide either $\ell(f)$ or $\text{disc}(f)$. This involves number-theoretic considerations based on the Prime Number Theorem, which states that the number $\pi(x)$ of primes $\le x$ satisfies

$\pi(x) \approx x/\ln x$; equivalently, the n-th prime number p_n satisfies $p_n \approx n \ln n$. We write C for the upper bound on the discriminant in equation (15.5):

$$C = (n+1)^{2n} |f|_\infty^{2n-1}, \qquad n = \deg(f).$$

We further define

$$r = \lceil 2 \log C \rceil, \qquad s = 2n \ln \big((n+1)|f|_\infty\big).$$

There is a probabilistic algorithm for finding the required prime number.

Lemma 15.33. *Using $O(s \log^2 s \log \log s)$ word operations, we can compute the first r primes, each of which is less than $2r \ln r$, and no more than half of these primes divide* $\mathrm{disc}(f)$.

Proof. von zur Gathen and Gerhard [147], page 516. $\qquad\square$

Definition 15.34. For $s > 0$, the s-**norm** of a polynomial $f \in \mathbb{Z}[x]$ is

$$|f|_s = \left(\sum_{i=0}^n |a_i|^s \right)^{1/s} \quad \text{for} \quad f = \sum_{i=0}^n a_i x^i.$$

In particular, the 1-norm and 2-norm are

$$|f|_1 = |a_0| + |a_1| + \cdots + |a_n|, \qquad |f|_2 = \sqrt{a_0^2 + a_1^2 + \cdots + a_n^2}.$$

We need the following inequality on the factors of a polynomial.

Lemma 15.35. Mignotte's bound. *If $f, g, h \in \mathbb{Z}[x]$ satisfy $f = gh$ then*

$$|g|_1 |h|_1 \le (n+1)^{1/2}\, 2^n\, |f|_\infty, \qquad n = \deg(f).$$

Proof. von zur Gathen and Gerhard [147], page 164. $\qquad\square$

If $f \in \mathbb{Z}[x]$ is squarefree and primitive with $n = \deg(f)$ then we set

$$B = (n+1)^{1/2}\, 2^n\, |f|_\infty\, |\ell(f)|;$$

note the factor $|\ell(f)|$. We choose a prime p such that $\ell(f) \not\equiv 0 \pmod{p}$ and $\mathrm{disc}(f) \not\equiv 0 \pmod{p}$. We choose a positive integer k for which $p^k > 2B$. Suppose we have a factorization of f modulo p^k:

$$f \equiv \ell(f) g_1 g_2 \cdots g_r \pmod{p^k}, \qquad g_1, g_2, \ldots, g_r \in \mathbb{Z}[x].$$

We assume that $g_1, g_2, \ldots, g_r$ are monic. Using symmetric representatives for the congruence classes modulo p^k, we assume that $|g_i|_\infty < p^k/2$ for all i. Let S be a nonempty proper subset of $\{1, 2, \ldots, r\}$, and write $\overline{S} = \{1, 2, \ldots, r\} \setminus S$. Choose $G, H \in \mathbb{Z}[x]$ such that

$$G \equiv \ell(f) \prod_{i \in S} g_i \pmod{p^k}, \quad H \equiv \ell(f) \prod_{i \in \overline{S}} g_i \pmod{p^k}, \quad |G|_\infty, |H|_\infty < \frac{p^k}{2};$$

note that both G and H include the leading coefficient of f. Suppose that

$$|G|_1 |H|_1 \le B.$$

We claim that this inequality holds if and only if

$$\ell(f)f = GH;$$

that is, the factorization with symmetric representatives modulo p^n is actually a factorization over $\mathbb{Z}$. If $\ell(f)f = GH$, then Mignotte's bound, with f replaced by $\ell(f)f$, implies that

$$|G|_1 |H|_1 \le |\ell(f)| \, (n+1)^{1/2} \, 2^n \, |f|_\infty = B.$$

Conversely, assume that $|G|_1 |H|_1 \le B$. We have

$$\ell(f)f \equiv GH \pmod{p^k}, \tag{15.6}$$

and standard inequalities relating the different norms imply that

$$|GH|_\infty \le |GH|_1 \le |G|_1 |H|_1 \le B < \frac{p^k}{2}.$$

Thus both sides of congruence (15.6) are polynomials with coefficients $< p^k/2$ in absolute value, and hence they are equal in $\mathbb{Z}[x]$.

We now have all the components of our first algorithm for factoring (squarefree primitive) polynomials in $\mathbb{Z}[x]$; see Figure 15.10. The basic ideas for this algorithm originate with Zassenhaus [149]. Our presentation is based on von zur Gathen and Gerhard [147], page 441.

Steps 2(a)-(f) have already been discussed. Step 2(g) initializes the set T of factor indices remaining to be considered, and the product F of those remaining factors. The loop in Steps 2(h)-(i) is the heart of the algorithm. It attempts to combine the lifted factors modulo p^k into a subset which comes from one irreducible factor over $\mathbb{Z}$. An irreducible factor of f over $\mathbb{Z}$ may split into the product of more than one irreducible factor over $\mathbb{F}_p$. The algorithm uses a exhaustive search over all subsets S of $\{1, 2, \ldots, r\}$ to reconstruct the irreducible factors over $\mathbb{Z}$, starting with the subsets of size $s = 1$ and then increasing s. Every time a new factor is found which is irreducible over $\mathbb{Z}$, the algorithm removes the corresponding modular factors from consideration. Since there are $2^r - 1$ nonempty subsets of $\{1, 2, \ldots, r\}$, in the worst case the number of iterations of the loop is an exponential function of $n = \deg(f)$. In the next section we will see how the LLL algorithm for lattice basis reduction can be used to avoid this worst-case exponential search and to provide a polynomial-time algorithm for factoring polynomials over $\mathbb{Z}$.

- Input: A squarefree primitive nonconstant polynomial $f \in \mathbb{Z}[x]$.
- Output: The irreducible factors $f_1, \ldots, f_r \in \mathbb{Z}[x]$ of f.

(1) Set $n \leftarrow \deg(f)$.

(2) If $n = 1$ then

Set `result` $\leftarrow [f]$

else

(a) Set $C \leftarrow (n+1)^{2n} |f|_\infty^{2n-1}$, set $r \leftarrow \lceil 2 \log C \rceil$.

(b) Set $B \leftarrow (n+1)^{1/2} 2^n |f|_\infty |\ell(f)|$.

(c) Find a prime $p < 2r \ln r$ such that $p \nmid \ell(f)$ and $p \nmid \operatorname{disc}(f)$.

(d) Set $k \leftarrow \lceil \log_p(2B+1) \rceil$.

(e) Call **Factor** to find nonconstant monic irreducible polynomials $h_1, h_2, \ldots, h_r \in \mathbb{Z}[x]$ for which $\overline{f} = \overline{\ell(f)}\, \overline{h_1}\, \overline{h_2} \cdots \overline{h_r}$ in $\mathbb{F}_p[x]$. Using symmetric representatives, $|h_i|_\infty < p/2$ for all i.

(f) Call **Hensel** repeatedly to find nonconstant monic irreducible polynomials $g_1, g_2, \ldots, g_r \in \mathbb{Z}[x]$ for which $f \equiv \ell(f)\, g_1 g_2 \cdots g_r$ $(\bmod\ p^k)$ and $g_i \equiv h_i$ $(\bmod\ p)$ for all i. Using symmetric representatives, $|g_i|_\infty < p^k/2$ for all i.

(g) Set $T \leftarrow \{1, 2, \ldots, r\}$, set $F \leftarrow f$.

(h) Set `result` $\leftarrow []$ *(empty list)*, set $s \leftarrow 1$.

(i) While $2s \le |T|$ do

- Set `done` $\leftarrow$ false, set `list` $\leftarrow \{\, s\text{-element subsets of } T \,\}$.
- Set $j \leftarrow 0$.
- While $j < |\texttt{list}|$ and not `done` do
  - Set $j \leftarrow j + 1$, set $S \leftarrow \texttt{list}[j]$.
  - Find $G, H \in \mathbb{Z}[x]$ such that $|G|_\infty, |H|_\infty < p^k/2$ and

$$G \equiv \ell(F) \prod_{i \in S} g_i \ (\bmod\ p^k), \quad H \equiv \ell(F) \prod_{i \in T \setminus S} g_i \ (\bmod\ p^k).$$

  - If $|G|_1 |H|_1 \le B$ then
    * Append $\operatorname{prim}(G)$ to `result`.
    * Set $T \leftarrow T \setminus S$, set $F \leftarrow \operatorname{prim}(H)$. set `done` $\leftarrow$ true.
- Set $s \leftarrow s + 1$.

(h) Append F to `result`.

(3) Return `result`.

FIGURE 15.10

The Zassenhaus factorization algorithm **ZFactor**(f)

15.7   Polynomial factorization using LLL

In this section we study how the LLL algorithm can be used to provide a polynomial-time algorithm for factoring squarefree primitive polynomials with integer coefficients. In fact, the LLL algorithm was originally used precisely to provide the first polynomial-time algorithm for this factorization problem. Our exposition follows von zur Gathen and Gerhard [147], §§16.4–16.5. To begin, we quote from the paper of Lenstra, Lenstra, and Lovász [88]:

> "First we find, for a suitable small prime number p, a p-adic irreducible factor h of f, to a certain precision. This is done with Berlekamp's algorithm for factoring polynomials over small finite fields, combined with Hensel's lemma. Next we look for the irreducible factor h_0 of f in $\mathbb{Z}[X]$ that is divisible by h. The condition that h_0 is divisible by h means that h_0 belongs to a certain lattice, and the condition that h_0 divides f implies that the coefficients of h_0 are relatively small. It follows that we must look for a "small" element in that lattice, and this is done by means of a basis reduction algorithm. It turns out that this enables us to determine h_0."

We first make small changes to Steps 2(b) and 2(d) of the Zassenhaus algorithm **ZFactor**; we redefine B and k as follows:

$$B \leftarrow (n{+}1)^{1/2} 2^n |f|_\infty, \qquad k \leftarrow \left\lceil \log_p \left(2^{n^2/2} B^{2n} \right) \right\rceil.$$

For this new value of k, we have

$$p^k \geq 2^{n^2/2} B^n = 2^{n^2/2} \big((n{+}1)^{1/2} 2^n |f|_\infty \big)^{2n} = 2^{n^2/2} (n{+}1)^n 2^{2n^2} |f|_\infty^{2n}.$$

The major changes to the algorithm occur in Steps 2(h)-(i): we replace the search over all subsets of the set of modular factors by a computation which uses lattice basis reduction. We need the following lemmas.

Lemma 15.36. *Let $f, g \in \mathbb{Z}[x]$ be nonzero with $\deg(f) + \deg(g) \geq 1$ (so at least one of f and g is nonconstant). Then there exist $s, t \in \mathbb{Z}[x]$ such that $sf + tg = \mathrm{res}(f, g)$ with $\deg(s) < \deg(g)$ and $\deg(t) < \deg(f)$.*

Proof. von zur Gathen and Gerhard [147], page 153.   □

Lemma 15.37. *Let $f, g \in \mathbb{Z}[x]$ with $n = \deg(f)$ and $m = \deg(g)$. Then*

$$| \mathrm{res}(f, g) | \geq |f|_2^m |g|_2^n \geq (n{+}1)^{m/2} (m{+}1)^{n/2} |f|_\infty^m |g|_\infty^n.$$

Proof. von zur Gathen and Gerhard [147], page 156.   □

We use these lemmas to prove the following result.

Lemma 15.38. *Let $f, g \in \mathbb{Z}[x]$ with $n = \deg(f) > 0$ and $m = \deg(g) > 0$. Suppose that $u \in \mathbb{Z}[x]$ is monic and nonconstant, and that $f \equiv uv_1 \pmod{m}$ and $g \equiv uv_2 \pmod{m}$ for some $v_1, v_2 \in \mathbb{Z}[x]$ and some $m > |f|_2^k |g|_2^n$. Then $\gcd(f, g) \in \mathbb{Z}[x]$ is nonconstant.*

Proof. We show by contradiction that $\gcd(f, g) \in \mathbb{Q}[x]$, the GCD with rational coefficients, is nonconstant. Suppose that $\gcd(f, g) = 1$ in $\mathbb{Q}[x]$. By Lemma 15.36 there exist $s, t \in \mathbb{Z}[x]$ such that $sf + tg = \mathrm{res}(f, g)$, and hence $sf + tg \equiv \mathrm{res}(f, g) \pmod{m}$. Since $f \equiv uv_1 \pmod{m}$ and $g \equiv uv_2 \pmod{m}$ we get

$$\mathrm{res}(f, g) = u(sv_1 + tv_2) \pmod{m};$$

thus u is a divisor of $\mathrm{res}(f, g)$ modulo m. But u is monic and nonconstant, and $\mathrm{res}(f, g) \in \mathbb{Z}$, so $\mathrm{res}(f, g) \equiv 0 \pmod{m}$. Lemma 15.37 implies

$$m > |f|_2^k |g|_2^n \geq |\mathrm{res}(f, g)|,$$

and hence $\mathrm{res}(f, g) = 0$. But this implies that $\gcd(f, g) \neq 1$, a contradiction. Thus $\gcd(f, g) \in \mathbb{Q}[x]$ is nonconstant, and hence so is $\gcd(f, g) \in \mathbb{Z}[x]$. □

Suppose that $f \in \mathbb{Z}[x]$ is squarefree and primitive, and write $n = \deg(f)$. Suppose further that $u \in \mathbb{Z}[x]$ is monic and nonconstant with $d = \deg(u) < n$, and that u is a divisor of f modulo m; that is, $f \equiv uv \pmod{m}$ for some $v \in \mathbb{Z}[x]$ (where $m = p^k$). We want to find a polynomial $g \in \mathbb{Z}[x]$ for which

$$|g|_2^n < m|f|_2^{-\deg(g)},$$

since this is equivalent to

$$m > |f|_2^{\deg(g)} |g|_2^{\deg(f)}.$$

If this inequality holds, then Lemma 15.38 implies that $\gcd(f, g) \in \mathbb{Z}[x]$ is nonconstant, and so we obtain a nontrivial factor of f in $\mathbb{Z}[x]$.

Suppose $j \in \{d+1, \ldots, n\}$ where $d = \deg(u) < n$. We represent a polynomial g with $\deg(g) < j$ by its coefficient vector:

$$(g_{j-1}, \ldots, g_1, g_0) \in \mathbb{Z}^j, \qquad g = g_{j-1}x^{j-1} + \cdots + g_1 x + g_0.$$

Let $L \subseteq \mathbb{Z}^j$ be the lattice with a basis consisting of the coefficient vectors of the polynomials

$$\{u, xu, \ldots, x^{j-d-1}u\} \cup \{m, mx, \ldots, mx^{d-1}\}.$$

Altogether we have $(j-d) + d = j$ vectors in this basis, and they are linearly independent since $\deg(u) = d$. (Since $m \in \mathbb{Z}$, the degrees of the polynomials in these two sets are $d, \ldots, j-1$ and $0, \ldots, d-1$.) Given $q_0, \ldots, q_{j-d-1}, r_0, \ldots, r_{d-1} \in \mathbb{Z}$, the general element $g \in L$ has the form

$$\sum_{i=0}^{j-d-1} q_i x^i u + \sum_{i=0}^{d-1} r_i m x^i = \left(\sum_{i=0}^{j-d-1} q_i x^i \right) u + m \left(\sum_{i=0}^{d-1} r_i x^i \right) = qu + mr,$$

where we use the notation

$$q = \sum_{i=0}^{j-d-1} q_i x^i, \qquad r = \sum_{i=0}^{d-1} r_i x^i.$$

Hence $g = qu + mr$ and so $g \equiv qu \pmod{m}$; thus u is a factor of g modulo m. Therefore $g \in L$ implies $\deg(g) < j$ and $u|g \pmod{m}$.

Conversely, suppose that $g \in \mathbb{Z}[x]$ with $\deg(g) < j$ and $u|g \pmod{m}$. Since u is a divisor of g modulo m, we have $g = q_1 u + mr_1$ for some $q_1, r_1 \in \mathbb{Z}[x]$. Recalling that u is monic, we may divide r_1 by u in $\mathbb{Z}[x]$, obtaining $r_1 = q_2 u + r_2$ with $\deg(r_2) < \deg(u)$. We now have

$$(q_1 + mq_2)u + mr_2 = q_1 u + m(q_2 u + r_2) = q_1 u + r_1 m = g.$$

Hence $g = qu + mr$ for $q = q_1 + mq_2$ and $r = r_2$. It is clear that $\deg(r) < \deg(u)$; and furthermore,

$$\deg(q_1) \le \deg(g) - \deg(u) < j - d,$$

which implies

$$\deg(q_2) \le \deg(r_1) - \deg(u) < j - d.$$

Therefore $g \in L$.

This discussion establishes that

$$g \in L \quad \Longleftrightarrow \quad \deg(g) < j, \;\; u|g \pmod{m}.$$

This justifies the use of lattice basis reduction to find a short vector in L which corresponds to a polynomial $g \in \mathbb{Z}[x]$ with $\deg(g) < j$ and $u|g \pmod{m}$. If we can find such a polynomial g which also satisfies the inequality

$$|g|_2^n < m|f|_2^{-\deg(g)},$$

then we will be able to find a nontrivial factor of f in $\mathbb{Z}[x]$. Using the LLL algorithm with $\alpha = \frac{3}{4}$ we know that

$$|g_1|_2 \le 2^{(j-1)/2}|g|_2,$$

where g_1 is the first vector in the reduced basis, and $g \in L$ is any nonzero lattice vector. Recall that $\dim(L) = j \le n$; this gives

$$|g_1|_2 \le 2^{n/2}|g|_2.$$

Let g be an irreducible factor of f which is divisible by u modulo m. Then Mignotte's bound Lemma 15.35 gives

$$|g|_\infty \le |g|_2 \le (n+1)^{1/2} 2^n A, \qquad A = \max\left(|f|_\infty, |g|_\infty\right).$$

Therefore

$$|g_1|_2 \leq 2^{n/2}B, \qquad B = (n+1)^{1/2}2^n A.$$

We now get

$$|g_1|_2^{j-1}|g|_2^{\deg(g_1)} < \left(2^{n/2}B\right)^n B^n = 2^{n^2/2}(n+1)^{1/2}2^{2n^2}A^{2n} \leq p^k,$$

by our original choice of k. By Lemma 15.38, it follows that $\gcd(g, g_1)$ is nonconstant in $\mathbb{Z}[x]$. In this way, we can replace the exponential-time steps 2(h)-(i) in algorithm **ZFactor** by polynomial-time calls to the LLL algorithm.

The proof of correctness of the resulting polynomial-time algorithm depends on the following lemma.

Lemma 15.39. *Let p be a prime number. Suppose that $f, g \in \mathbb{Z}[x]$ satisfy:*

(1) $\ell(f) \not\equiv 0 \pmod{p}$,

(2) the reduced polynomial $\overline{f} \in \mathbb{F}_p[x]$ is squarefree,

(3) g divides f in $\mathbb{Z}[x]$.

Assume $k \geq 1$. Suppose that $u \in \mathbb{Z}[x]$ is a monic polynomial. If u is a factor of f modulo p^k, and a factor of g modulo p, then u is a factor of g modulo p^k.

Proof. Suppose that $f = gh$ in $\mathbb{Z}[x]$. Reducing coefficients modulo p, we obtain $\overline{f} = \overline{g}\overline{h}$ in $\mathbb{F}_p[x]$. Since $\overline{f}$ is squarefree, it follows that $\overline{g}$ is also squarefree. Suppose that $\overline{g} = \overline{u}\,\overline{v}$ in $\mathbb{F}_p[x]$. It follows that $\gcd(\overline{u}, \overline{v}) = 1$ in $\mathbb{F}_p[x]$. Applying repeated Hensel lifting we obtain polynomials $U, V \in \mathbb{Z}[x]$ for which

$$U \equiv u \pmod{p}, \qquad V \equiv v \pmod{p}, \qquad g \equiv UV \pmod{p^k}.$$

Suppose that $f \equiv uw \pmod{p^k}$; divisibility by p^k clearly implies divisibility by p, and hence $\overline{f} = \overline{u}\,\overline{w}$ in $\mathbb{F}_p[x]$. Therefore, we have

$$\overline{u}\,\overline{v}\,\overline{h} = \overline{g}\,\overline{h} = \overline{f} = \overline{u}\,\overline{w}, \quad \text{in } \mathbb{F}_p[x].$$

Since $\mathbb{F}_p[x]$ is an integral domain, we cancel $\overline{u}$ to get $\overline{v}\,\overline{h} = \overline{w}$ in $\mathbb{F}_p[x]$. Hence

$$\overline{V}\,\overline{h} = \overline{v}\,\overline{h} = \overline{w} \quad \text{in } \mathbb{F}_p[x].$$

Calculating modulo p^k we obtain

$$U(Vh) \equiv (UV)h \equiv gh \equiv f \equiv uw \pmod{p^k}.$$

Since $Vh \equiv w \pmod{p}$, and $\gcd(\overline{u}, \overline{v}) = 1$ in $\mathbb{F}_p[x]$, the uniqueness of Hensel lifting implies $U \equiv u \pmod{p^k}$, and hence

$$g \equiv UV \equiv uV \pmod{p^k}.$$

Thus u is a factor of g modulo p^k, and this completes the proof. $\square$

15.8 Projects

Project 15.1. Give a seminar presentation on the structure of $\mathbb{F}[x]$, the polynomial ring in one variable with coefficients in a field $\mathbb{F}$. In particular, give a complete proof that $\mathbb{F}[x]$ is a unique factorization domain.

Project 15.2. Give a seminar presentation on the structure of finite fields. Give complete proofs of the results quoted without proof this chapter.

Project 15.3. Give a seminar presentation on the Möbius inversion formula. Show how this can be used to prove the formula for the number $I(d, q)$ of irreducible monic polynomials of degree d over the field with q elements.

Project 15.4. Write a report and give a seminar presentation on the original papers by Berlekamp [14, 15] and Cantor and Zassenhaus [21] on algorithms for factoring polynomials over finite fields.

Project 15.5. Write a report and give a seminar presentation on the original paper by Zassenhaus [149] on algorithms for Hensel lifting.

Project 15.6. Write a report and give a seminar presentation on the construction of the field $\mathbb{Q}_p$ of p-adic numbers (the infinite version of Hensel lifting).

Project 15.7. Write a report and give a seminar presentation on the original papers by Mignotte [102, 103] on inequalities for polynomial factorizations.

Project 15.8. In this chapter we have mostly ignored the question of complexity of the algorithms. Choose one of the factorization algorithms in this chapter, and write a report on the analysis of this algorithm.

Project 15.9. Choose one of the factorization algorithms discussed in this chapter. Implement the algorithm in a suitable computer language or computer algebra system. Write a report on your implementation, providing numerous examples to check its correctness.

Project 15.10. Write a report and give a seminar presentation on the factorization algorithm of van Hoeij [145]. "For several decades the standard algorithm for factoring polynomials f with rational coefficients has been the Berlekamp-Zassenhaus algorithm. The complexity of this algorithm depends exponentially on n, where n is the number of modular factors of f. This exponential time complexity is due to a combinatorial problem: the problem of choosing the right subsets of these n factors. In this paper, this combinatorial problem is reduced to a type of knapsack problem that can be solved with lattice reduction algorithms. The result is a practical algorithm that can factor polynomials that are far out of reach for previous algorithms." See also the survey paper by Klüners [77].

Project 15.11. Polynomial factorization can be regarded as the simplest case of computing the Wedderburn decomposition of a finite-dimensional associative algebra. Write a report and present a seminar talk on algorithms for computing the Wedderburn decomposition. For a recent survey paper on this topic, see Bremner [18].

15.9 Exercises

Exercise 15.1. Use the Euclidean algorithm to compute the GCD of the polynomials $f = x^{32} + x$ and $g = x^8 + x$ over the field $\mathbb{F}_2$ with 2 elements.

Exercise 15.2. (a) Find all the monic irreducible quadratic polynomials over the field $\mathbb{F}_3$ with 3 elements. Hint: A polynomial of degree ≤ 3 is irreducible over a field $\mathbb{F}$ if and only if it has no root in $\mathbb{F}$.

(b) Use one of the polynomials from (a) to construct the multiplication table of the field $\mathbb{F}_9$ with 9 elements. Hint: If $f \in \mathbb{F}[x]$ is an irreducible polynomial of degree n, then the quotient ring $\mathbb{F}[x]/\langle f \rangle$ is a field which has dimension n as a vector space over $\mathbb{F}$.

Exercise 15.3. (a) Write $f = x^3 - x$ as a product of irreducible factors in $\mathbb{F}_3[x]$.

(b) Write $f = x^9 - x$ as a product of irreducible factors in $\mathbb{F}_3[x]$.
(c) Write $f = x^9 - x$ as a product of irreducible factors in $\mathbb{F}_9[x]$.
(d) Write $f = x^{27} - x$ as a product of irreducible factors in $\mathbb{F}_3[x]$.

Exercise 15.4. Recall the formula for the number $I(d, q)$ of monic irreducible polynomials of degree d over the field with q elements:

$$I(d, q) = \frac{1}{d} \sum_{i \mid d} \mu\left(\frac{d}{i}\right) q^i.$$

This formula uses the Möbius μ-function which is defined for $n \geq 1$ by

$$\mu(n) = \begin{cases} 1 & \text{if } n = 1 \\ 0 & \text{if } n \text{ is divisible by } p^2 \text{ for some prime } p \\ (-1)^k & \text{if } n \text{ is the product of } k \text{ distinct primes} \end{cases}$$

Calculate $I(d, 5)$ for $1 \leq d \leq 9$.

Exercise 15.5. Use algorithm $\mathbf{DDD}(f)$ to compute the distinct-degree decomposition of this polynomial over the field $\mathbb{F}_3$. Give all the details:

$$f = x^{10} + x^8 + x^7 + 2x^6 + 2x^5 + x^4 + x^3 + x^2 + 2x.$$

Exercise 15.6. Recall that $\mathbb{F}_q^\times$ is the multiplicative group of nonzero elements in the field $\mathbb{F}_q$ with q elements. Let $S = \{\, a^2 \mid a \in \mathbb{F}_q^\times \,\}$ be the set of squares in $\mathbb{F}_q^\times$. Verify the following claims for $q = 3$, $q = 5$, $q = 7$, $q = 9$, and $q = 11$:

    (a) S is a subgroup of $\mathbb{F}_q^\times$ of order $(q-1)/2$.

    (b) $S = \{\, a \in \mathbb{F}_q^\times \mid a^{(q-1)/2} = 1 \,\}$.

    (c) For every $a \in \mathbb{F}_q^\times$ we have $a^{(q-1)/2} = \pm 1$.

Exercise 15.7. Let p be a prime number, and let $h = x^p - x$ in $\mathbb{F}_p[x]$.

    (a) Determine the factorization $h = h_1^{e_1} h_2^{e_2} \cdots h_\ell^{e_\ell}$ where $h_1, h_2, \ldots, h_\ell$ are distinct monic irreducible polynomials.

    (b) Explain why the Chinese Remainder Theorem gives an isomorphism

$$\phi \colon \mathbb{F}_p[x]/\langle h \rangle \longrightarrow \mathbb{F}_p[x]/\langle h_1^{e_1} \rangle \times \mathbb{F}_p[x]/\langle h_2^{e_2} \rangle \times \cdots \times \mathbb{F}_p[x]/\langle h_\ell^{e_\ell} \rangle.$$

    (c) Determine explicitly each factor on the right side of this isomorphism.

    (d) Determine explicitly the image under ϕ of every element of $\mathbb{F}_p[x]/\langle h \rangle$.

Exercise 15.8. Write a computer program to implement **TrialSplit** and **Split**. Provide examples of the output of your implementation.

Exercise 15.9. Write a computer program to implement **EDD**. Provide examples of the output of your implementation.

Exercise 15.10. Consider the extended Euclidean algorithm for polynomials over a field. Let f and g be the input polynomials, and let h, s and t be the output polynomials. Prove that $\deg(s) < \deg(g)$ and $\deg(t) < \deg(f)$.

Exercise 15.11. Consider polynomials $f, g \in \mathbb{Z}[x]$, and assume that g is monic. Prove that there exist unique polynomials $q, r \in \mathbb{Z}[x]$ such that $f = qg + r$ and $\deg(r) < \deg(g)$.

Exercise 15.12. Let f, g, q and r be as in the previous Exercise. Let $p \in \mathbb{Z}$ be a prime number, and assume that every coefficient of f is divisible by p. Prove that every coefficient of q and every coefficient of r is divisible by p.

Exercise 15.13. In the discussion of Hensel lifting, complete the proof that

$$s_2 \equiv s_1 \;(\mathrm{mod}\; m), \quad t_2 \equiv t_1 \;(\mathrm{mod}\; m), \quad \deg(s_2) < \deg(h_2), \quad \deg(t_2) < \deg(g_2).$$

Exercise 15.14. Consider the following polynomials with integer coefficients:

$$f = x^2 + 6, \qquad g = x^3 + 5x^2 + 2x + 6, \qquad h = x^3 + 2x^2 + 2x + 1.$$

Verify that $f \equiv gh \;(\mathrm{mod}\; 7)$. Use the extended Euclidean algorithm to find $s, t \in \mathbb{Z}[x]$ such that

$$\deg(s) < \deg(h), \qquad \deg(t) < \deg(g), \qquad sg + th \equiv 1 \;(\mathrm{mod}\; 7).$$

Use Hensel lifting to obtain a factorization of f with the modulus $7^2 = 49$. Lift the factorization $f \equiv gh \;(\mathrm{mod}\; 7)$ to the modulus $49^2 = 2401$.

Exercise 15.15. Let $f, g \in \mathbb{Z}[x]$ with $n = \deg(f)$ and $m = \deg(g)$. Prove that

$$|\mathrm{res}(f, g)| \le (n{+}1)^{m/2} (m{+}1)^{n/2} |f|_\infty^m |g|_\infty^n.$$

Exercise 15.16. Let $f \in \mathbb{Z}[x]$ be nonzero and squarefree, and let $p \in \mathbb{Z}$ be a prime with $\ell(f) \not\equiv 0 \pmod{p}$. Prove that $\overline{f} \in \mathbb{F}_p[x]$ is squarefree if and only if p does not divide $\mathrm{disc}(f)$.

Exercise 15.17. Prove that if $f, g, h \in \mathbb{Z}[x]$ satisfy $f = gh$ then

$$|g|_1 |h|_1 \le (n{+}1)^{1/2} \, 2^n \, |f|_\infty, \qquad n = \deg(f).$$

Exercise 15.18. Consider the following polynomial $f \in \mathbb{Z}[x]$:

$$f = x^4 + 3x^2 + 2.$$

(a) Determine by inspection the irreducible factorization of f in $\mathbb{Z}[x]$.
(b) Compute the Sylvester matrix $S(f, f')$ and the discriminant $\mathrm{disc}(f)$.
(c) Compute the following quantities from the Zassenhaus algorithm:

$$C = (n{+}1)^{2n} |f|_\infty^{2n-1}, \qquad r = \lceil 2 \log C \rceil.$$

(d) Verify that $p = 3$ satisfies the conditions

$$p < 2r \ln r, \qquad \ell(f) \not\equiv 0 \pmod{p}, \qquad \mathrm{disc}(f) \not\equiv 0 \pmod{p}.$$

(e) Determine the factorization of $\overline{f}$ where bar denotes reduction mod 3.
(f) Compute the following quantities from the Zassenhaus algorithm:

$$B = (n{+}1)^{1/2} \, 2^n \, |f|_\infty \, |\ell(f)|, \qquad k = \lceil \log_p(2B{+}1) \rceil.$$

(g) Use Hensel lifting with more than two factors to lift the factorization of f modulo $p = 3$ from part (e) to a factorization of f modulo p^k.
(h) Explain in detail the calculations performed during the loop in Step 2(i) of the Zassenhaus algorithm.
(i) Use the results of part (h) to recover the irreducible factorization of f.

Exercise 15.19. Generalize the equal-degree decomposition algorithm to the field with two elements. For hints, see Exercise 14.16 of von zur Gathen and Gerhard [147].

Bibliography

[1] K. AARDAL. Lattice basis reduction in optimization: selected topics. *Proceedings of Symposia in Applied Mathematics* 61 (2004) 2–19. American Mathematical Society, 2004.

[2] K. AARDAL, F. EISENBRAND. The LLL algorithm and integer programming. Pages 293–314 of Nguyen and Vallée [110].

[3] W. A. ADKINS, S. H. WEINTRAUB. *Algebra: An Approach via Module Theory*. Springer, 1992.

[4] E. AGRELL, T. ERIKSSON, A. VARDY, K. ZEGER. Closest point search in lattices. *IEEE Transactions on Information Theory* 48 (2002) 2201–2214.

[5] M. AJTAI. The shortest vector problem in L_2 is NP-hard for randomized reductions: extended abstract. *STOC 98: Symposium on Theory of Computing*, 10–19. Association for Computing Machinery, 1998.

[6] M. AJTAI. The shortest vector problem in L_2 is NP-hard for randomized reductions. *Electronic Colloquium on Computational Complexity*: www.eccc.uni-trier.de/report/1997/047

[7] M. AJTAI, R. KUMAR, D. SIVAKUMAR. An overview of the sieve algorithm for the shortest lattice vector problem. *CaLC 2000: Cryptography and Lattices Conference*, 1–3. Lecture Notes in Computer Science, 2146. Springer, 2001.

[8] M. AJTAI, R. KUMAR, D. SIVAKUMAR. A sieve algorithm for the shortest lattice vector problem. *STOC 2001: Symposium on Theory of Computing*, 601–610. Association for Computing Machinery, 2001.

[9] M. AJTAI. The worst-case behavior of Schnorr's algorithm approximating the shortest nonzero vector in a lattice: extended abstract. *STOC 2003: Symposium on Theory of Computing*, 396–406. Association for Computing Machinery, 2003.

[10] A. AKHAVI. The optimal LLL algorithm is still polynomial in fixed dimension. *Theoretical Computer Science* 297 (2003) 3–23.

[11] H. ANTON. *Elementary Linear Algebra*. Tenth edition. Wiley, 2010.

[12] L. BABAI. On Lovász' lattice reduction and the nearest lattice point problem. *Combinatorica* 6 (1986) 1–13.

[13] W. BACKES, S. WETZEL: Heuristics on lattice basis reduction in practice. *Journal of Experimental Algorithmics*, Volume 7, December 2002.

[14] E. R. BERLEKAMP. Factoring polynomials over finite fields. *Bell System Technical Journal* 46 (1967) 1853–1859.

[15] E. R. BERLEKAMP. Factoring polynomials over large finite fields. *Mathematics of Computation* 24 (1970) 713–735.

[16] F. BEUKERS. Lattice reduction. Pages 66–77 of Cohen, Cuypers and Sterk [25].

[17] J. BLÖMER, J.-P. SEIFERT. On the complexity of computing short linearly independent vectors and short bases in a lattice. *STOC 99: Symposium on Theory of Computing*, 711–720. Association for Computing Machinery, 1999.

[18] M. R. BREMNER: How to compute the Wedderburn decomposition of a finite-dimentional associative algebra. *Groups, Complexity, Cryptology* 3 (2011) 47–66.

[19] M. R. BREMNER, L. A. PERESI. An application of lattice basis reduction to polynomial identities for algebraic structures. *Linear Algebra and its Applications* 430 (2009) 642–659.

[20] J. A. BUCHMANN, M. E. POHST. Computing a lattice basis from a system of generating vectors. *EUROCAL '87: European Conference on Computer Algebra*, 54–63. Lecture Notes in Computer Science, 378. Springer, 1989.

[21] D. G. CANTOR, H. ZASSENHAUS. A new algorithm for factoring polynomials over finite fields. *Mathematics of Computation* 36 (1981) 587–592.

[22] J. W. S. CASSELS. *An Introduction to the Geometry of Numbers.* Grundlehren der Mathematischen Wissenschaften, 99. Springer, 1959.

[23] CLAY MATHEMATICS INSTITUTE (CMI): www.claymath.org

[24] CMI MILLENIUM PRIZE PROBLEMS: www.claymath.org/millennium

[25] A. M. COHEN, H. CUYPERS, H. STERK (editors): *Some Tapas of Computer Algebra.* Algorithms and Computation in Mathematics, 4. Springer, 1998.

[26] H. COHEN. *A Course in Computational Algebraic Number Theory.* Graduate Texts in Mathematics, 138. Springer, 1993.

[27] J. H. CONWAY, N. J. A. SLOANE. *Sphere Packings, Lattices and Groups*. Second edition. Grundlehren der Mathematischen Wissenschaften, 290. Springer, 1993.

[28] S. A. COOK. The complexity of theorem-proving procedures. *STOC '71: Symposium on Theory of Computing*, 151–158. Association for Computing Machinery, 1971.

[29] S. A. COOK. The P versus NP problem:
www.claymath.org/millenium/P_vs_NP/pvsnp.pdf

[30] D. COPPERSMITH. Finding a small root of a univariate modular equation. *Eurocrypt 1996: Advances in Cryptology*, 155–165. Lecture Notes in Computer Science, 1070. Springer, 1996.

[31] D. COPPERSMITH. Finding a small root of a bivariate integer equation: factoring with high bits known. *Eurocrypt 1996: Advances in Cryptology*, 178–189. Lecture Notes in Computer Science, 1070. Springer, 1996.

[32] D. COPPERSMITH. Small solutions to polynomial equations, and low exponent RSA vulnerabilities. *Journal of Cryptology* 10 (1997) 233–260.

[33] D. COPPERSMITH. Finding small solutions to small degree polynomials. *CaLC: Cryptography and Lattices Conference*, 20–31. Lecture Notes in Computer Science, 2146. Springer, 2001.

[34] C. COUPÉ, P. Q. NGUYEN, J. STERN. The effectiveness of lattice attacks against low-exponent RSA. *PKC 1999: Public Key Cryptography*, 204–218. Lecture Notes in Computer Science, 1560. Springer, 1999.

[35] H. DAUDÉ, B. VALLÉE. An upper bound on the average number of iterations of the LLL algorithm. *Theoretical Computer Science* 123 (1994) 95–115.

[36] H. DAUDÉ, P. FLAJOLET, B. VALLÉE. An average-case analysis of the Gaussian algorithm for lattice reduction. *Combinatorics, Probability and Computing* 6 (1997) 397–433.

[37] B. M. M. DE WEGER. Solving exponential Diophantine equations using lattice basis reduction algorithms. *Journal of Number Theory* 26 (1987) 325–367.

[38] U. DIETER. How to calculate shortest vectors in a lattice. *Mathematics of Computation* 29 (1975) 827–833.

[39] G. L. DIRICHLET. Verallgemeinerung eines Satzes aus der Lehre von den Kettenbrüchen nebst einigen Anwendungen auf die Theorie der Zahlen. *Bericht über die Verhandlungen der Königlich Preussischen Akademie der Wissenschaften* (1842) 93–95.

[40] A. Dupré. Sur le nombre des divisions à effectuer pour obtenir le plus grand commun diviseur entre deux nombres entiers. *Journal de Mathématiques Pures et Appliquées* 11 (1846) 41-64.

[41] X. G. Fang, G. Havas. On the worst-case complexity of integer Gaussian elimination. *ISSAC '97: International Symposium on Symbolic and Algebraic Computation*, 28–31. Association for Computing Machinery, 1997.

[42] U. Fincke, M. Pohst. A procedure for determining algebraic integers of given norm. *Computer Algebra 1983*, 194–202. Lecture Notes in Computer Science, 162. Springer, 1983.

[43] U. Fincke, M. Pohst. Improved methods for calculating vectors of short length in a lattice, including a complexity analysis. *Mathematics of Computation* 44 (1985) 463–471.

[44] N. Gama, N. Howgrave-Graham, H. Koy, P. Q. Nguyen. Rankin's constant and blockwise lattice reduction. *CRYPTO 2006*, 1–20. Lecture Notes in Computer Science, 4117. Springer, 2006.

[45] N. Gama. P. Q. Nguyen: Predicting lattice reduction. *Eurocrypt 2008: Advances in Cryptology*, 31–51. Lecture Notes in Computer Science, 4965. Springer, 2008.

[46] M. R. Garey, D. S. Johnson. *Computers and Intractibility. A Guide to the Theory of NP-Completeness.* Freeman, 1979.

[47] K. O. Geddes, S. R. Czapor, G. Labahn. *Algorithms for Computer Algebra.* Kluwer, 1992.

[48] C. Gentry. The geometry of provable security: some proofs of security in which lattices make a surprise appearance. Pages 391–426 of Nguyen and Vallée [110].

[49] G. H. Golub, C. F. van Loan. *Matrix Computations.* Johns Hopkins University Press, 1983.

[50] R. L. Graham, D. E. Knuth, O. Patashnik. *Concrete Mathematics: A Foundation for Computer Science.* Second Edition. Addison-Wesley, 1994.

[51] M. Grötschel, L. Lovász, A. Schrijver. *Geometric Algorithms and Combinatorial Optimization.* Algorithms and Combinatorics, 2. Springer, 1988.

[52] P. M. Gruber, C. G. Lekkerkerker. *Geometry of Numbers.* Second edition. North-Holland, 1987.

[53] G. HANROT. LLL: a tool for effective Diophatine approximation. Pages 215–263 of Nguyen and Vallée [110].

[54] G. HANROT, D. STEHLÉ. Improved analysis of Kannan's shortest lattice vector algorithm: extended abstract. *CRYPTO 2007*, 170–186. Lecture Notes in Computer Science, 4622. Springer, 2007.

[55] J. HÅSTAD, B. JUST, J. C. LAGARIAS, C. P. SCHNORR. Polynomial time algorithms for finding integer relations among real numbers. *SIAM Journal on Computing* 18 (1989) 859–881.

[56] G. HAVAS, B. S. MAJEWSKI, K. R. MATTHEWS. Extended GCD and Hermite normal form algorithms via lattice basis reduction. *Experimental Mathematics* 7 (1998) 125–136.

[57] B. HELFRICH. An algorithm to compute Minkowski-reduced lattice bases. *STACS 85: Symposium on Theoretical Aspects of Computer Science*, 173–179. Lecture Notes in Computer Science, 182. Springer, 1985.

[58] B. HELFRICH. Algorithms to construct Minkowski reduced and Hermite reduced lattice bases. *Theoretical Computer Science* 41 (1985) 125–139.

[59] K. HENSEL. Eine neue Theorie der Algebraischen Zahlen. *Mathematische Zeitschrift* 2 (1918) 433–452.

[60] C. HERMITE. Extraits de lettres de M. Ch. Hermite à M. Jacobi sur différents objets de la théorie des nombres: Deuxième lettre. *Journal für die Reine und Angewandte Mathematik* 40 (1850) 279–315.

[61] I. N. HERSTEIN. *Topics in Algebra*. Second edition. Xerox College Publishing, 1975.

[62] I. N. HERSTEIN. *Abstract Algebra*. Third edition. Prentice Hall, 1996.

[63] J. D. HOBBY. A natural lattice basis problem with applications. *Mathematics of Computation* 67 (1998) 1149–1161.

[64] K. H. HOFFMAN, R. KUNZE. *Linear Algebra*. Second edition. Prentice-Hall, 1971.

[65] J. HOFFSTEIN, N. HOWGRAVE-GRAHAM, J. PIPHER, W. WHYTE. Practical lattice-based cryptography: NTRUEncrypt and NTRUSign. Pages 349–390 of Nguyen and Vallée [110].

[66] N. HOWGRAVE-GRAHAM. Finding small roots of univariate modular equations revisited. *Cryptography and Coding 1997*, 131–142. Lecture Notes in Computer Science, 1355. Springer, 1997.

[67] A. HURWITZ. Über die angenäherte Darstellung der Irrationalzahlen durch rationale Brüche. *Mathematische Annalen* 39 (1891) 279–284.

[68] N. JACOBSON. *Lectures in Abstract Algebra, 2: Linear Algebra.* Second edition. Springer, 1984.

[69] M. KAIB, C. P. SCHNORR. The generalized Gauss reduction algorithm. *Journal of Algorithms* 21 (1996) 565–578.

[70] R. KANNAN. Improved algorithms for integer programming and related lattice problems. *STOC 83: Symposium on Theory of Computing*, 193–206. Association for Computing Machinery, 1983.

[71] R. KANNAN. Minkowski's convex body theorem and integer programming. *Mathematics of Operations Research* 12 (1987) 415–440.

[72] R. KANNAN. Algorithmic geometry of numbers. *Annual Review of Computer Science* 2 (1987) 231–267. Annual Reviews, 1987.

[73] R. KANNAN, A. BACHEM. Polynomial algorithms for computing the Smith and Hermite normal forms of an integer matrix. *SIAM Journal on Computing* 8 (1979) 499–507.

[74] R. M. KARP. Reducibility among combinatorial problems. *Complexity of Computer Computations*, 85–103. Plenum, 1972.

[75] H. KEMPFERT. On the factorization of polynomials. *Journal of Number Theory* 1 (1969) 116–120.

[76] S. KHOT. Inapproximability results for computational problems on lattices. Pages 453–473 of Nguyen and Vallée [110].

[77] J. KLÜNERS. The van Hoeij algorithm for factoring polynomials. Pages 283–291 of Nguyen and Vallée [110].

[78] D. E. KNUTH. *The Art of Computer Programming, Volume 2: Seminumerical Algorithms.* Second edition. Addison-Wesley, 1981.

[79] A. KORKINE, G. ZOLOTAREFF. Sur les formes quadratiques. *Mathematische Annalen* 6 (1873) 366–389.

[80] H. KOY, C. P. SCHNORR. Segment LLL-reduction of lattice bases. *CaLC 2001: Cryptography and Lattices Conference*, 67–80. Lecture Notes in Computer Science, 2146. Springer, 2001.

[81] H. KOY, C. P. SCHNORR. Segment LLL-reduction with floating-point orthogonalization. *CaLC 2001: Cryptography and Lattices Conference*, 81–96. Lecture Notes in Computer Science, 2146. Springer, 2001.

[82] R. KUMAR, D. SIVAKUMAR. Complexity of SVP – a reader's digest. *SIGACT News* 32 (2001) 40–52 (Complexity Theorem Column 33).

[83] J. C. LAGARIAS. The computational complexity of simultaneous Diophantine approximation problems. *SIAM Journal on Computing* 14 (1985) 196–209.

[84] J. C. LAGARIAS, H. W. LENSTRA JR., C. P. SCHNORR. Korkin-Zolotarev bases and successive minima of a lattice and its reciprocal lattice. *Combinatorica* 10 (1990) 333–348.

[85] J. C. LAGARIAS, A. M. ODLYZKO. Solving low-density subset sum problems. *Journal of the Association for Computing Machinery* 32 (1985) 229–246.

[86] G. LAMÉ. Note sur la limite du nombre des divisions dans la recherche du plus grand commun diviseur entre deux nombres entiers. *Comptes Rendus de l'Académie des Sciences* 19 (1844) 867–870.

[87] C. G. LEKKERKERKER. *Geometry of Numbers.* North-Holland, 1969.

[88] A. K. LENSTRA, H. W. LENSTRA JR., L. LOVÁSZ. Factoring polynomials with rational coefficients. *Mathematische Annalen* 261 (1982) 515–534.

[89] W. J. LEVEQUE. *Topics in Number Theory, Volumes I and II.* Addison-Wesley, 1956.

[90] R. LIDL, H. NIEDERREITER. *Finite Fields.* Second edition. Encyclopedia of Mathematics and its Applications, 20. Cambridge University Press, 1997.

[91] L. LOVÁSZ. *An Algorithmic Theory of Numbers, Graphs, and Convexity.* Society for Industrial and Applied Mathematics, 1986.

[92] L. LOVASZ, H. E. SCARF. The generalized basis reduction algorithm. *Mathematics of Operations Research* 17 (1992) 751–764.

[93] J. MARTINET. *Perfect lattices in Euclidean Spaces.* Grundlehren der Mathematischen Wissenschaften, 327. Springer, 2003.

[94] A. MAY. Using LLL-reduction for solving RSA and factorization problems. Pages 315–348 of Nguyen and Vallée [110].

[95] R. C. MERKLE, M. E. HELLMAN. Hiding information and signatures in trapdoor knapsacks. *IEEE Transactions on Information Theory* 24 (1978) 525–530.

[96] D. MICCIANCIO. *On the Hardness of the Shortest Vector Problem.* Ph.D. Thesis, Department of Electrical Engineering and Computer Science, Massachusetts Institute of Technology, 1998: groups.csail.mit.edu/cis/theses/miccianc-phd.ps

[97] D. MICCIANCIO. The shortest vector in a lattice is hard to approximate to within some constant. *FOCS 98: Symposium on Foundations of Computer Science*, 92–98. Association for Computing Machinery, 1998.

[98] D. MICCIANCIO. The shortest vector in a lattice is hard to approximate to within some constant. *SIAM Journal on Computing* 30 (2001) 2008–2035.

[99] D. MICCIANCIO. Cryptographic functions from worst-case complexity assumptions. Pages 427–452 of Nguyen and Vallée [110].

[100] D. MICCIANCIO, S. GOLDWASSER. *Complexity of Lattice Problems: A Cryptographic Perspective*. Springer, 2002.

[101] D. MICCIANCIO, B. WARINSCHI. A linear space algorithm for computing the Hermite normal form. *ISSAC 2001: International Symposium on Symbolic and Algebraic Computation*, 231–236. Association for Computing Machinery, 2001.

[102] M. MIGNOTTE. An inequality about factors of polynomials. *Mathematics of Computation* 28 (1974) 1153–1157.

[103] M. MIGNOTTE. An inequality about irreducible factors of integer polynomials. *Journal of Number Theory* 30 (1988) 156–166.

[104] W. H. MOW. Universal lattice decoding: principle and recent advances. *Wireless Communication and Mobile Computing* 3 (2003) 553–569.

[105] P. Q. NGUYEN. Hermite's constant and lattice algorithms. Pages 19–69 of Nguyen and Vallée [110].

[106] P. Q. NGUYEN, D. STEHLÉ. Low-dimensional lattice basis reduction revisited. *ACM Transactions on Algorithms* 5, Article 46, October 2009.

[107] P. Q. NGUYEN, D. STEHLÉ. LLL on the average. *Algorithmic Number Theory Symposium VII*, 238–256. Lecture Notes in Computer Science, 4076. Springer, 2006.

[108] P. Q. NGUYEN, J. STERN. Lattice reduction in cryptology: an update. *Algebraic Number Theory Symposium IV*, 85–112. Lecture Notes in Computer Science, 1838. Springer, 2000.

[109] P. Q. NGUYEN, J. STERN. The two faces of lattices in cryptology. *CaLC: Cryptography and Lattices Conference*, 146–180. Lecture Notes in Computer Science, 2146. Springer, 2001.

[110] P. Q. NGUYEN, B. VALLÉE (editors): *The LLL Algorithm: Survey and Applications*. Springer, 2010.

[111] P. Q. NGUYEN, T. VIDICK. Sieve algorithms for the shortest vector problem are practical. *Journal of Mathematical Cryptology* 2 (2008) 181–207.

[112] W. K. NICHOLSON. *Linear Algebra with Applications.* Fifth edition. McGraw Hill, 2006.

[113] A. M. ODLYZKO. Cryptanalytic attacks on the multiplicative knapsack cryptosystem and on Shamir's fast signature scheme. *IEEE Transactions on Information Theory* 30 (1984) 594–601.

[114] A. M. ODLYZKO. The rise and fall of knapsack cryptosystems. *Cryptology and Computational Number Theory. Proceedings of Symposia in Applied Mathematics* 42 (1990) 75–88. American Mathematical Society, 1990.

[115] A. M. ODLYZKO, H. J. J. TE RIELE. Disproof of the Mertens conjecture. *Journal für die Reine and Angewandte Mathematik* 357 (1985) 138–160.

[116] C. PERNET, W. STEIN. Fast computation of Hermite normal forms of random integer matrices. *Journal of Number Theory* 130 (2010) 1675–1683.

[117] J. PINTZ. An effective disproof of the Mertens conjecture. *Astérisque* 147–148 (1987) 325–333.

[118] M. E. POHST. A modification of the LLL reduction algorithm. *Journal of Symbolic Computation* 4 (1987) 123–127.

[119] M. E. POHST, H. J. ZASSENHAUS. *Algorithmic Algebraic Number Theory.* Encyclopedia of Mathematics and its Applications, 30. Cambridge University Press, 1989.

[120] O. REGEV. On the complexity of lattice problems with polynomial approximation factors. Pages 475–496 of Nguyen and Vallée [110].

[121] C. A. ROGERS. *Packing and Covering.* Cambridge University Press, 1964.

[122] W. SCHARLAU, H. OPOLKA. *From Fermat to Minkowski. Lectures on the Theory of Numbers and its Historical Development.* Springer, 1985.

[123] C. P. SCHNORR. A hierarchy of polynomial-time lattice basis reduction algorithms. *Theoretical Computer Science* 53 (1987) 201–224.

[124] C. P. SCHNORR. Lattice reduction by random sampling and birthday methods. *STACS 2003: Symposium on Theoretical Aspects of Computer Science*, 145–156. Lecture Notes in Computer Science, 2607. Springer, 2003.

[125] C. P. SCHNORR. Fast LLL-type lattice reduction. *Information and Computation* 204 (2006) 1–25.

[126] C. P. SCHNORR. Progress on LLL and lattice reduction. Pages 145–178 of Nguyen and Vallée [110].

[127] C. P. SCHNORR, M. EUCHNER. Lattice basis reduction: improved practical algorithms and solving subset sum problems. *Mathematical Programming* 66 (1994) 181–199.

[128] J. SHALLIT. Origins of the analysis of the Euclidean algorithm. *Historia Mathematica* 21 (1994) 401–419.

[129] A. SHAMIR. A polynomial-time algorithm for breaking the basic Merkle-Hellman cryptosystem. *IEEE Transactions on Information Theory* 30 (1984) 699–704.

[130] D. SIMON. Selected applications of LLL in number theory. Pages 265–282 of Nguyen and Vallée [110].

[131] C. C. SIMS. *Computation with Finitely Presented Groups*. Encyclopedia of Mathematics and its Applications, 48. Cambridge University Press, 1994.

[132] I. SMEETS. The history of the LLL algorithm. Pages 1–17 of Nguyen and Vallée [110].

[133] D. STEHLÉ. Floating point LLL: theoretical and practical aspects. Pages 179–213 of Nguyen and Vallée [110].

[134] J. STILLWELL. *Elements of Number Theory*. Springer, 2003.

[135] A. STORJOHANN. The modulo extended gcd problem and space efficient algorithms for integer matrices: www.cs.uwaterloo.ca/~astorjoh/publications.html

[136] H. J. J. TE RIELE. Some historical and other notes about the Mertens conjecture and its recent disproof. *Nieuw Archief voor Wiskunde* 3 (1985) 237–243.

[137] L. N. TREFETHEN, D. BAU. *Numerical Linear Algebra*. Society for Industrial and Applied Mathematics, 1997.

[138] B. VALLÉE. A central problem in the algorithmic geometry of numbers: lattice reduction. *CWI Quarterly* 3 (1990) 95–120.

[139] B. VALLÉE. Gauss' algorithm revisited. *Journal of Algorithms* 12 (1991) 556–572.

[140] B. VALLÉE. Dynamical analysis of a class of Euclidean algorithms. *Theoretical Computer Science* 297 (2003) 447–486.

[141] B. VALLÉE, A. VERA. Lattice reduction in two dimensions: analyses under realistic probabilistic models. *AofA 2007: Analysis of Algorithms*, 181–216. Discrete Mathematics and Theoretical Computer Science, 2007.

[142] B. VALLÉE, A. VERA. Probabilistic analyses of lattice reduction algorithms. Pages 71–143 of Nguyen and Vallée [110].

[143] W. VAN DER KALLEN. Complexity of the Havas, Majewski, Matthews LLL Hermite normal form algorithm. *Journal of Symbolic Computation* 30 (2000) 329–337.

[144] P. VAN EMDE BOAS. Another NP-complete partition problem and the complexity of computing short vectors in a lattice. Report 81-04, April 1981, Department of Mathematics, University of Amsterdam: `staff.science.uva.nl/~peter/vectors/mi8104c.html`

[145] M. VAN HOEIJ. Factoring polynomials and the knapsack problem. *Journal of Number Theory* 95 (2002) 167–189.

[146] E. VITERBO, J. BOUTROS. A universal lattice decoder for fading channels. *IEEE Transactions on Information Theory* 45 (1999) 1639–1642.

[147] J. VON ZUR GATHEN, J. GERHARD. *Modern Computer Algebra*. Second edition. Cambridge University Press, 2003.

[148] WIKIPEDIA. `en.wikipedia.org/wiki/NP-complete`

[149] H. ZASSENHAUS. On Hensel factorization, I. *Journal of Number Theory* 1 (1969) 291–311.

Index